Quantitative Assessment of Environmental/Human Health Risks

Quantitative Assessment of Environmental/Human Health Risks

Editor

Cheng Yan

MDPI • Basel • Beijing • Wuhan • Barcelona • Belgrade • Manchester • Tokyo • Cluj • Tianjin

Editor
Cheng Yan
China University of
Geosciences (Wuhan)
China

Editorial Office
MDPI
St. Alban-Anlage 66
4052 Basel, Switzerland

This is a reprint of articles from the Special Issue published online in the open access journal *International Journal of Environmental Research and Public Health* (ISSN 1660-4601) (available at: https://www.mdpi.com/journal/ijerph/special_issues/environmental_human_risk).

For citation purposes, cite each article independently as indicated on the article page online and as indicated below:

LastName, A.A.; LastName, B.B.; LastName, C.C. Article Title. *Journal Name* **Year**, *Volume Number*, Page Range.

ISBN 978-3-0365-5851-6 (Hbk)
ISBN 978-3-0365-5852-3 (PDF)

Contents

Riccardo Petrini, Lisa Ghezzi, Simone Arrighi, Lisa Genovesi, Chiara Frassi and Luca Pandolfi
Trace Elements in Soil and Urban Groundwater in an Area Impacted by Metallurgical Activity: Health Risk Assessment in the Historical Barga Municipality (Tuscany, Italy)
Reprinted from: *Int. J. Environ. Res. Public Health* **2022**, *19*, 13419, doi:10.3390/ijerph192013419 . **1**

Qianlei Yang, Rui Yan, Yuemei Mo, Haixuan Xia, Hanyi Deng, Xiaojuan Wang, Chunchun Li, Koichi Kato, Hengdong Zhang, Tingxu Jin, Jie Zhang and Yan An
The Potential Key Role of the NRF2/NQO1 Pathway in the Health Effects of Arsenic Pollution on SCC
Reprinted from: *Int. J. Environ. Res. Public Health* **2022**, *19*, 8118, doi:10.3390/ijerph19138118 . . . **19**

Bin Zhang, Ying Yang, Wenjia Xie, Wei He, Jia Xie and Wei Liu
Identifying Algicides of *Enterobacter hormaechei* F2 for Control of the Harmful Alga *Microcystis aeruginosa*
Reprinted from: *Int. J. Environ. Res. Public Health* **2022**, *19*, 7556, doi:10.3390/ijerph19137556 . . . **35**

Yan Yang, Mengdi Zhang, Haojia Chen, Zenghua Qi, Chengcheng Liu, Qiang Chen and Tao Long
Estimation of Children's Soil and Dust Ingestion Rates and Health Risk at E-Waste Dismantling Area
Reprinted from: *Int. J. Environ. Res. Public Health* **2022**, *19*, 7332, doi:10.3390/ijerph19127332 . . . **51**

Yue Lang, Yanan Yu, Hongtao Zou, Jiexu Ye and Shihan Zhang
Performance and Mechanisms of Sulfidated Nanoscale Zero-Valent Iron Materials for Toxic TCE Removal from the Groundwater
Reprinted from: *Int. J. Environ. Res. Public Health* **2022**, *19*, 6299, doi:10.3390/ijerph19106299 . . . **65**

Ruonan Wang, Aoyuan Yu, Tianlei Qiu, Yajie Guo, Haoze Gao, Xingbin Sun, Min Gao and Xuming Wang
Aerosolization Behaviour of Fungi and Its Potential Health Effects during the Composting of Animal Manure
Reprinted from: *Int. J. Environ. Res. Public Health* **2022**, *19*, 5644, doi:10.3390/ijerph19095644 . . . **77**

Junzhi Zhang, Xiao He, Huixin Zhang, Yu Liao, Qi Wang, Luwei Li and Jianwei Yu
Factors Driving Microbial Community Dynamics and Potential Health Effects of Bacterial Pathogen on Landscape Lakes with Reclaimed Water Replenishment in Beijing, PR China
Reprinted from: *Int. J. Environ. Res. Public Health* **2022**, *19*, 5127, doi:10.3390/ijerph19095127 . . . **93**

Qingman Li, Jie Liang, Qun Wang, Yuntong Chen, Hongyu Yang, Hong Ling, Zhiwen Luo and Jian Hang
Numerical Investigations of Urban Pollutant Dispersion and Building Intake Fraction with Various 3D Building Configurations and Tree Plantings
Reprinted from: *Int. J. Environ. Res. Public Health* **2022**, *19*, 3524, doi:10.3390/ijerph19063524 . . . **105**

Xin Kang, Dongpeng Li, Xiaoxiang Zhao, Yanfeng Lv, Xi Chen, Xinshan Song, Xiangyu Liu, Chengrong Chen and Xin Cao
Long-Term Exposure to Phenanthrene Induced Gene Expressions and Enzyme Activities of *Cyprinus carpio* below the Safe Concentration
Reprinted from: *Int. J. Environ. Res. Public Health* **2022**, *19*, 2129, doi:10.3390/ijerph19042129 . . . **139**

Zhihong Zheng and Xiaohan Duan
Mitigating the Health Effects of Aqueous Cr(VI) with Iron-Modified Biochar
Reprinted from: *Int. J. Environ. Res. Public Health* **2022**, *19*, 1481, doi:10.3390/ijerph19031481 . . . **153**

International Journal of
Environmental Research and Public Health

MDPI

Article

Trace Elements in Soil and Urban Groundwater in an Area Impacted by Metallurgical Activity: Health Risk Assessment in the Historical Barga Municipality (Tuscany, Italy)

Riccardo Petrini, Lisa Ghezzi *, Simone Arrighi, Lisa Genovesi, Chiara Frassi and Luca Pandolfi

Department of Earth Sciences, University of Pisa, Via S. Maria 53, 56126 Pisa, Italy
* Correspondence: lisa.ghezzi@unipi.it; Tel.: +39-050-2215787

Abstract: Trace elements were measured in soil and groundwater collected within the Fornaci di Barga urban area (Serchio River Valley, Tuscany, Italy), a territory that integrates natural assets with touristic vocation, impacted by long-lasting metallurgical activity. Epidemiological studies highlighted that the area surrounding the industrial plants is characterized by a persistent excess of diseases, attributed to heavy metal pollution. Soils were taken in school gardens, public parks, sport grounds and roadsides. The results indicate that Cu, Zn and Cd represent the main contaminants in surface soil, likely originated by deposition of airborne particulate matter from metallurgical activity. Risk assessment considering soil ingestion and dermal contact exposure routes revealed that the cadmium Hazard Quotient approaches unity for children, and the cadmium risk-based concentration obtained by combining exposure information with toxicity data is only slightly lower compared with the cadmium maximum concentration actually measured in soil. Groundwater does not show evidence of trace metal contamination, suggesting that the migration of contaminants from soil to subsurface is a slow process. However, assessment of the possible interconnections between shallow and deep-seated aquifers requires monitoring to be continued. The obtained results highlight the possible link between space clusters of diseases and metal concentration in soil.

Keywords: trace elements; health risk assessment; heavy metals; Tuscany (Italy); soil contamination; urban environment

Citation: Petrini, R.; Ghezzi, L.; Arrighi, S.; Genovesi, L.; Frassi, C.; Pandolfi, L. Trace Elements in Soil and Urban Groundwater in an Area Impacted by Metallurgical Activity: Health Risk Assessment in the Historical Barga Municipality (Tuscany, Italy). *Int. J. Environ. Res. Public Health* **2022**, *19*, 13419. https://doi.org/10.3390/ijerph192013419

Academic Editor: Cheng Yan

Received: 20 September 2022
Accepted: 13 October 2022
Published: 17 October 2022

1. Introduction

The quantitative assessment of environmental soil quality in urban green spaces is a priority to enhance resilience and to promote high standard of living [1]. Soils in metropolitan areas are variably affected by human activities [2–4]; indeed, urban soils contain different materials related to the urban development distributed in anthropogenic layers and act as sinks for pollutants from localized and/or diffused manifold anthropogenic sources. These include automotive traffic, industry, fossil fuel combustion, emissions from municipal incinerators and waste disposal [5]. The urban soil structure ranges from relatively undisturbed to completely man-made, lacking the established horizons that characterize natural soil [6] and influencing the retention and transport of contaminants through the subsurface. Potentially toxic elements in soils have different mobility and can be transported by rainwater during runoff and/or leached from soil by infiltrating water [7], possibly extending the pollution out of the urban area. Once soil is contaminated by PTE such as heavy metals it is difficult to recover, yielding accumulation gradients and long-term effects along urbanization [8,9]. The prolonged exposure to contaminated soil represents a potential hazard for inhabitants, due to both direct pathways that include dermal contact, ingestion and dust inhalation [10], and through indirect routes of exposure such as the food chain [11]. Potentially toxic elements in urban soils may exert chronic effects of toxicity, especially on preschool children that have the tendency to place objects and their fingers in their mouths, unintentionally eating soil while spending their time in urban parks and kindergartens [12,13]. In Italy, the Code of Environment (Legislative Decree 152/2006 which regroups in a single legislative text the environmental laws previously contained in several decrees) recognizes the importance of soil in public green spaces and for residential use, posing restrictive concentration thresholds for PTE and health risk models and tools, to estimate the probability of adverse health effects in humans exposed.

The present study aims at depicting the distribution of PTE in soil and groundwater in the urban setting of Fornaci di Barga (Lucca), within the historical Barga municipality (Middle Serchio River Valley, Tuscany, Italy). In this area, metallurgical activities situated directly adjacent the residential area of Fornaci di Barga have lasted more than 100 years, creating a potential ecotoxicological hazard by PTE primarily delivered to soil through industrial dust varying in age of settlement. Indeed, health statistics and epidemiological evidence highlighted the prevalence of lung cancer and chronic kidney and cardiovascular disease in the Middle Valley [14,15], with a significantly higher incidence compared with the average of Tuscany. In particular, the incidence of kidney diseases increases approaching the site of current and historical metallurgical manufacturing, as revealed by an ongoing cohort study, and the cause has been postulated as due to environmental pollution by PTE, particularly cadmium [15]. Fornaci di Barga has been included among five pilot cities in Europe for industrial pollution and health impact [16] (CitieS-Health project: https://citieshealth.eu/ (accessed on 1 June 2022)).

The results of this study provide valuable insights to policy makers and urban planners into how to mitigate potential health risks in industrial threatened urban areas.

2. Environmental and Geological Setting

The Fornaci di Barga village belongs to the Barga municipality (Province of Lucca, Tuscany, Italy) in the Serchio River Middle Valley (Figure 1) between the Apuan Alps mountain range and Apennine chain, a territory with cultural and historical heritage, naturalistic assets, and a great touristic vocation. From a geological point of view, the Serchio River Valley is a NW-SE-trending tectonic depression dissecting the architecture of the Neogenic Northern Apennine orogenic belt made up of tectonic units derived from both oceanic (Ligurian Units, composed by ophiolitic rocks and their sedimentary cover) and continental domains (Tuscan Nappe and Tuscan metamorphic Units) (Figure 1). The metamorphic and the non-metamorphic Tuscan units are separated by thick carbonate cataclasites mainly originated during the nappe stacking.

Whatever is the trigger enhancing the exhumation of the Tuscan metamorphic rocks currently exposed in the Apuan Alps [17,18], the tectonic nappe stack of Northern Apennines was affected by an extensional tectonic regime at earlier than 11 Ma [19] with the activation of low- and subsequently of NW-SE trending high-angle faults [20]. The latter are responsible for the last uplift stages of the Apuan Alps region (2–6 Ma) [19] and the development of the tectonic depression of the Serchio River Valley.

The depression is bounded by high-angle normal faults dipping toward east on the west side and toward west on the east side of the Valley. Their activity allows the exposure of the Ligurian units s.l. (i.e., the units at the top of the Apennine nappe stack) at the bottom of the Valley (Figure 1). It also strongly controls the landscape, which shows landforms typical of tectonically active regions with high rates of uplift producing high relief topography with steep mountain slopes (i.e., the Apuan Alps). From early Pliocene to early Pleistocene the tectonic depression was filled by 80 to 320 m-thick continental sequence deposited in fluvial environment associated to debris flows sourced from both sides of the tectonic depression in rapid uplift [21–23]. Currently, the best outcrops of these deposits are in the areas of Barga-Fornaci di Barga and Castelnuovo di Garfagnana (Figure 1). These deposits consist of a basal sequence of sands and silty sands interbedded by unsorted to poorly sorted medium to very coarse-grained conglomerates, 20–40 cm thick organic-rich lenses and paleosols [22]. Clasts of conglomerates are mainly represented by Tuscan non-metamorphic lithotypes with a 15–30% of clasts deriving from the Tuscan metamorphic rocks [21]. Clasts are mostly sandstones referable to the youngest formation of the Tuscan

Nappe currently exposed in the Apennine main ridge. Moving upward, the sequence is characterized by a dominance of poorly- to well-sorted, medium to coarse-grained and mainly clast-supported conglomerates with interbedded sandy material. Clasts mainly consist of carbonate lithotypes equally derived from metamorphic and non-metamorphic Tuscan units whereas sandstones clasts are subordinates [21].

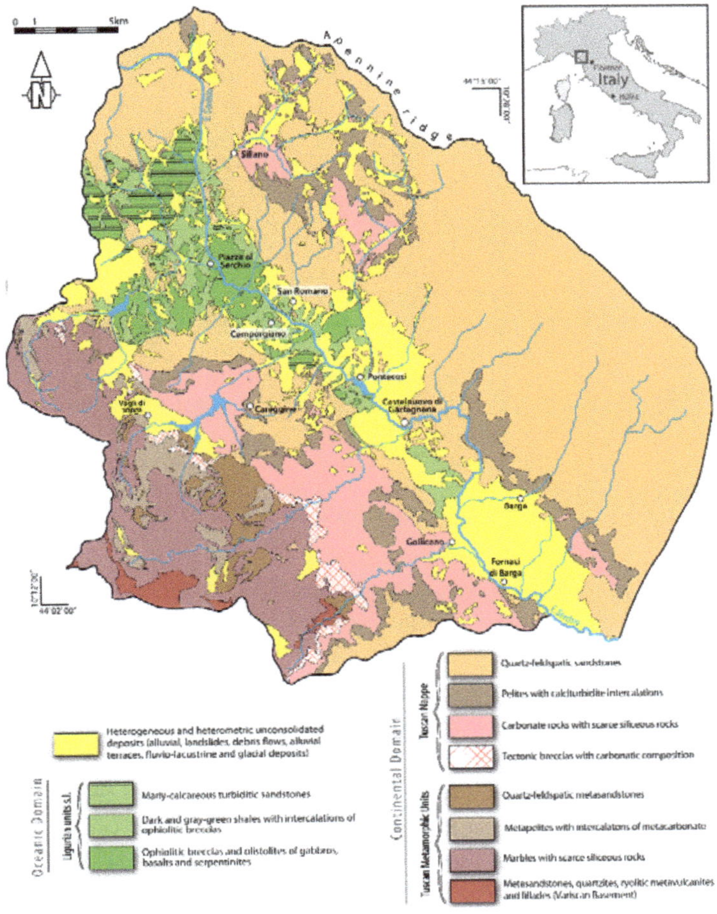

Figure 1. Lithological map of the northern hydrographic basin of Serchio River. The location of the study area, Fornaci di Barga, is shown.

During middle to late Pleistocene, these sediments were cut by the Serchio River and its tributaries forming different orders of alluvial terraces [22].

Fornaci di Barga develops on one of these fluvial terraces. In the area the thickness of these sediments commonly ranges from 5 to 30 m. From a hydrogeological point of view these deposits have a medium-high to high permeability associated with their primary porosity. They lie on poorly permeable or almost impermeable deposits, formed by Pleistocene clays and sandy clays, which form the bedrock of the unconfined aquifer.

Metallurgical activity in the area started in 1915, and Fornaci di Barga is now one of the three main industrial sites of KME (Engineering Copper Solutions) one of the largest manufacturers in the world of copper and copper alloy preliminary and semi-finished product. Copper represents the second most used non-ferrous metal in industry. Excesses above the regional average of cardiovascular diseases, chronic kidney diseases and malignant tumors such as lung cancer have been reported in the area. A strong tendency to associate environmental pollution with the perceived health profile by inhabitants has been demonstrated [15].

3. Materials and Methods

3.1. Sampling

Soils were collected in 14 stations during surveys between February and June 2020 (samples FB_1S to FB_14S). Sampling was adapted to the Fornaci di Barga urban pattern, considering different urban land uses including school gardens (FB_1S, FB_2S, FB_9S, FB_12S); sports grounds (FB_3S, FB_11S); private gardens (FB_8S, FB_13S); roadsides (FB_4S, FB_7S), public parks (FB_6S, FB_10S) and riverside fishing access (FB_14S) near KME. One station (FB_5S) was in a peri-urban area. The georeferenced distribution of sampling sites is shown in Figure 2a. Most soils were taken at 0–10 cm depth and in some cases at greater depth (Table 1) by using a hand auger. To quantify the geogenic fingerprint of trace elements in Serchio River Valley sediments, at the watershed scale of interest, fluvial sediments were collected within the reservoir system of the hydroelectric lake of Pontecosi, about 11 km upstream Fornaci di Barga (Figure 1), where the impact of anthropogenic activities is absent. In the lake, sediments were collected in 15 sampling stations during September 2019, both in subaerial domains, to have an archive that includes environmental conditions closed to exchanges with the free water column, using a manual auger (Matest manufacturer, Treviolo, Italy) (composite samples PC1 to PC9 combining all the cores from top up to 1 m depth) and dredged from subaquatic lake floor using a Van Veen grab sampler (samples PC10 to PC15). To compare PTE concentration against Italian guidelines soil and sediment samples were quartered and sieved to separate the <2 mm size fraction for multi-element chemical analysis.

3.2. Analytical Procedures

For all instrumental analysis calibration curves were performed prior samples analysis. The minimal number of standard calibration solutions was four and correlation coefficients for calibration curves were better than 0.99, otherwise the calibration curve was repeated. Quality control samples were analyzed one in every ten samples. Quality Control samples included blanks (to control the purity of reagents, the other possible contamination in all work procedures and the memory effect) and check solutions (or check solid matrices) whose elements concentration is known. Certified reference materials (see below) were used to evaluate accuracy and precision of the method.

Soil samples were digested by using the Ethos Easy microwave platform (Milestone Srl, Sorisole, Italy) (US EPA Method 3052, reversed aqua regia). The concentration of a set of trace elements was determined by ICP-MS using the NexION 300X (Perkin Elmer Inc., Waltham, MA, USA). The analytical uncertainty was evaluated by the analysis of soil reference material NIST SRM 2711a (Montana soil). RSD was 5% for Li, Be, Mn, Ni, Ag, Sn, Cd, Tl, Pb, Fe and As, and 10% for Co, Cu, Zn, Sr, Sb, Ba, Th, U, V, and Cr. In general, the accuracy was better than 10%. Soil pH was also measured (ISO 10390). The total mercury content was determined (US EPA Method 7473) by the DMA-80 (Milestone Srl, Sorisole, Italy). NIST 2711a, ERM-CC018 (sandy soil) and MESS-3 (marine sediment) were used as reference materials. RSD was 5% and accuracy was within 10%. Leaching tests on soil samples were performed following the procedure for granular waste EN 12457-2:2002. Briefly, one stage batch test under stirring for 24 h at a liquid to solid ratio of 10 L/kg with particle size below 4 mm and deionized water as liquid was applied.

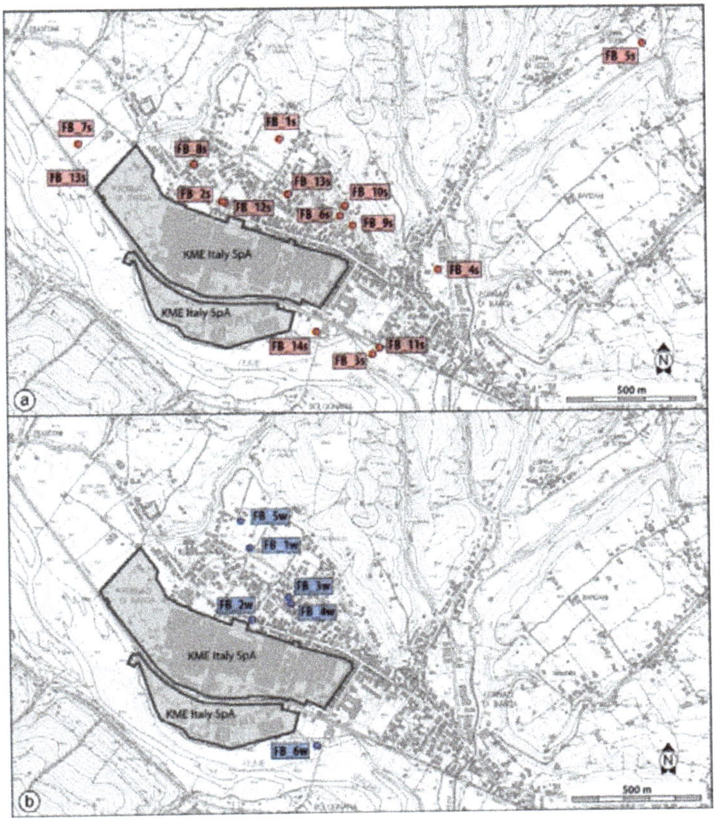

Figure 2. Location of soil (**a**) and groundwater (**b**) sampling stations. Groundwater was sampled in the unconfined aquifer of Fornaci di Barga on July, 2020 through five wells tapping groundwater bodies at different depth (FB_1W, depth 50 m; FB_2W, depth 8 m; FB_3W, depth 7 m; FB_4W, depth 18 m; FB_5W, depth 21 m) and in a piezometer nearby the station FB_14S, using a bailer at depth of 3.5 and 8.7 m below ground level (samples FB_6Wa and FB_6Wb, respectively). The georeferenced distribution of groundwater sampling stations is shown in Figure 2b. Water samples were filtered in the field through 0.45 μm filters and refrigerated in pre-cleaned polyethylene bottles. Ultrapure nitric acid was used as a preservative for major cation and trace element analysis.

Water temperature, pH, electrical conductivity (EC), redox potential (Eh), and dissolved oxygen (DO) were measured in the field. Alkalinity (expressed as HCO_3^-) was also determined in the field by acidimetric titration. Major ions were determined by ion chromatography using the ICS 900 (Thermo Fischer Scientific, Waltham, MA, USA). RSD was less than 5%. Trace elements in waters and leachates were determined by ICP-MS using a PerkinElmer NexION 300X. Deviations from the certified values of water standards NIST SRM 1640a (trace elements in natural water) and 1643f (trace elements in water) (20 replicates) were less than 5%, except for Cu, Fe, Li and Zn (5–10%). RSD was 10% except for Sr, Mo, Ag, Cd, Sb, Tl, Pb, U and As (5%).

Table 1. Trace element concentration in Fornaci di Barga soils (mg/kg) and in soil leachates (μg/L) from FB_13S and FB_14S samples) together with the maximum concentration level (MCL, mg/kg) imposed by Italian regulations (Legislative Decree No. 152/2006 approving the Code on the Environment). Values in bold exceed the MCL for residential soil use or groundwater in case of leachates.

Sample	Depth	Li	Be	Mn	Co	Ni	Cu	Zn	Sr	Mo	Ag	Sn	Cd	Sb	Ba	Tl	Pb	Th	U	V	Cr	Fe	As	Hg
FB_1Sa	0–10	43	1.70	892	16.1	80	39	84	41	0.72	0.12	2.3	0.28	0.27	121	0.33	22.1	7.1	1.42	66	123	32,900	6.6	0.05
FB_1Sb	50–60	41	1.63	1120	16.3	75	41	84	43	0.70	0.16	2.5	0.30	0.42	127	0.36	24.2	5.8	1.22	70	121	34,900	7.5	0.07
FB_2Sa	0–10	22	0.91	707	8.0	47	**216**	**439**	92	<0.5	0.16	4.1	**5.3**	0.57	113	0.25	51	5.1	1.00	42	84	20,900	4.5	0.31
FB_2Sb	20–30	20	0.88	719	8.0	51	**233**	**465**	94	<0.5	0.15	4.9	**5.6**	0.84	112	0.25	60	4.4	1.00	41	85	20,700	4.5	0.34
FB_3S	0–10	28	1.27	951	15.2	63	**234**	**270**	26	<0.5	0.08	2.2	**2.6**	0.55	77	0.28	27	5.0	0.9	50	90	25,700	10.6	0.10
FB_4S	0–10	22	1.06	519	8.2	45	43	96	23	<0.5	0.07	2.1	0.61	0.42	75	0.22	21	4.8	1.00	41	80	20,400	4.3	0.07
FB_5S	0–10	31	1.49	754	11.5	66	42	74	31	<0.5	0.05	2.0	0.43	0.42	97	0.29	25	6.1	1.27	53	107	27,200	5.3	0.06
FB_6S	0–10	31	1.20	768	10.1	61	103	**213**	29	<0.5	0.08	2.3	**3.3**	0.51	85	0.27	29	5.6	1.38	50	107	25,500	5.6	0.09
FB_7S	0–10	32	1.20	506	11.2	66	37	80	31	<0.5	0.05	1.50	0.57	0.22	58	0.19	13.2	5.4	1.06	48	106	27,100	4.6	0.04
FB_8S	0–10	23	1.07	377	9.4	55	38	62	26	<0.5	0.05	1.39	0.27	0.20	53	0.15	10.6	5.2	1.06	37	89	22,900	3.0	0.06
FB_9S	0–10	19.9	0.90	427	8.1	49	91	**200**	79	0.22	0.06	1.81	**3.1**	0.29	73	0.20	17.9	4.4	0.92	36	74	18,500	3.5	0.04
FB10S	0–10	27	1.15	607	10.9	68	35	81	33	0.22	0.03	1.43	0.50	0.23	61	0.19	12.0	6.3	1.08	47	100	23,800	4.7	0.05
FB_11S	0–10	26	1.09	1017	12.0	56	**155**	**248**	25.6	0.36	0.09	2.3	**2.2**	0.51	73	0.29	29	4.7	0.86	47	84	23,100	5.5	0.16
FB_12Sa	35–50	24	0.99	701	9.7	47	78	**189**	56	0.75	0.13	2.3	1.67	0.34	99	0.33	25	4.8	1.09	45	87	21,900	4.0	0.11
FB_12Sb	50–60	14.4	0.60	636	6.3	37	**134**	**296**	78	0.28	0.13	2.9	**2.8**	0.5	65	0.21	42	3.0	0.67	28.6	55	14,300	3.4	0.38
FB_13Sa	20–30	19.5	0.88	455	7.6	46	**129**	**445**	23.5	0.70	0.26	4.5	**6.7**	0.61	104	0.23	41	5.6	1.26	36	84	20,000	4.5	0.23
FB_13Sb	40–50	18.2	0.84	381	7.0	42	107	**353**	20.1	0.42	0.18	3.7	**4.8**	0.49	83	0.22	36	5.1	1.27	34	72	17,300	4.1	0.22
FB_13Sc	60–70	23	0.98	401	8.4	48	41	126	17.9	0.36	0.11	2.0	1.20	0.36	72	0.24	18.6	6.6	1.36	38	82	21,400	4.4	0.16
FB_14S	15–22	32	1.24	1530	16.5	**265**	**5280**	**27,400**	46	1.37	8.5	22.1	**15.8**	**11.7**	169	0.23	**221**	5.3	1.29	55	146	34,800	**255**	**2.2**
MCL			2		20	120	120	150					2	10		1	100			90	150		20	1
FB13a$_{leach}$		<4	<0.2	9.0	<0.2	<4	58	153	<36	<5	<0.4	<2	1.41	1.15	<14	<0.15	<4	0.16		<2.5	<2.5	**479**	1.10	
FB14$_{leach}$		<4	<0.2	<4	<0.2	<4	47	178	40	<5	<0.4	<2	1.20	**6.9**	183	<0.15	<4	<0.06		<2.5	<2.5	46	0.94	
MCL			4	50	50	20	1000	3000					5	5		2	10	<0.06			50	200	10	

3.3. Risk Assessment

Risk analysis was performed following the approaches described both by ASTM standards [24] and United States Environmental Protection Agency (US EPA) guidelines [25] and using the Risk-net software (version 3.1.1pro, September 2019. http://www.reconnet.net/Software.htm (accessed on 1 June 2022)).

The selected exposure routes were surface soil ingestion and dermal absorption in unpaved outdoor areas of residential zones. The soil dust inhalation was not included in risk analysis due to the lack of site-specific data (e.g., ambient air velocity in the mixing zone); it has to be mentioned that inhalation pathways can result in much higher exposure than dermal exposure in many circumstances depending on the size distribution of the dust. In addition, fugitive particulate emission reflects the occurrence of different categories of open dust sources, including re-entrained dust due to traffic traveling over the paved roadway, that would be necessary to estimate a realistic risk assessment through inhalation exposure pathway. Human receptors have been identified with both adults and children. The Hazard Quotient (HQ) for non-carcinogenic chronic (long term) effects in humans was calculated for each exposure pathway [26]. In case of direct ingestion and dermal contact, HQ (i.e $HQ_{ingestion}$ and HQ_{derma}l) was calculated by dividing the chronic daily intake (CDI, mg/kg/day) by the corresponding reference dose (RfD, mg/kg/day) [27,28], defined as the maximum daily exposure to a toxic agent that would not produce any appreciable deleterious effects on human health:

$$HQ = \frac{CDI}{RfD} \quad (1)$$

where CDI represents the exposure to a toxic agent, averaged over a long period of time, through ingestion (CDI_{ing}) or dermal contact (CDI_{derm}), given by:

$$CDI_{ing} = C_{POE} * \frac{R_{ing} * EF * ED}{BW * AT} * 10^{-6} \quad (2)$$

$$CDI_{derm} = C_{POE} * \frac{SA * SAF * ABS * EF * ED}{BW * AT} * 10^{-6} \quad (3)$$

where C_{POE} is the exposure point concentration of contaminant in soil (mg/kg), equal to the concentration at the source (Cs) for direct exposure pathways. For the remaining parameters, recommend values were used [29–31]: R_{ing} is the ingestion rate (100 mg/day for adult, 200 mg/day for children); EF is the exposure frequency (350 day/year, maximum value suitable for a residential setting), ED is the exposure duration (24 years for adults, 6 years for children); SA is the exposed skin area (5700 cm^2 for adults, 2800 cm^2 for children); SAF is the skin adherence factor (0.07 mg/cm^2 day for adults, 0.2 mg/cm^2 day for children); ABS is the dermal absorption factor (chemical specific, unitless); BW is the average body weight (70 kg for adults,15 kg for children); AT is the average time of exposure to non-carcinogens (ED × 365 day/year).

The total risk for a single substance defines a screening level individual Hazard Index (HI_i), such as:

$$HI_i = HQ_{ingestion} + HQ_{dermal} \quad (4)$$

A HI_i value less than unity indicates that risk is acceptable [30]. Hazard Index resulting from simultaneous exposure to n non-carcinogens (HI_{cum}), is given by:

$$HI_{cum} = \sum_{i=1}^{n} HI_i \quad (5)$$

The maximum allowed concentration of contaminants in soil, intended to be protective of human health (soil screening levels–SSLs, according to US EPA guidelines), was obtained by following the Risk Based Corrective Action procedure [24,25]. In this approach, the exposure equations and pathway models are run in reverse to back-calculate the acceptable level of a contaminant concentration in soil corresponding to the target risk. Risk based

SSLs for the different outdoor exposure pathways and for residential setting were derived from standardized sets of equations that are based on the updated U.S. Environmental Protection Agency's human health risk assessment methods [32]. Furthermore, individual SSLs were reduced iteratively dividing each SSL by a corrective factor (f) until an acceptable cumulative risk was obtained. The reduced SSL (SSL/f), ensuring the respect of both the individual and cumulative risks for each contaminant, represents the cumulative soil screening level (SSL_{cum}) to be compared with the measured concentrations in soil.

4. Results

4.1. Soil and Lake Sediments

Soils represent natural soil reworked artificially; unattached pieces of millimetric to centimetric well rounded or more angular rock fragments are dispersed in a loamy sand matrix. Soil pH is in the range between 4 and 5. Trace elements concentration in Fornaci di Barga soils, including PTE, are given in Table 1. It is observed that in most stations Cu, Zn and Cd exceed the maximum concentration level (MCL) for residential soil use imposed by Italian regulations. In particular, the Ni and Cr content range from 37 to 265 mg/kg and from 55 to 146 mg/kg, respectively; Cu, Zn and Pb range from 35 to 5275 mg/kg, 62 to 27405 mg/kg and 10.6 to 221 mg/kg, respectively; Cd and Sb range from 0.28 mg/kg to 15.8 mg/kg and from 0.20 to 11.7 mg/kg, respectively; As and Hg are in the range from 3.0 mg/kg to 24.5 mg/kg and 0.04 to 2.20 mg/kg, respectively. The areal distribution of Cu, Zn and Cd concentration in soil is shown in Figure 3. Exceedingly high concentrations of pollutants characterize soil in station FB_14S, near the industrial site, suggesting a point source of pollution. The data indicate that most of the collected soils are being polluted at least for one potentially toxic element.

To assess the potential for solubility and release of inorganic pollutant from soil to groundwater, leaching tests were performed on FB_13S and FB_14S soils. The obtained results (Table 1), indicate that soil leachates are characterized by trace metal concentrations below the MCL reported for groundwater by Italian regulations, except Sb and Fe.

Trace element data on Pontecosi lake sediments are reported in Table 2. Sediments are characterized by relatively high Mn content (between 440 and 1090 mg/kg), as observed in basalt-hosted deposits in Apennine ophiolites, and Fe (between 18.8 and 31.2 g/kg); the lack of Fe vs. Mn correlation (not shown) likely reflects the complex cycling of these redox-sensitive trace metals in lacustrine environment [33]. Vanadium, Cr and Ni concentrations are in the range from 23 to 58 mg/kg, 55 to 111 mg/kg and 55 to 96 mg/kg, respectively. Vanadium vs. Cr and Ni vs. Cr show a positive covariance and plot along single, well-defined regression lines (not shown; $R^2 = 0.82$ and $R^2 = 0.86$, respectively) with no distinction between subaerial and lake floor deposits. This suggests that hydraulic processes and grain-size do not significantly influence the sediment composition. Copper, Zn and Pb concentration ranges from 18 to 51 mg/kg, 48 to 59 mg/kg and 6.7 to 20 mg/kg, respectively. Despite some scatter, Pb vs. Zn and Pb vs. Cu correlate linearly (not shown; $R^2 = 0.89$ and $R^2 = 0.73$, respectively), a feature associated to the (Fe)–Cu–Zn–Pb sulfides that characterize ophiolites of the Ligurian Units in northern Apennine, currently exposed north of Castelnuovo di Garfagnana at the bottom of the Serchio Valley [34] (Figure 1).

The Co/Ni ratio is quite constant (Co/Ni = 0.20 $\pm$ 0.01), also resembling what reported for pyrites in serpentinite-hosted deposits of northern Apennine ophiolites. Arsenic ranges between 2.9 and 5.5 mg/kg, likely reflecting a source from pyrite in sulfide ores. These observations indicate that trace metals are supplied to sediments by ophiolite-rich sources. Strontium ranges from 41 to 350 mg/kg, reflecting inputs from the carbonate rocks. Potentially toxic elements in Pontecosi lake sediments are all below the MCL for soils by Italian regulations (see Tables 1 and 2), and can be used to establish reliable natural trace elements (including PTE) concentration in downstream Serchio River sediments and related soils.

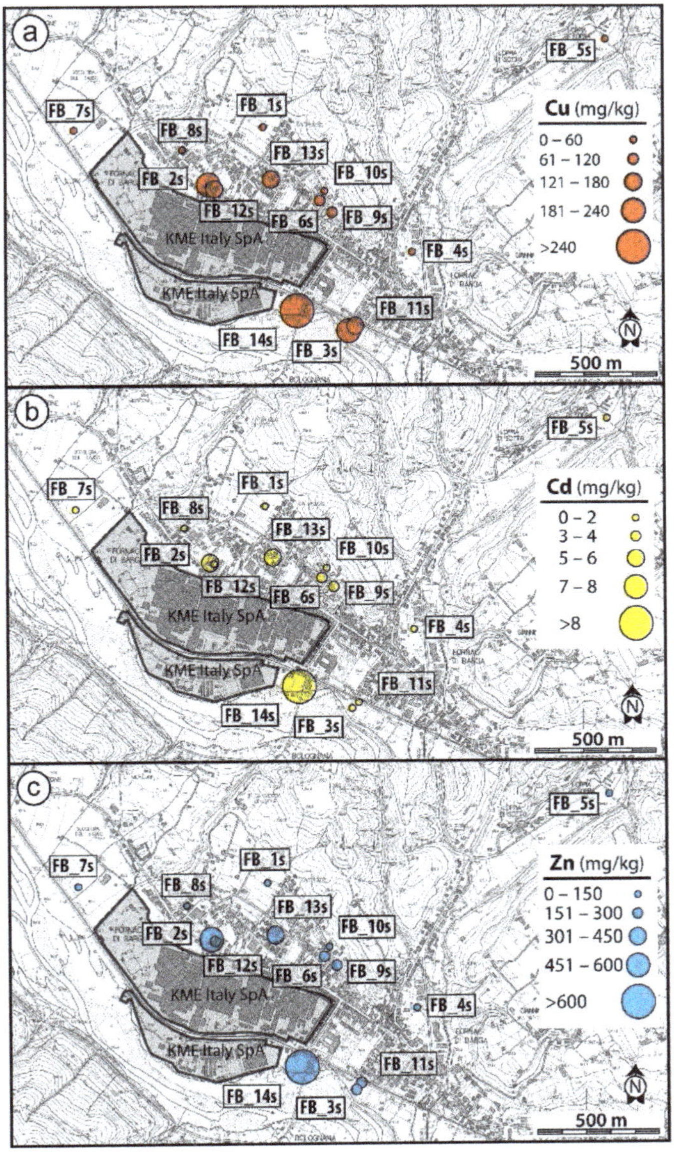

Figure 3. Copper (**a**), Zn (**b**), Cd (**c**) concentration in Fornaci di Barga soil.

4.2. Groundwater

The physico-chemical parameters and major element chemistry measured on groundwater are given in Table 3; trace elements analyses are reported in Table 4. The major ions chemistry is graphically shown in the Piper diagram of Figure 4.

Table 2. Trace elements on Pontecosi lake sediments (mg/kg).

Sample	Li	Be	Mn	Co	Ni	Cu	Zn	Sr	Mo	Ag	Sn	Cd	Sb	Ba	Tl	Pb	Th	U	V	Cr	Fe	As	Hg
PC1	25	0.77	1090	13.4	69	22	50	350	<0.4	0.08	0.19	0.08	<0.07	253	0.08	6.7	3.5	0.48	42	85	29,500	2.6	<0.07
PC2	29	0.76	1050	13.6	68	51	77	162	<0.4	0.08	0.33	0.10	<0.07	115	0.08	8.7	3.4	0.49	42	76	29,600	2.8	<0.07
PC3	28	0.84	710	12.3	66	25	53	131	<0.4	0.06	0.21	0.09	<0.07	65	0.10	9.5	4.0	0.58	38	71	24,900	2.9	<0.07
PC4	27	0.76	930	14.4	78	24	56	173	<0.4	0.08	0.19	0.19	<0.07	81	0.08	8.9	3.5	0.48	47	85	30,400	2.9	<0.07
PC5	27	0.81	630	13.0	64	30	55	129	<0.4	0.08	0.17	0.10	<0.07	72	0.11	9.4	4.3	0.62	41	83	26,800	2.8	<0.07
PC6	29	0.94	710	13.3	72	28	59	131	<0.4	0.06	0.27	0.09	<0.07	78	0.12	9.9	5.7	0.85	47	92	29,600	3.4	<0.07
PC7	30	0.90	670	13.9	70	29	63	125	<0.4	0.07	0.19	0.12	<0.07	81	0.12	11.6	4.8	0.80	44	81	27,300	3.3	<0.07
PC8	36	1.12	810	16.7	84	35	74	149	<0.4	0.08	0.24	0.15	<0.07	97	0.15	14.4	6.4	1.00	51	95	30,900	4.2	<0.07
PC9	30	0.88	720	12.9	70	26	56	123	<0.4	0.12	0.21	0.09	<0.07	60	0.09	9.4	4.5	0.64	38	76	26,400	3.3	<0.07
PC10	28	0.83	600	10.6	55	27	50	115	<0.4	0.08	0.15	0.11	<0.07	75	0.13	10.1	3.6	0.62	36	62	19,900	3.2	<0.07
PC11	35	1.09	820	14.8	74	37	69	166	<0.4	0.09	0.37	0.15	<0.07	94	0.16	13.9	5.1	0.86	49	88	28,400	3.8	<0.07
PC12	42	1.34	650	16.1	75	37	89	100	<0.4	0.07	0.21	0.16	<0.07	104	0.21	16.9	5.6	0.80	43	86	30,900	4.3	<0.07
PC13	25	0.76	440	10.1	59	18	48	41	<0.4	<0.03	0.24	0.08	<0.07	51	0.09	8.3	5.1	0.82	29	70	18,800	2.7	<0.07
PC14	41	1.43	830	17.0	96	37	86	88	<0.4	0.06	0.49	0.23	<0.07	137	0.23	20.0	7.0	1.48	58	111	31,200	5.5	<0.07
PC15	27	0.84	730	12.6	65	26	53	136	<0.4	0.06	0.19	0.09	<0.07	62	0.09	7.5	4.5	0.55	42	76	27,900	2.9	<0.07

Table 3. The physico-chemical parameters and major ions chemistry (mg/L) measured on groundwater.

Sample	Depth	T_w (°C)	T_{air} (°C)	DO (mg/L)	EC (mS)	pH	Na^+	K^+	Mg^{2+}	Ca^{2+}	F^-	Cl^-	NO_3^-	SO_4^{2-}	HCO_3^-
FB_1W	50	14.0	30.2	3.5	293	7.3	11	1.2	6.4	60	<0.2	8.4	5.8	9.1	210
FB_2W	8	16.6	30.2	2.4	391	6.6	15	1.1	9.6	58	0.3	16	32	17	183
FB_3W	7	16.4	30.2	2.0	296	6.9	9.9	1.0	16	40	0.3	9.7	3.2	14	207
FB_4W	18	16.2	30.2	1.0	432	7.3	11	1.0	16	84	<0.2	11	1.5	18	332
FB_5W	21	14.9	30.2	3.8	266	6.9	9.0	1.0	5.6	43	<0.2	8.4	7.5	8.8	143
FB_6Wa	2	17.0	15.4	6.1	520	7.6	39	3.5	10	68	n.d.	42	n.d.	132	159

n.d.: not determined.

Table 4. Trace elements in groundwater (μg/L) together with the maximum concentration level (MCL, μg/L) imposed by Italian regulations (Legislative Decree No. 152/2006 approving the Code on the Environment). Values in bold exceed the concentration threshold for groundwater.

Samples	Li	Be	Mn	Co	Ni	Cu	Zn	Sr	Mo	Ag	Sn	Cd	Sb	Ba	Tl	Pb	Th	U	V	Cr	Fe	As
FB_1W	1.46	<0.01	5.5	0.12	3.8	2.0	<7	166	0.20	<0.02	<0.1	0.02	0.16	39	0.01	<0.2	0.006	0.17	0.23	0.23	53	0.14
FB_2W	0.26	<0.01	1.06	0.18	3.7	2.4	9	156	<0.2	<0.02	<0.1	0.02	0.04	28	<0.006	0.34	<0.006	0.11	0.58	0.55	50	0.10
FB_3W	4.1	<0.01	**306**	0.11	2.6	2.2	14	130	<0.2	<0.02	<0.1	0.03	0.04	4.7	0.01	<0.2	<0.006	0.19	0.21	<0.1	60	0.10
FB_4W	2.7	<0.01	10.6	0.18	4.8	<1	41	189	<0.2	<0.02	<0.1	0.01	0.04	13.2	<0.006	<0.2	<0.006	0.27	<0.1	0.11	123	0.09
FB_5W	1.15	<0.01	0.50	0.08	2.4	3.4	77	131	<0.2	<0.02	<0.1	0.02	0.05	22	<0.006	<0.2	<0.006	0.14	0.20	0.21	34	0.05
FB_6Wa	35	<0.01	**199**	0.20	4.4	2.4	<7	570	1.59	0.03	0.11	0.04	0.18	48	0.14	<0.18	<0.006	0.33	0.48	<0.12	7.0	0.39
FB_6Wb	35	<0.01	**155**	0.35	4.4	1.10	<7	552	1.76	<0.02	<0.1	0.03	0.20	52	0.01	<0.18	<0.006	0.39	0.51	<0.12	9.0	0.47
MCL		4	**50**	50	20	1000	3000					5	5		2	10				50	200	10

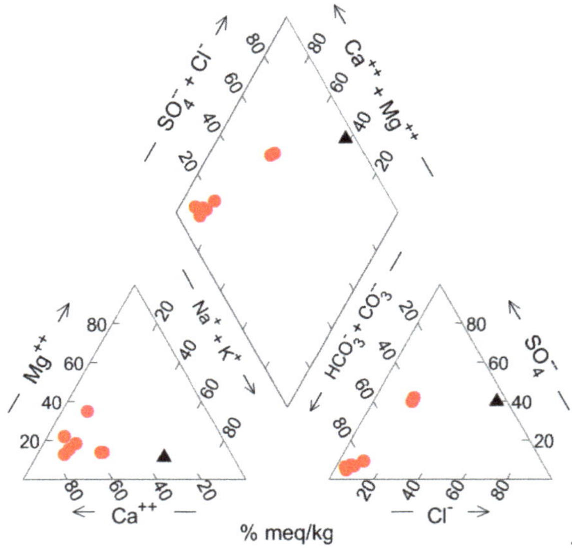

Figure 4. Piper diagram for groundwater samples. Red circles: FB_1W to FB_6W wells and piezometer; black filled triangle: deep water (see text).

The diamond-shaped field in the Piper diagram classifies groundwater as belonging to the Ca-HCO_3 type; it has however to be noted that the water samples FB_6Wa,b are shifted toward SO_4^{2-}, Cl^- and Na^+ enrichments, a feature that characterizes the end-member thermal waters outflowing in the area and interpreted as reflecting a deep circulation in Triassic carbonate-evaporite aquifers [35].

Trace element contaminants in groundwater have concentration below the MCL reported by Italian regulations, except Mn that exceed the maximum concentration value of 50 μg/L in FB_3W and FB_6W.

4.3. Risk Analysis

Health risk assessment was calculated for Cu, Zn and Cd (Table 5), exceeding the MCL for soil (Table 1) imposed by Italian regulations. The highest measured concentrations were precautionary used (Table 1) due to the small sample size [36]. Sample FB_14S, interpreted as a primary source of contamination, was not included in calculations.

Table 5. Summary of Hazard Quotient (HQ) for non-carcinogenic risk calculated for each contaminant and for direct exposure routes, for children and adults receptor (in brackets values for adult when different). The individual (HI_i) and cumulative (HI_{cum}) Hazard Index are also reported.

HQ	Soil Ingestion	Dermal Contact	HI_i
Cadmium $RfD_{oral} = 1 \times 10^{-4}$ mg/kg/day	0.857 (9.18×10^{-2})	2.40×10^{-3} (3.66×10^{-4})	0.859 (9.22×10^{-2})
Copper $RfD_{oral} = 4 \times 10^{-2}$ mg/kg/day	7.48×10^{-2} (8.01×10^{-3})	2.09×10^{-3} (3.20×10^{-4})	7.69×10^{-2} (8.33×10^{-3})
Zinc $RfD_{oral} = 3 \times 10^{-1}$ mg/kg/day	1.98×10^{-2} (2.12×10^{-3})	5.55×10^{-4} (8.47×10^{-5})	2.04×10^{-2} (2.21×10^{-3})
Sum of individual hazard quotients for each route considered	0.952 (0.102)	5.05×10^{-3} (7.71×10^{-4})	HI_{cum} (all contaminants) 0.956 (0.103)

The obtained HI_{cum} (0.956, Table 5) was only slightly lower than the acceptance threshold. The calculated individual and cumulative soil screening levels (SSLs,) are given in Table 6. It can be observed that the obtained screening values are very close to the maximum concentrations measured in soil.

Table 6. Summary of individual SSLs for Cadmium, Copper and Zinc respectively calculated for direct exposure routes, for children and adults receptor (in brackets values for adult). * Calculated according to USEPA [32]; ** SSLs recalculated taking into account the simultaneous presence of all contaminants (applying a corrective factor, *f*).

Individual SSL	Soil Ingestion	Dermal Contact	Sum of Direct Exposures *	*f*	Cumulative SSL **
Cadmium C_{max} = 6.7 mg/kg	7.82 (7.30×10^1)	2.79×10^3 (1.83×10^4)	7.80 (7.27×10^1)	1.11	7.03 (6.55×10^1)
Copper C_{max} = 234 mg/kg	$3{,}13 \times 10^3$ (2.92×10^4)	$1{,}12 \times 10^5$ (7.32×10^5)	3.04×10^3 (2.81×10^4)	13	2.34×10^2 (2.16×10^3)
Zinc C_{max} = 465 mg/kg	2.35×10^4 (2.19×10^5)	8.38×10^5 ($>10^6$)	2.28×10^4 (2.11×10^5)	49	4.66×10^2 (4.30×10^3)

5. Discussion

Potentially toxic elements data indicate that Fornaci di Barga soils are variably contaminated by Cu, Zn, Cd, exceeding the thresholds for residential use. It must be noted that these elements in sample FB_5S, a station in a rural setting, do not exceed the MCL for soil, suggesting that contamination is confined to the urbanized environment, in particular in proximity of the metallurgical plant. Soil in station FB_14S, near the industry (Figure 2a), deviates towards the highest level of pollution, including Ni, Sb, Pb, As and Hg in addition to Cu, Zn and Cd. These data suggest that contaminants enter the environment from a confined space, possibly representing the legacy of waste storage. The extent of this polluted area suspected to be hazardous must be investigated and clean-up procedures for remediation should be planned.

In order to characterize natural background concentrations of PTE in soils, the Pontecosi lake sediments unaffected by pollution have been used as reference conditions. In particular, vanadium has been intended as an indicator for ophiolitic sources. It is observed that Fornaci di Barga soils and Pontecosi lake sediments have similar patterns for Cr, Fe, Be, Ba, Ni, As vs. V (Figure 5), suggesting common lithogenic sources. On the contrary, deviations towards higher concentrations compared with natural values are observed for Cu, Zn, Cd and Pb vs. V (Figure 5), even if Pb does not exceed the maximum concentration level for residential soils.

The observed trends confirm the input of Cu, Zn and Cd from anthropogenic sources. To assess the contribution of anthropogenic deposition on surface soil, enrichment factor (EF) for Cu, Zn and Cd, exceeding the maximum concentration level, has been calculated using iron as a reference element and the relationship:

$$EF = \left(\frac{Me}{Fe}\right)_{sample} \Big/ \left(\frac{Me}{Fe}\right)_{UCC}$$

where *Me* represent the metal concentration and *UCC* the upper continental crust [37]. The results indicate an extremely high enrichment for Cd (EF > 40) and moderate to significant enrichments for Zn and Cu (2 < EF < 5 and 5 < EF < 10, respectively). Soil pollution has been also evaluated by using the geo-accumulation index (I_{geo}) calculated by using the Müller relationship [38] given by:

$$I_{geo} = log_2\left(\frac{Cn}{1.5Bn}\right)$$

where *Cn* represents the element concentration in soil and Bn the geogenic background level for individual element taken as the concentration measured in the Pontecosi lake sediments. The obtained results highlight that most soil samples belong to Class 6 ($I_{geo} > 5$, extremely contaminated soil) for Cu, Zn, Cd. Similar results are obtained by using background concentration values for sediments representative of the Earth's crust [37,39].

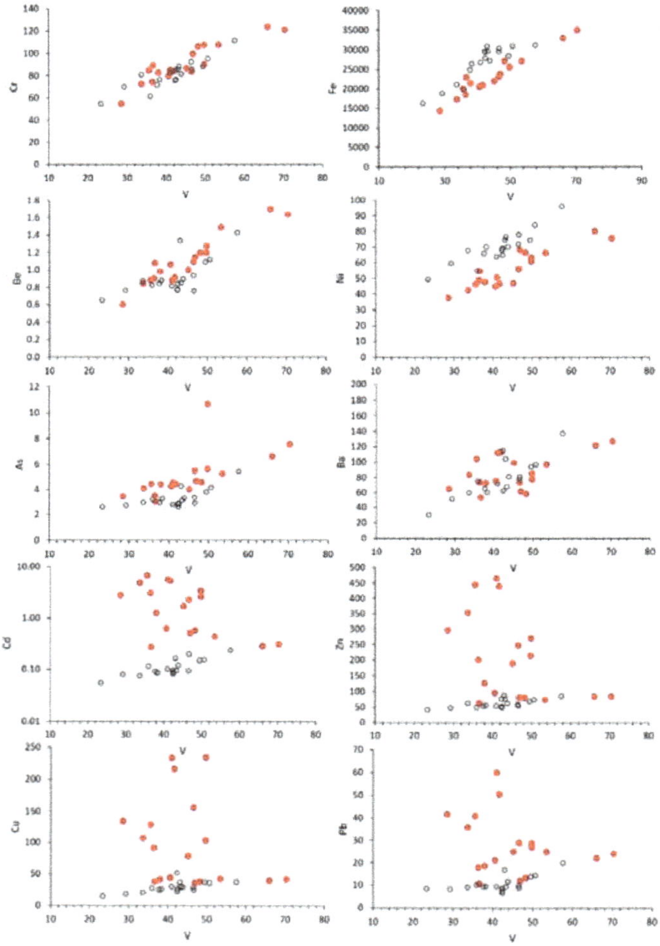

Figure 5. Cr, Fe, Be, Ba, Ni, As, Cu, Zn, Cd and Pb concentrations (mg/kg) vs. V (mg/kg) in Fornaci di Barga soils (filled circles) and Pontecosi lake sediments (open circles).

The calculated SSL concentration values, that are deemed protective of receptors that come into contact with soil, reveal that the highest measured Cu, Zn and Cd concentration in Fornaci di Barga soil (FB_3S, FB_2S and FB_13S stations, respectively) should be considered of potential toxicological concern. Risk assessment indicates that the major threat for health is associated to Cd through soil ingestion pathway by children. Dermal contact pathway for Cu, Zn and Cd does not rise potential adverse health effects.

The local, long-lasting metallurgical activity might be responsible for Cu, Zn, Pb and Cd emission; in particular, gaseous Cd volatilizes and condense forming particulate [40,41]. Road traffic also represents an emission source for trace metal loading on soil [42,43]; in particular, Cu is released during vehicle brake abrasion, Zn by automotive tire wear, Pb mostly represents the legacy of burned leaded gasoline [44]. It has to be noted that the highest concentration of PTE has not been measured in roadside environments, suggesting that the major contribution is from industrial activity. However, additional studies are necessary to unequivocally identify the main anthropogenic sources of pollutant emission in the area.

Soil profiles show that Cu, Zn and Cd concentration gradually decreases with increasing depth, confirming that the anthropogenic contribution correlates with local air active emission sources. Furthermore, experiments indicate that Cu, Zn and Cd concentrations in soil leachates (Table 1) were significantly higher than those measured in groundwater samples (Table 4), suggesting that soils have the potential to release contaminants according to the principle that elements of accumulated anthropogenic nature have the tendency to be mobile in soil when compared with those originated from lithogenic sources [45]. Once below the water table, contaminants may be subject to dispersion and diffusion, with different flowpaths and velocities [46], possibly extending groundwater contamination. The geochemical data on samples FB_6Wa,b show a relatively high Li, Ba, Cl, Na and SO_4 concentration (Tables 3 and 4), interpreted as the evidence of a deep-water component in evaporitic-carbonatic hydrogeological structures rising to the surface through fault systems [35]. The linear correlation in the binary Na vs. Cl diagram (Figure 6) highlights the contribution of such deep component to the shallower groundwater in alluvial deposits. In particular, the FB_6W water chemistry may be simulated using the PHREEQC Code [47] by 4% mixing of the deep water (represented by Pieve di Fosciana outflows [35] (Figure 6) with modelled Ca-HCO_3 water in equilibrium with calcite and undersaturated with dolomite (saturation index = −0.5) in a system closed to CO_2. These observations suggest that mixing-induced transport might occur in cases when soil leachates reach saturated strata, and contaminants might in this case travel over variable distances from the discharge point. Groundwater monitoring should hence be carried out regularly.

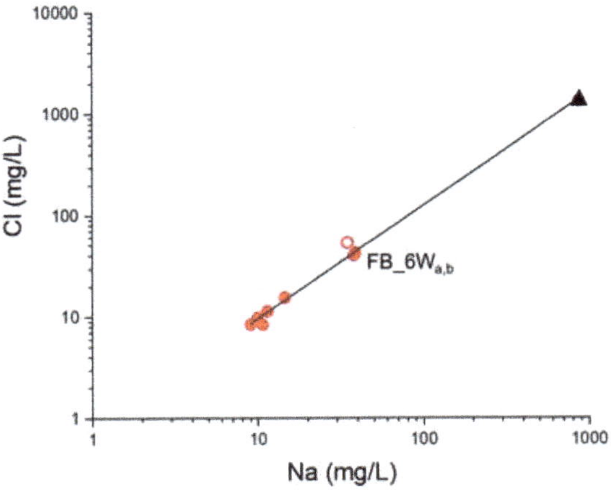

Figure 6. Na (mg/L) vs. Cl (mg/L) diagram with the linear fitting (black line). Red filled circles: groundwater from FB_1W to FB_6W; red open circle: modelled FB_6W water (see text); black filled triangle: deep water (see text).

6. Conclusions

The results obtained in this study indicate that Cu, Zn and Cd, likely originated by deposition of airborne particulate matter released during the historical and long-lasting metallurgical activity that characterized the Fornaci di Barga area, represent the main contaminants in surface soil.. Even low-level environmental exposure to cadmium is a risk factor for lung cancer [48,49] and may potentiate the effects diabetes on the kidney and the chronic kidney disease [50–52]. Risk assessment indicates that cadmium may contribute to adverse health outcomes for children by direct soil ingestion exposure routes. The calculated general soil screening level closely approaches the highest measured contaminant concentration in soil. Urban groundwater does not show evidence of trace element pollution, suggesting that the migration of contaminants from soil to subsurface is a slow process. However, the hydrologic interconnection between shallow and deep aquifers through fault zones requires a network monitoring.

These results represent the first evidence of the links between space clusters of diseases, soil contamination and risk assessment in the densely populated area of the Middle Serchio River Valley, and provide the scientific insights to policy makers and stakeholders for future-oriented solutions.

Author Contributions: All the authors contributed to the investigation. Conceptualization, R.P. and L.G. (Lisa Ghezzi); formal analysis, L.G. (Lisa Ghezzi) and L.G. (Lisa Genovesi); software, S.A.; writing—original draft preparation, R.P., L.G. (Lisa Ghezzi), S.A., C.F. and L.P.; supervision, R.P. All authors have read and agreed to the published version of the manuscript.

Funding: This research received no external funding.

Institutional Review Board Statement: Not applicable.

Informed Consent Statement: Not applicable.

Data Availability Statement: Not applicable.

Acknowledgments: The Mayor of Barga municipality and the City Hall staff provided the logistic support for site selection and sampling; we also received considerable help from the citizen's Committee "La Libellula"; Annibale Biggeri and co-workers introduced us to the research agenda on environmental epidemiology in the study area. Roberto Balatri and Roberto Giannecchini contributed with helpful discussions and are sincerely acknowledged.

Conflicts of Interest: The authors declare no conflict of interest.

References

1. Gómez-Baggethun, E.; Barton, D.N. Classifying and valuing ecosystem services for urban planning. *Ecol. Econ.* **2013**, *86*, 235–245. [CrossRef]
2. Novák, T.J.; Balla, D.; Kamp, J. Changes in anthropogenic influence on soils across Europe 1990-2018. *Appl. Geogr.* **2020**, *124*, 102294. [CrossRef]
3. De Kimpe, C.R.; Morel, J.-L. Urban soil management: A growing concern. *Soil Sci.* **2000**, *165*, 31–40. [CrossRef]
4. Yang, J.L.; Zhang, G.L. Formation, characteristics and eco-environmental implications of urban soils–A review. *Soil Sci. Plant Nutr.* **2015**, *61*, 30–46. [CrossRef]
5. Sager, M. Urban soils and road dust—Civilization effects and metal pollution—A review. *Environments* **2020**, *7*, 98. [CrossRef]
6. Pickett, S.T.A.; Cadenasso, M.L. Altered resources, disturbance, and heterogeneity: A framework for comparing urban and non-urban soils. *Urban Ecosyst.* **2009**, *12*, 23–44. [CrossRef]
7. Müller, A.; Österlund, H.; Marsalek, J.; Viklander, M. The pollution conveyed by urban runoff: A review of sources. *Sci. Total Environ.* **2020**, *709*, 136125. [CrossRef] [PubMed]
8. Ajmone-Marsan, F.; Biasioli, M. Trace elements in soils of urban areas. *Water Air Soil Pollut.* **2010**, *213*, 121–143. [CrossRef]
9. Simon, E.; Vidic, A.; Braun, M.; Fábián, I.; Tóthmérész, B. Trace element concentrations in soils along urbanization gradients in the city of Wien, Austria. *Environ. Sci. Poll. Res.* **2013**, *20*, 917–924. [CrossRef] [PubMed]
10. Poggio, L.; Vrščaj, B.; Schulin, R.; Hepperle, E.; Ajmone Marsan, F. Metals pollution and human bioaccessibility of topsoils in Grugliasco (Italy). *Environ. Pollut.* **2009**, *157*, 680–689. [CrossRef]
11. Li, G.; Sun, G.-X.; Ren, Y.; Luo, X.-S.; Zhu, Y.-G. Urban soil and human health: A review. *Eur. J. Soil. Sci.* **2018**, *69*, 196–215. [CrossRef]
12. Davis, S.; Mirik, D. Soil ingestion in children and adults in the same family. *J. Expo. Sci. Environ. Epidemiol.* **2006**, *16*, 63–75. [CrossRef]
13. Różański, S.L.; Peñas Castejón, J.M.; McGahan, D.G. Child risk assessment of selected melal(loid)s from urban soils using in vitro UBM procedure. *Ecol. Indic.* **2021**, *127*, 107726. [CrossRef]
14. Biggeri, A.; Marchi, M. Case-control designs for the detection of space clusters of diseases. *Environmetrics* **1995**, *6*, 385–393. [CrossRef]
15. De Marchi, B.; Ficorilli, A.; Biggeri, A. Research in the air in Valle del Serchio. *Futures* **2022**, *137*, 102906. [CrossRef]
16. Froeling, F.; Gignac, F.; Hoek, G.; Vermeulen, R.; Nieuwenhuijsen, M.; Ficorilli, A.; De Marchi, B.; Biggeri, D.; Robinson, J.; Righi, V.; et al. Narrative review of citizen science in environmental epidemiology: Setting the stage for co-created research projects in environmental epidemiology. *Environ. Int.* **2021**, *152*, 106470. [CrossRef] [PubMed]

17. Carmignani, L.; Kligfield, R. Crustal extension in the northern Apennines: The transition from compression to extension in the Alpi Apuane core complex. *Tectonics* **1990**, *9*, 1275–1303. [CrossRef]
18. Carosi, R.; Frassi, C.; Montomoli, C.; Pertusati, P.C. Structural evolution of the Tuscan Nappe in the southeastern sector of the Apuan Alps metamorphic dome (Northern Apennines, Italy). *Geol. J.* **2005**, *40*, 103–119. [CrossRef]
19. Fellin, M.G.; Reiners, P.W.; Brandon, M.T.; Wuthrich, E.; Balestrieri, M.L.; Molli, G. Thermo-chronologic evidence of exhumational history of the Alpi Apuane metamorphic core complex, northern Apennines, Italy. *Tectonics* **2007**, *26*, TC6015. [CrossRef]
20. Frassi, C.; Ottria, G.; Ferdeghini, A. Fault tectonics of the Tuscan Nappe in the eastern sector of the Apuan Alps (Italy). *J. Maps* **2020**, *16*, 745–754. [CrossRef]
21. Landi, E.; Ravani, S.; Sarti, G.; Sodini, M. The Villafranchian deposits of the Castelnuovo Garfagnana and Barga Basins (Lucca, Tuscany, Italy): Facies analysis and paleoenvironmental reconstruction. *Atti Soc. Tosc. Sci. Nat. Mem.* **2003**, *108*, 81–93.
22. Perilli, N.; Puccinelli, A.; Sarti, G.; Avanzi, G.D.A. Villafranchian deposits of the Barga and Castelnuovo Garfagnana basins (Tuscany, Italy): Lithostratigraphy and sedimentary features. *Quaternario* **2004**, *17*, 313–322.
23. Coltorti, M.; Pieruccini, P.; Rustioni, M. The Barga basin (Tuscany): A record of Plio-Pleistocene mountain building of the northern Apennines, Italy. *Quat. Intern.* **2008**, *189*, 56–70. [CrossRef]
24. ASTM. *Standard Provisional Guide for Risk-Based Corrective Action*; Report E 2081-00; American Society for Testing Materials: West Conshohocken, PA, USA, 2000.
25. US EPA. *Soil Screening Guidance: Technical Background Document*; EPA/540/R-95/128; Office of Emergency and Remedial Response, US Environmental Protection Agency: Washington, DC, USA, 1996.
26. Ghezzi, L.; Arrighi, S.; Giannecchini, R.; Bini, M.; Valerio, M.; Petrini, R. The legacy of mercury contamination from a past leather manufacturer and health risk assessment in an urban area (Pisa Municipality, Italy). *Sustainability* **2022**, *14*, 4367. [CrossRef]
27. USDoE. RAIS: Risk Assessment Information System. 2010. University of Tennessee. Available online: http://rais.ornl.gov (accessed on 1 June 2022).
28. US EPA. Regional Screening level (RSLs)—Generic Tables. US Environmental Protection Agency 2022. Washington, DC. Available online: https://www.epa.gov/risk/regional-screening-levels-rsls-generic-tables (accessed on 1 June 2022).
29. USTCEQ. Texas Commission of Environmental Quality. 2022. Current PCL Tables. Available online: www.tceq.texas.gov/remediation/trrp/trrppcls.html (accessed on 1 June 2022).
30. US EPA. *Risk Assessment Guidance for Superfund. Volume I. Human Health Evaluation Manual, Part A*; EPA/540/1-89/002; Office of Emergency and Remediation Response, US Environmental Protection Agency: Washington, DC, USA, 1989.
31. US EPA. *Supplemental Guidance for Developing Soil Screening Levels for Superfund Sites*; OSWER 9355.4-24; Office of Emergency and Remediation Response, US Environmental Protection Agency: Washington, DC, USA, 2002.
32. US EPA. Regional Screening level (RSLs)—Equations. US Environmental Protection Agency. 2022. Washington, DC. Available online: https://www.epa.gov/risk/regional-screening-levels-rsls-equations#res (accessed on 1 June 2022).
33. Kuznetsov, S.I. The Iron and Manganese Cycles in Lakes. In *The Microflora of Lakes and Its Geochemical Activity*; Oppenheimer, C.H., Ed.; University of Texas Press: Austin, TX, USA, 2021; pp. 365–407.
34. Zaccarini, F.; Garuti, G. Mineralogy and chemical composition of VMS deposits on northern Apennine ophiolites, Italy: Evidence for the influence of country rock type on ore composition. *Mineral. Petrol.* **2008**, *94*, 61–83. [CrossRef]
35. Gherardi, F.; Pierotti, L. The suitability of the Pieve di Fosciana hydrothermal system (Italy) as a detection site for geochemical seismic precursors. *Appl. Geochem.* **2018**, *92*, 166–179. [CrossRef]
36. US EPA. *Supplemental Guidance to RAGS: Calculating the Concentration Term*; Office of Solid Waste and Emergency Response. US Environmental Protection Agency: Washington, DC, USA, 1992.
37. Rudnick, R.ç.; Gao, S. The Composition of the Continental Crust. In *Treatise on Geochemistry*; Holland, H.D., Turekian, K.K., Eds.; The Crust; Elsevier-Pergamon: Oxford, UK, 2003; Volume 3, pp. 1–64.
38. Müller, G. Die Schwermetallbelastung der sedimente des Neckars und seiner Nebenflusse: Eine Bestandsaufnahme. *Chem. Ztg.* **1981**, *105*, 157–164.
39. Turekian, K.K.; Wedepohl, K.H. Distribution of the elements in some major units of the Earth's crust. *Geol. Soc. Am. Bull.* **1961**, *72*, 175–192. [CrossRef]
40. Zhu, Z.; Huang, Y.; Zha, J.; Yu, M.; Liu, X.; Li, H.; Zhu, X. Emission and retention of cadmium during combustion of contaminated biomass with mineral additives. *Energy Fuels* **2019**, *33*, 12508–12517. [CrossRef]
41. Lau, L.L.; Strezov, V.; Gonçalves, M.V.B.; Bagatini, M.V. Trace elements emission in iron ore sintering: A review. *Environ. Adv.* **2021**, *6*, 100123. [CrossRef]
42. Davis, A.P.; Shokouhian, M.; Ni, S. Loading of lead, copper, cadmium, and zinc in urban runoff from specific sources. *Chemosphere* **2001**, *44*, 997–1009. [CrossRef]
43. Nabulo, G.; Oryem-Origa, H.; Diamong, M. Assessment of lead, cdmium, and zinc contamination of roadside soils, surface filma, and vegetables in Kampala City, Uganda. *Environ. Res.* **2006**, *101*, 42–52. [CrossRef] [PubMed]
44. Wang, Z.; Wade, A.M.; Richter, D.D.; Stapleton, H.M.; Kaste, J.M.; Vengosh, A. Legacy of anthropogenic lead in urban soils: Co-occurrence with matal(loids) and fallout radionuclides, isotopic fingerprinting, and in vitro bioaccessibility. *Sci. Total Environ.* **2022**, *806*, 151276. [CrossRef] [PubMed]
45. Kuo, S.; Heilman, P.E.; Baker, A.S. Distribution and forms of copper, zinc, cadmium, iron, and manganese in soils near a copper smelter. *Soil Sci.* **1983**, *135*, 101–109. [CrossRef]

46. Violante, A.; Cozzolino, V.; Perelemov, L.; Caporale, A.G.; Pigna, M. Mobility and bioavailability of heavy metals and metalloids in soil environments. *J. Soil Sci. Plant. Nutr.* **2010**, *10*, 268–292. [CrossRef]
47. Parkhurst, D.L.; Appelo, C.A.J. *User's Guide to PHREEQC—A Computer Program for Speciation, Batch-Reaction, One-Dimensional Transport, and Inverse Geochemical Calculations*; U.G. Geological Survey Report 99-4250; USGS: Washington DC, USA, 1999.
48. Waalkes, M.P. Cadmium Carcinogenesis. *Mutat. Res.* **2003**, *533*, 107–120. [CrossRef] [PubMed]
49. Nawrot, T.S.; Martens, D.S.; Hara, A.; Plusquin, M.; Vangronsveld, J.; Roels, H.S.; Staessen, J.A. Association of total cancer and lung cancer with environmental exposure to cadmium: The meta-analytical evidence. *Cancer Causes Control* **2015**, *26*, 1281–1288. [CrossRef]
50. Ferraro, P.M.; Costanzi, S.; Naticchia, A.; Sturniolo, A.; Gambaro, G. Low level exposure to cadmium increases the risk of chronic kidney disease: Analysis of the NHANES 1999-2006. *BMC Public Heath* **2010**, *10*, 304. Available online: http://www.biomedcentral.com/1471-2458/10/304 (accessed on 1 June 2022). [CrossRef]
51. Nordberg, G.F.; Bernard, A.; Diamond, G.L.; Duffus, J.H.; Illing, P.; Nordberg, M.; Bergdahl, I.A.; Skerfving, S. Risk assessment of effects of cadmium on human heakth (IUPAC Thechnical Report). *Pure Appl. Chem.* **2018**, *90*, 755–808. [CrossRef]
52. Wang, X.; Cui, W.; Wang, M.; Liang, Y.; Zhu, G.; Jin, T.; Chen, X. The association between life-time dietary cadmium intake from rice and chronic kidney disease. *Ecotoxicol. Environ. Saf.* **2021**, *211*, 111933. [CrossRef]

International Journal of
Environmental Research and Public Health

MDPI

Article

The Potential Key Role of the NRF2/NQO1 Pathway in the Health Effects of Arsenic Pollution on SCC

Qianlei Yang [1,†], Rui Yan [1,†], Yuemei Mo [2,†], Haixuan Xia [1], Hanyi Deng [3], Xiaojuan Wang [1], Chunchun Li [4], Koichi Kato [5], Hengdong Zhang [6,7], Tingxu Jin [1,8,*], Jie Zhang [1] and Yan An [1,*]

1 Department of Toxicology, School of Public Health, Jiangsu Key Laboratory of Preventive and Translational Medicine for Geriatric Diseases, Medical College of Soochow University, Suzhou 215123, China; qlyang@suda.edu.cn (Q.Y.); 15895572995@163.com (R.Y.); 20184047006@stu.suda.edu.cn (H.X.); 20214047009@stu.suda.edu.cn (X.W.); zhangjie_78@suda.edu.cn (J.Z.)
2 Physical Examination Department, Center for Disease Control and Prevention of Suzhou Industrial Park, Suzhou 215100, China; 0219501041@163.com
3 Shanghai Municipal Center for Disease Control and Prevention, Shanghai 200336, China; dhysuda@126.com
4 Changzhou Wujin District Center for Disease Control and Prevention, Changzhou 213164, China; 20144247028@stu.suda.edu.cn
5 Laboratory of Environmental Toxicology and Carcinogenesis, School of Pharmacy, Nihon University, Chiba 274-8555, Japan; kato.koichi@nihon-u.ac.jp
6 Department of Occupational Disease Prevention, Jiangsu Provincial Center for Disease Control and Prevention, Nanjing 210028, China; hdzhangjscdc@yeah.net
7 Jiangsu Preventive Medicine Association, Nanjing 210009, China
8 School of Public Health, The Key Laboratory of Environmental Pollution Monitoring and Disease Control, Ministry of Education, Guizhou Medical University, Guiyang 550025, China
* Correspondence: lanting8310@sina.com (T.J.); dranyan@126.com (Y.A.)
† These authors contributed equally to this work.

Citation: Yang, Q.; Yan, R.; Mo, Y.; Xia, H.; Deng, H.; Wang, X.; Li, C.; Kato, K.; Zhang, H.; Jin, T.; et al. The Potential Key Role of the NRF2/NQO1 Pathway in the Health Effects of Arsenic Pollution on SCC. *Int. J. Environ. Res. Public Health* **2022**, *19*, 8118. https://doi.org/10.3390/ijerph19138118

Academic Editor: Cheng Yan

Received: 27 May 2022
Accepted: 29 June 2022
Published: 1 July 2022

Abstract: Arsenic is widely present in nature and is a common environmental poison that seriously damages human health. Chronic exposure to arsenic is a major environmental poisoning factor that promotes cell proliferation and leads to malignant transformation. However, its molecular mechanism remains unclear. In this study, we found that arsenite can promote the transformation of immortalized human keratinocyte cells (HaCaT) from the G0/G1 phase to S phase and demonstrated malignant phenotypes. This phenomenon is accompanied by obviously elevated levels of NRF2, NQO1, Cyclin E, and Cyclin-dependent kinase 2 (CDK2). Silencing the NRF2 expression with small interfering RNA (siRNA) in arsenite-transformed (T-HaCaT) cells was shown to reverse the malignant phenotype. Furthermore, the siRNA silencing of *NQO1* significantly decreased the levels of the cyclin E-CDK2 complex, inhibiting the G0/G1 to S phase cell cycle progression and transformation to the T-HaCaT phenotypes. Thus, we hypothesized that the NRF2/NQO1 pathway played a key role in the arsenite-induced malignancy of HaCaT cells. By increasing the expression of Cyclin E-CDK2, the NRF2/NQO1 pathway can affect cell cycle progression and cell proliferation. A new common health effect mechanism of arsenic carcinogenesis has been identified; thus, it would contribute to the development of novel treatments to prevent and treat skin cancer caused by arsenic.

Keywords: arsenite pollution; health effect; squamous cell carcinoma; NRF2/NQO1 pathway; cell proliferation; malignant transformation

1. Introduction

As the twentieth element of the Earth's crust, arsenic is most commonly found in the form of sulfides in nature. Arsenic is released into the environment from natural sources as a result of natural phenomena, such as dust storms, volcanic eruptions, geothermal/hydrothermal activity, and forest fires [1]. Arsenic can interact with oxygen and other molecules of nature to form different compounds. Therefore, it is extremely commonplace in the daily life of humans [2]. Arsenic is widely used in the manufacture of industrial and agricultural products such as semiconductors, glassware, alloys, wood preservatives, and pesticides and in medical treatments for diseases such as syphilis, rickets, amoebic dysentery, and leukemia [3,4]. These human activities also have the potential to cause severe arsenic pollution in the environment. Arsenic contamination of drinking water and soil has become an international and regional problem over the past two decades. A survey of arsenic concentrations in groundwater in eastern Wisconsin, United States, found naturally occurring arsenic concentrations in groundwater of more than 12,000 μg/L, which is much higher than the U.S. Environmental Protection Agency (EPA) and WHO standard/baseline concentrations for arsenic in drinking water [5]. In other countries and regions, such as Argentina, Bangladesh, Chile, China, Mexico, India, Thailand, and Taiwan, the health effects of arsenic contamination in groundwater have become one of the major public health problems [3,6].

Arsenic is a common environmental toxin with toxic and teratogenic effects in many metazoans [7–9]. As water is the primary polluting medium for arsenic, aquatic organisms bear the brunt of arsenic [10–12]. In addition, humans have the highest risk of arsenic contamination due to the food chain. It is estimated that nearly hundreds of millions of people are exposed to the threat of arsenic compounds in drinking water and food worldwide [13–15]. The main source of human arsenic intake comes from water and food, with an average daily intake of approximately 50 μg [3]. The human arsenic intake from air, water, and soil is usually much smaller. However, these environmental media also deserve particular attention in areas contaminated with arsenic. In certain workplaces where arsenic compounds are produced and used, workers may be exposed to greater concentrations of arsenic than in the natural environment [16–18]. Thus, arsenic exposure is considered the leading cause of global health problems today. In the Indian subcontinent, for instance, exposure to arsenic and having related diseases have reached alarming levels, where nearly 359 million people may be at risk [19]. Long-term exposure to arsenic can lead to a number of adverse health effects, including hyperkeratosis, jaundice, and vascular diseases [20]. However, the carcinogenic role of arsenic has brought more significant concerns in recent years, as arsenic-related cancers can be diagnosed in the skin, lungs, liver, and other malignant tumors [21–24]. Arsenic is now classified as a Group A carcinogen by the EPA and a Group I carcinogen by the International Agency for Research on Cancer (IARC). Among the cancers caused by arsenic, skin cancer was the first one that has been confirmed. As early as the late 1880s, White et al. reported one case of skin cancer caused by psoriasis treated with potassium arsenite [25]. Currently, skin cancer is commonly referred to as skin melanoma (MC) and non-melanoma skin cancer (NMSC), the latter being composed of basal cell carcinoma (BCC) and squamous cell carcinoma (SCC) [26]. Numerous epidemiological studies have shown that apart from ultraviolet (UV) radiation, arsenic is the most significant chemical damage factor related to SCC and BCC [27–30]. Unfortunately, the mechanics of arsenic-induced skin cancer are unclear, making treatment and control extremely difficult.

In most cases, the final outcome of cellular differentiation is antiproliferative and constitutes a barrier to the continued proliferation required for tumorigenesis. Therefore, the hallmarks and characteristics of cancer currently include the acquired capabilities for sustaining proliferative signaling, evading growth suppressors, resisting cell death, and enabling replicative immortality [31]. As the mechanisms of cell proliferation, malignant transformation, and carcinogenesis are complex, the Nrf2/NQO1 pathway is considered one of them. Nuclear erythroid factor 2 p45 (NRF2) is of primary importance in the alteration of cell homeostasis to respond in an adaptive manner to xenobiotic and oxidative stress [32]. The accumulation of studies has shown that NRF2 plays a significant role in cell proliferation, except for the regulation of intracellular redox homeostasis [33,34]. NRF2 has long been considered a cancer inhibitor, providing the rationale for cancer prevention strategies using NRF2 activators [35]. However, increased NRF2 activity has recently been observed in a wide range of cancers [36]. This deregulated Nrf2 condition provides several growth advantages to cancer cells [35]. Therefore, NRF2 has attracted great interest for its use in cancer therapy and diagnosis. NAD(P)H: quinine oxidoreductase 1 (*NQO1*) is one of the target genes regulated by the transcription factor NRF2 [37,38]. The function of NQO1 is widely regarded as a "cytoprotective agent" that protects cells from oxidative damage by inducing responses to various noxious stimuli [39,40]. As with NRF2, the accumulation of studies confirmed the high expression of the *NQO1* gene in solid tumor tissue in recent years [41–43]. The upregulation of NQO1 may contribute to the growth of cancer cells, especially in oxidative stress environments [44–46]. Furthermore, the negative expression of NRF2 and NQO1 provides a better prognosis for patients with non-small cell lung cancer (NSCLC) [47]. In fact, we have reported that the long-term exposure to arsenite induced NRF2 accumulation and enhanced proliferation in human bronchial epithelial cells (HBE) [48] and immortalized human keratinocyte cells (HaCaT) [44], and we have found that the upregulation of NRF2/NQO1 caused the proliferation and transformation of arsenite-induced HBE cells by increasing cyclin E-CDK2 and then affecting the cell cycle [48]. It is suggested that a novel mechanistic pathway for arsenite carcinogenesis, the NRF2/NQO1 pathway, is involved in the dislocation of cell cycle progression in arsenite-induced malignancies. Nevertheless, this mechanism has not been confirmed in the study of arsenic-induced skin cancer.

In this study, we aimed to elucidate the mechanism of the NRF2/NQO1 pathway in mediating arsenite-induced malignant phenomena in HaCaT cells. To achieve our study objective, we analyzed the mechanism of dynamic changes in cell proliferation, NRF2, NQO1, cell cycle-related proteins, and cyclin-dependent kinases with arsenic-induced neoplastic transformation in HaCaT cells. In addition, we examined the actions of NRF2/NQO1 in arsenite-transformed HaCaT cells (T-HaCaT). We hope this study will lead to a greater understanding of arsenite skin toxicity, which may provide more possibilities for clinical applications for SCC.

2. Materials and Methods

2.1. Cell Culture and Exposure to Arsenic

HaCaT cells were developed by Boukamp et al. as a line of spontaneously immortalized human epithelial cells. They are immortal, nontumorigenic, and retain most of the characteristics of normal human keratinocytes (NHKs). HaCaT cells offer a suitable and stable model for keratinization studies. Moreover, this line could be reproducibly transfected with the activated human Ha-ras oncogene. Selected clones gave rise to highly differentiated benign epidermal cysts and/or squamous cell carcinomas in nude mice [49]. Thus, the HaCaT cell line is ideal for SCC studies. In this study, HaCaT cells were generously supplied by Prof M.Y (Dalian Medical University, Dalian, China). The cells were incubated at 37 °C in a humidified 5% atmosphere in Dulbecco's modified Eagle's medium (DMEM; Life Technologies/Gibco, Grand Island, NY, USA) with 10% fetal bovine serum (FBS; Hyclone, Thermo Scientific, Waltham, MA, USA), 100 U penicillin/mL, and 100 µg of streptomycin/mL. To create cellular patterns for the chronic arsenic exposure model with final concentrations of DMEM with 0.0 (control) or 1.0 µM sodium arsenite ($NaAsO_2$; Merck Drugs & Biotechnology, Darmstadt, Germany, purity 99.0%), HaCaT cells were cultured for 35 passages (approximately 18 weeks). The experimental control cells were HaCaT cells exposed to 0.0 µM $NaAsO_2$ for passage 0, while the transient matched control cells were exposed to 0.0 µM $NaAsO_2$ at other times. T-HaCaT cells were defined as 35-passage HaCaT cells exposed to arsenite based on our previous study [44]. Analytical and quality-assured reagents were used in all experiments.

2.2. Proliferation of Cells Assay

To determine whether exposure to 0.0 or 1.0 µM $NaAsO_2$ at various passages (0, 1, 7, 14, 21, 28, and 35 passages) of HaCaT cells would induce transformation, cellular proliferation changes were measured. As instructed by the manufacturer, the cell proliferation assay was measured by a 3-(4,5-dimethyl-2-thiazolyl)-2,5-diphenyl-2-H tetrazolium bromide

(MTT) kit. In brief, before 10 μL of MTT labeling reagent was added, 50,000 cells/mL were washed with phosphate-buffered saline (PBS), seeded onto 96-well plates, and incubated at 37 °C with 5% CO_2 for 24 h. Afterward, 100 μL of solubilization solution was added 4 h later. Finally, a microplate scanner was used to measure blue formazan from viable cells at 570 nm in accordance with the manufacturer's protocol. The experiment was repeated three times.

2.3. Flow Cytometry Analysis of the Cell Cycle

The cell cycle was detected by the cell cycle test kit. As per the manufacturer's instructions, synchronizing the cells before exposure required 48 h of culture in DMEM with 0.5% FBS. After continuous exposure to arsenite over a period of time, HaCaT cells were harvested at different passages, collected with 0.25% trypsin without EDTA, and washed twice with 4 °C pre-cooled PBS. The cells were left at −20 °C overnight with the addition of 1 mL of 70% pre-cooled ethanol. After two rounds of washing in cold PBS, the cells were stained for 30 min with 20 μg/mL of PI (Beyotime Institute of Biotechnology, Haimen, China) and 200 μg/mL of ribonuclease A (RNaseA, Beyotime Institute of Biotechnology, Haimen, China) diluted in PBS. Following the above processing, the data were analyzed using cell cycle analysis software based on flow cytometry (Beckman Coulter FC500, Brea, CA, USA) for detection. The experiment was repeated three times.

2.4. Western Blots

The experimental process was as follows: After three washes with ice-cold PBS and centrifugation at 16,000× *g* for 10 min at 4 °C, whole cell extracts were obtained using a cell lysis buffer for Western blot and Immunol precipitation (IP) (Beyotime Institute of Biotechnology, Haimen, China) with 1.0% phenylmethanesulfonyl fluoride (PMSF; Beyotime Institute of Biotechnology, Haimen, China), and the protein fractions were quantified using a BCA Protein Assay Kit (Beyotime). Protein lysates were subjected to sodium polyacrylamide dodecyl sulphate gel electrophoresis (SDS-PAGE) and transferred to polyvinylidene difluoride membranes (Millipore, Bedford, MA, USA) in equal amounts. The membranes were blocked with 5% skim milk (Yili Industrial Group Limited by Share Ltd., Hohhot, China) for 1 h and then incubated with the appropriate primary antibodies overnight at 4 °C. The following primary antibodies were used in this research: NRF2 (sc-13032) (dilution 1:500), NQO1 (sc-32793) (dilution 1:500), and CDK2 (sc-6248) (dilution 1:500) were purchased from Santa Cruz Biotech (Santa Cruz, CA, USA); Cyclin E (WL01072) (WB 1:750) and Cyclin D1 (WL01435a) (WB 1:750) were acquired from Wanlei Organic Company (Shenyang, China); Cycline A (RLT1167) (dilution 1:500) was acquired from Suzhou Ruiying Organic Company. The loading control was GAPDH (Beyotime Institute of Biotechnology, Haimen, China) (dilution 1:1000) or β-actin (sc-47778) (dilution 1:500) (Santa Cruz, CA, USA). Following three washes in TBST (Tween 20 Tris-buffered saline) saline solution, membranes were incubated with secondary antibodies conjugated with equine radish peroxidase (Beyotime). An improved chemiluminescence kit (Millipore, Bedford, MA, USA) and G: BOX Chemi XRQ (Syngene, Cambridge, UK) were used for the detection of immunoreactive proteins. To correct for changes in protein load between the different test groups, densitometry was used to quantify the plots and standardize them with GAPDH or β-actin. The results obtained relate to three autonomous experiments.

2.5. SiRNA Interference Assays

According to the manufacturer's protocol, *NRF2* siRNA transfection and *NQO1* siRNA transfection were performed on the 35 passages (T-HaCaT) cells exposed to 1.0 μM arsenite to silence the *NRF2* and *NQO1* genes to further understand the roles of NRF2/NQO1 in our research. First, 2×10^5 cells were seeded in 2 mL of antibiotic-free DMEM containing FBS and then added to each well of a 6-well plate 24 h before transfection. Next, the biphasic siRNA solution (solution A) and diluted transfection reagent (solution B) (sc-29528, Santa Cruz Biotechnology) were mixed with siRNA transfection medium (sc-36868, Santa Cruz

Biotechnology) prior to siRNA in preparing a mix of transfection reagents. The above mixtures were incubated at room temperature for 15–45 min and then added to the cells and stirred evenly. The cells were incubated in a 37 °C 5% CO_2 incubator for 5–7 h, and then 1 mL of DMEM containing 20% fetal bovine serum and 2% antibiotics was added without removing the transfection mixture. After incubation for 18–24 h, the cells were collected for analysis. Control siRNA, *NRF2* siRNA, and *NQO1* siRNA were purchased from Santa Cruz Biotechnology (Santa Cruz, CA, USA). HaCaT cells (passage 35), not exposed to arsenite, served as a negative control.

2.6. Wound-Healing (Cell Migration) Assay

The wound-healing assay is a method for studying the role of various experimental conditions in cellular migration and proliferation. It is simple and cost-effective and is one of the earliest methods to be used for targeted in vitro cell migration studies. To determine cell migration, wound healing experiments were conducted. The objective of the wound-healing test was to analyze the impact of NRF2 and NQO1 inhibition on cell migration. In short, HaCaT and T-HaCaT cells with or without *NRF2* siRNA and *NQO1* siRNA transfection were seeded in 6 cm dishes to form confluent monolayers. The cell monolayer was wounded using a 200 μL pipette tip and washed twice with PBS. Then, the wound width (0 h and the subsequent 24 h incubation period were photographed immediately after scratching. The tests were conducted in three petri dishes in triplicate, and the results are expressed as the relative scratch width based on the distance migrated relative to the original scratched distance. Cell migration rate = (the width of the wound at 0 h-the width of the wound at 48 h)/the width of the wound at 48 h.

2.7. Anchor-Independent Growth Experiment

Normal cell proliferation requires the costimulation of growth factor signaling and cell adhesion, but an important feature of phenotypically transformed cells is anchor-independent growth. To determine the growth of cells independent of the anchor, a soft agar cloning assay was used. Sweet AGAR plates were prepared with 0.70% agarose underlay in DMEM supplemented with 10% FBS in a 35 mm (diameter) container. To observe the growth capacity of colonies on soft AGAR, cells were cultivated in three copies at a density of 5×10^3 in 2 mL of 0.35% agarose overlaid with 0.70% agarose in the same medium. Cells were incubated at 37 °C in a humidified atmosphere at 5% with a routing replenished of medium every three days. After 3 weeks, colonies of more than 30 cells were examined under a microscope and photographed under a dissective microscope.

2.8. Statistical Analysis

Data analysis was performed using SPSS software (version 20.0, SPSS Inc., Chicago, IL, USA). All experimental data were tested for homoscedasticity, and differences between groups were assessed using one-way analysis of variance (ANOVA). Pairwise comparisons of sample means with uniform variance were performed by ANOVA using the SNK test and Tamhane's T2 test for nonuniform variance. The test results were statistically significant at $p < 0.05$. Data from the triplicate experiments reported as the mean ± standard deviation (SD) were analyzed.

3. Results

3.1. Malignant Phenotypes Were Induced in HaCaT Cells with Continuous Arsenic Exposure

To drive cells toward a malignant phenotype, normal HaCaT cells were exposed to 0.0 or 1.0 μM $NaAsO_2$ for 18 weeks (35 passes). As a result, the proliferation rate in the treatment group (1.0 μM $NaAsO_2$) was significantly higher at passages 21, 28, and 35 than that in the passage-matched control group (0.0 μM $NaAsO_2$) and showed a trend of increasing passage by passage ($p < 0.05$) (Figure 1a). Furthermore, the ratio of G1 phase cells was significantly decreased, and the ratio of the S phase was significantly increased in the treatment group from passage 1 to 35 ($p < 0.05$) (Figure 1b), which indicated that arsenite can promote the transformation of HaCaT cells from the G0/G1 phase to S phase. The data demonstrated malignant phenotypes (T-HaCaT) in cells processed by $NaAsO_2$.

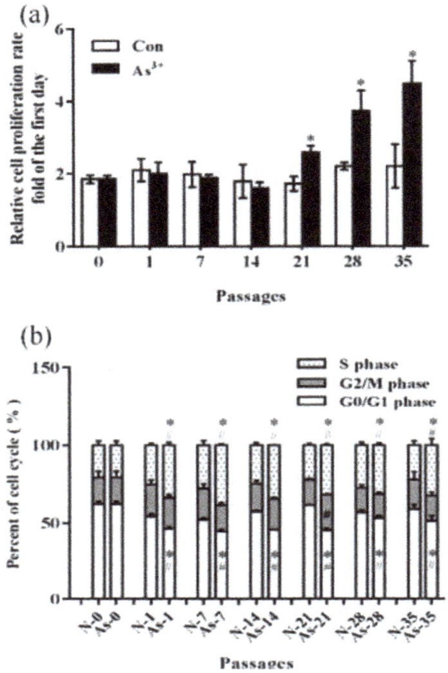

Figure 1. Changes in cell proliferation and cell cycle in HaCaT cells exposed to 1.0 μM arsenite or without arsenite exposure for different passages. (**a**) Change in the cell proliferation rate in HaCaT cells. (**b**) Change in the cell cycle in HaCaT cells. Error bars are the mean ± SD (n = 3). * Compared with the passage-matched control group, $p < 0.05$; # Compared with the experimental control group, $p < 0.05$.

3.2. Decreased Cell Cycle-Related Proteins and NRF2 and NQO1 in HaCaT Cells Exposed to Arsenite

Cyclin proteins, including A, B, D, E, G, and H, collaborate with their cyclin-dependent kinases. They act as the nucleus of the cell cycle process [50]. Cyclin and cyclin kinases were measured in HaCaT cells as indicators to assess changes in the cell cycle in arsenite-induced malignant transformation in this study. Compared with the passage-matched control groups, Cyclin E was significantly increased after passage 7 and showed an upward trend after continuous exposure to 1.0 μM $NaAsO_2$ ($p < 0.05$) (Figure 2b–d). Moreover, the protein expression of cyclin-dependent kinase CDK2 showed a significant increase after 21 passages in HaCaT cells and increased with each passage; however, similar changes were not observed in the control groups ($p < 0.05$) (Figure 2e). No significant differences were observed between the other cyclin and the cyclin-dependent kinases with the control groups. These results suggest that arsenite-exposed cell proliferation promotes the progression of the cell cycle from the G0/G1 phase to S phase through an increase in cyclin E-CDK2, thereby causing the malignant transformation of HaCaT. Changes in NRF2 and NQO1 were also measured. During the 1.0 μM $NaAsO_2$-induced malignant transformation in HaCaT cells, the elevated expression levels of NRF2 and NQO1 increased. As shown in Figure 2f,g, compared to the control groups, the protein level of NRF2 expression increased significantly after seven passages and displayed an upward trend. As a target of NRF2, the protein expression of NQO1 was upregulated after passage 21 ($p < 0.05$). These results indicate that the protein level of NRF2 and its target gene *NQO1* were increased by arsenite exposure.

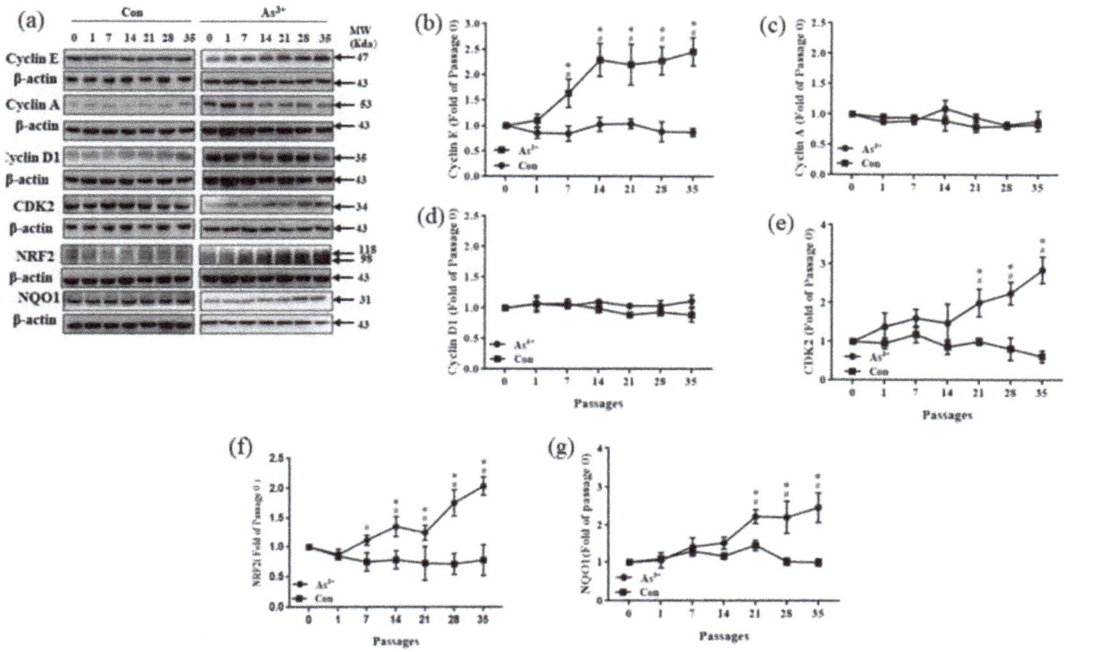

Figure 2. Relative protein expression and quantification of changes in NRF2, NQO1, and cell cycle-related proteins in HaCaT cells exposed to 1.0 μM arsenite or without arsenite exposure for different passages. (**a**) Protein bands. (**b**) Relative protein levels of Cyclin E. (**c**) Relative protein levels of Cyclin A. (**d**) Relative protein levels of Cyclin D1. (**e**) Relative protein levels of CDK2. (**f**) Relative protein levels of NRF2. (**g**) Relative protein levels of NQO1. Error bars are the mean ± SD (n = 3). * Compared with the passage-matched control group, $p < 0.05$; # Compared with the experimental control group, $p < 0.05$.

3.3. NRF2-Mediated Cellular Proliferation in Arsenite-Exposed HaCaT Cells

Considering that the levels of NRF2 and NQO1 proteins in T-HaCaT cells were higher than those in the control, to understand more about the role in the arsenite-induced malignant transformation of HaCaT cells, *NRF2* was silenced by *NRF2* siRNA in arsenite-transformed T-HaCaT cells. After T-HaCaT cells were transfected with *NRF2* siRNA, the protein levels of NRF2 and NQO1 were significantly decreased in T-HaCaT cells compared with the con-siRNA transfected group (T-HaCaT (−)) ($p < 0.05$, Figure 3a). In addition, compared to the T-HaCaT (−) cells, the G0/G1 phase of the cells increased, whereas the S phase decreased after the T-HaCaT cells were transformed by *NRF2* siRNA (Figure 3b). It is suggested that silencing *NRF2* resulted in cell cycle stoppage. Levels of cyclic E-CDK2 proteins were reduced in transfected T-HaCaT cells compared to T-HaCaT (−) cells ($p < 0.05$, Figure 3d). Meanwhile, *NRF2* silencing blocked the proliferation of T-HaCaT cells ($p < 0.05$, Figure 3c).

More importantly, silencing *NRF2* activity inhibited HaCaT cell migration, as detected by the wound-healing assay ($p < 0.05$, Figure 4a), and the anchorage-independent growth capacity was suppressed, as detected by colony formation in soft agar ($p < 0.05$, Figure 4b), which suggests that the malignant phenotype in HaCaT cells was relieved. These results demonstrated that NRF2 deficiency blocked the proliferation of T-HaCaT cells by blocking the change in the cell cycle, which suggested that NRF2 is a key factor in mediating cellular proliferation in arsenite-exposed HaCaT cells.

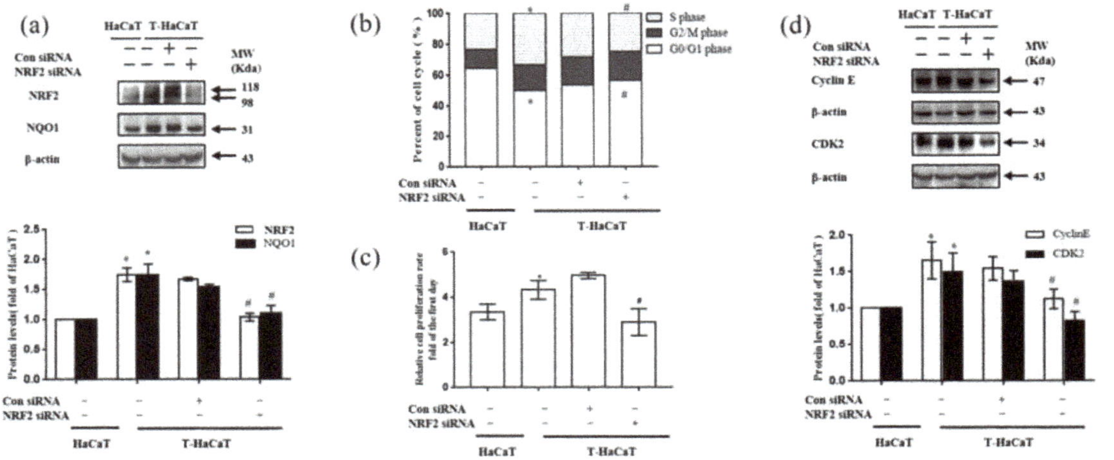

Figure 3. Effects of NRF2 on arsenite-induced malignant transformation of HaCaT cells. (a) Protein bands and relative protein levels of NRF2 and NQO1 in HaCaT cells and T-HaCaT cells with or without *NRF2* siRNA transfection. (b) Changes in the cell cycle. (c) Change in the cell proliferation rate. (d) Protein bands and relative protein levels of the cell-cycle-related proteins Cyclin E and CDK2. Error bars are the mean ± SD (n = 3). * Compared with HaCaT cells, $p < 0.05$; # Compared with control siRNA cells, $p < 0.05$.

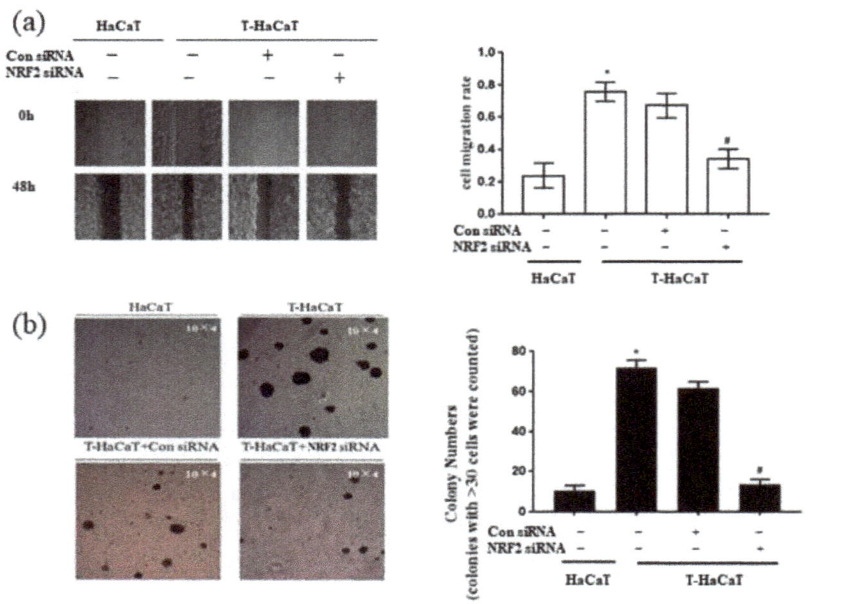

Figure 4. The cell migration rate and the colony number in T-HaCaT cells after transfection *NRF2*. (a) Wound-healing assay (Scheme 10 × 4) and the rate of migration. (b) Colony formation assay and colony numbers. Error bars are the mean ± SD (n = 3). * Compared with HaCaT cells, $p < 0.05$; # Compared with control siRNA cells, $p < 0.05$.

3.4. *NQO1-Mediated Cellular Proliferation in Arsenite-Exposed HaCaT Cells*

To learn more about the role of NQO1 in the arsenide-induced malignant transformation of HaCaT cells, *NQO1* was silenced by *NQO1* siRNA in arsenite-transformed T-HaCaT cells. After *NQO1* siRNA transfection of T-HaCaT cells, the NQO1 protein level was significantly decreased in T-HaCaT cells, while the NRF2 protein level was not changed ($p < 0.05$, Figure 5a). The proportion of G0/G1 phase cells was significantly increased and S phase cells were decreased significantly after silencing *NQO1* ($p < 0.05$, Figure 5b), and the expression of the cell cycle-related proteins cyclin E and CDK2 was reduced after transfection ($p < 0.05$, Figure 5d). Furthermore, the proliferation of T-HaCaT cells was blocked ($p < 0.05$, Figure 5c), HaCaT cell migration was suppressed ($p < 0.05$, Figure 6a), and the anchorage-independent growth capacity of T-HaCaT cells was decreased ($p < 0.05$, Figure 6b) after blocking NQO1 expression with siRNA, which suggested that the malignant phenotype in HaCaT cells was relieved. These results demonstrated that NQO1 deficiency blocked the proliferation of T-HaCaT cells by blocking the change in the cell cycle, which suggested that, in addition to NRF2, NQO1 is also a key factor in cell proliferation in HaCaT cells exposed to arsenite.

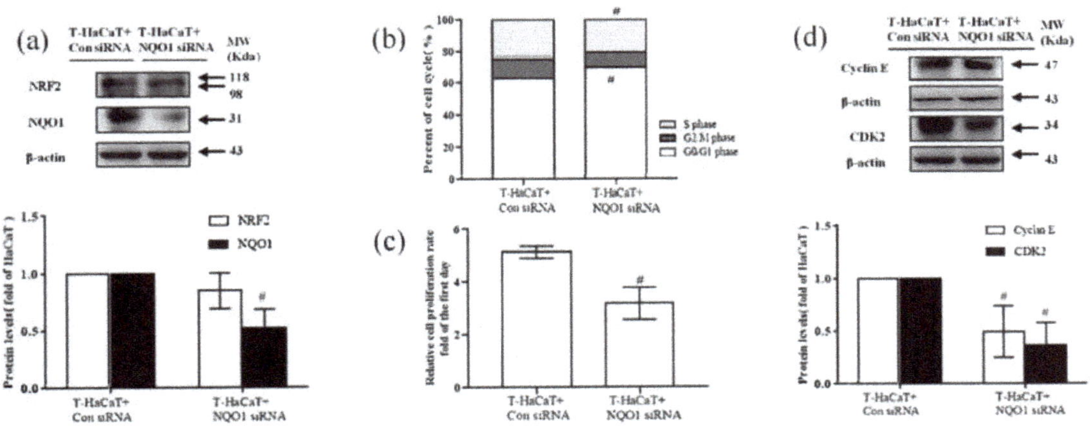

Figure 5. Effect of NQO1 on arsenite-induced malignant transformation of HaCaT cells. (**a**) Protein bands and relative protein levels of NRF2 and NQO1 in HaCaT cells and T-HaCaT cells with or without NQO1. (**b**) Changes in the cell cycle. (**c**) Change in the cell proliferation rate. (**d**) Protein bands and relative protein levels of cell cycle related protein Cyclin E and CDK2. Error bars are the mean ± SD ($n = 3$). # Compared with control siRNA cells, $p < 0.05$.

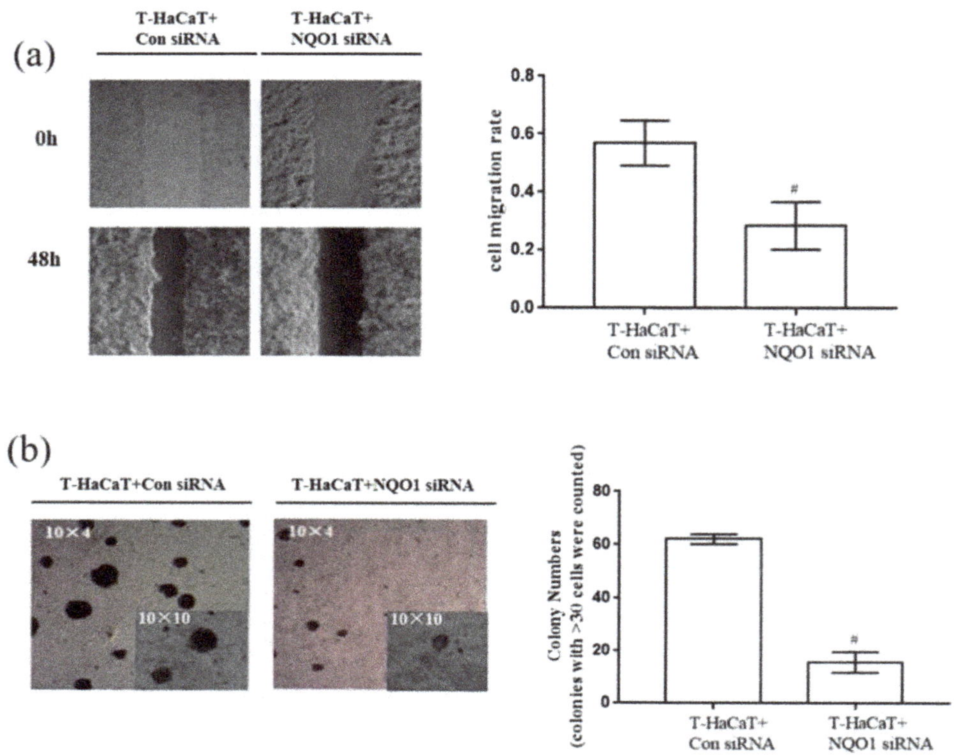

Figure 6. The cell migration rate and the colony number in T-HaCaT cells after transfection with *NQO1* siRNA. (**a**) Wound-healing assay (magnification factor: 10 × 4) and the rate of migration. (**b**) Colony formation assay and colony numbers. Error bars are the mean ± SD (n = 3). # Compared with control siRNA cells, $p < 0.05$.

4. Discussion

Long-term exposure to inorganic arsenic can lead to the occurrence of multisystem and multiorgan cancers [51,52]. Many studies have demonstrated the malignant transformation of normal cells after arsenite exposure and revealed the mechanism of arsenic-induced cancer [53–55]. For animal models of arsenic-induced skin cancer, which have not been constructed [56], HaCaT cells are widely used as an in vitro model to study the mechanism of arsenic-mediated skin cancer, especially in SCC [57–59]. Numerous studies have shown that exposure to a variety of exogenous chemicals may lead to the malignant transformation of HaCaT cells [60–62]. We found in an earlier study that HaCaT cells in soft agar exhibited high proliferation rates, increased MMP-9 secretion, and multicolony characteristics after being treated with 1.0 M sodium arsenite for more than 35 passages [44]. In the present study, we again verified that 1.0 μM sodium arsenite exposure can lead to a malignant phenotype in HaCaT cells by affecting the cell cycle of HaCaT cells (from the first passage) and increasing their proliferation rate from passage to passage (from passage 7) because sustaining proliferative signaling, evading growth suppressors, and replicative immortality are the hallmarks of cancer [31].

Furthermore, we explored the potential molecular mechanisms for promoting cellular proliferation for the carcinogenesis of arsenite. Many previous reports have suggested that the activation of NRF2 in tumor cells could affect proliferation [63,64] and accelerate the development of the disease [65–68], whose target gene *NQO1* also plays an important role in cellular proliferation and is overexpressed in many cancer tissues [41,69,70]. Indeed, our past studies have found that high levels of NRF2 may induce arsenite-induced tumors [44,46]; in this study, we observed that either high levels of NRF2 (after seven passages) or NQO1 (after 21 passages) were caused by arsenite. Notably, the in vitro UV irradiation of melanocytes has been demonstrated to induce the activation of NRF2 and its target genes, including *NQO1* [71]. Consistent with our results, NQO1 promotes an aggressive phenotype in hepatocellular carcinoma by amplifying ERK-NRF2 signaling [72]; immunohistochemical analysis showed that the protein levels of NRF2 and NQO1 were higher in carcinoma tissues than in benign follicular adenomas and hyperplastic nodules [73]. Overall, the NRF2/NQO1 pathway is important in response to environmental toxins and may be a tumor suppressor [74]. In addition, the results of the study also indicated that arsenite can regulate the protein expression levels of cyclin E and CDK2, which are also associated with cell proliferation and malignant transformation. This finding explains the cell cycle progression from the G0/G1 to S phase observed in this study. Cyclin E is an essential component of the nuclear cell cycle machinery. In mammalian cells, two types of cyclin E activate CDK2 and initiate cell cycle progression by phosphorylating various cellular proteins [75]. Cyclin E appears at the end of the G1 phase, which can facilitate and shorten the transition from the G/S phase [76]. Cyclin E binds to CDK2 at the end of stage G1, activates CDK2 activity, and promotes the expression of genes related to DNA replication, thereby initiating cellular DNA replication [77]. Lee et al. found that the fargesin-induced colony growth inhibition of colon cancer cells was mediated by suppression of the cyclin-dependent kinase 2 (CDK2)/cyclin E signaling axis by the upregulation of p21WAF1/Cip1, resulting in G1-phase cell cycle accumulation in a dose-dependent manner [78]. Liang et al. demonstrated that the administration of cyclin E siRNA could inhibit breast tumor growth in nude mice [79]. Therefore, we proposed that arsenite can induce the malignant transformation of HaCaT cells, affect cell cycle progression, and promote cell proliferation. The phase shift of the cell cycle can be adjusted by the G0/G1 and S phase structural ratio by increasing the expression of Cyclin E-CDK2.

In addition, the upstream–downstream relationship between NRF2/NQO1 and cyclin E-CDK2 in promoting cell proliferation and subsequent malignant transformation caused by arsenite drew our attention. To clarify this problem, we silenced the 35 passages of T-HaCaT cells by transfection with *NRF2* siRNA and *NQO1* siRNA, a kind of double-stranded RNA with a length of 20 to 25 nucleotides, by interfering with the post transcriptional degradation of mRNA expressing specific genes with complementary nucleotide sequences, thus preventing translation. We found that in addition to downregulating NQO1 expression, the silencing of *NRF2* by siRNA interference reduced the cell proliferation rate and cell cycle-related proteins Cyclin E and CDK2 and ultimately decreased the migration rate and colony-forming capacity of T-HaCaT cells. This result corresponds to Homma et al. [66]. They found that the knockout of *NRF2* in A549 and NCI-H292 cells caused cell cycle arrest in the G1 phase. Furthermore, other in vitro experiments showed that a lack of *NRF2* prevented cell cycle progression and slowed cell proliferation [80,81]. Moreover, there are few reports of the relationship between NQO1 and cyclin E-CDK2. We further observed the same situation as *NRF2* and *NQO1* silencing resulted in decreased cyclin E and CDK2 protein levels, arrested cell cycle progression from the G0/G1 to S phase, and restricted cell proliferation and malignant phenotypes in arsenite-transformed cells, which were characterized by reduced colonization and limited migration. This result is consistent with our previous report on HBC cells [48]. From this, we conclude that a new common pathway of arsenic carcinogenesis has been identified.

NRF2 and NQO1 are generally recognized to play important roles in cancer prevention and treatment [82–85], so they can be targeted for skin cancer. Researchers searched for molecules able to intentionally activate NRF2 because it has been shown to be a way to prevent skin cancer [86–88]. As the activation of NRF2 in cancer is two-sided [36,89], its functions are more far-reaching than originally envisioned, which presents new challenges and opportunities for targeting NRF2 and NQO1 in skin cancer prevention and treatment [90].

Unfortunately, we do not yet know the mechanism by which the NRF2/NQO1 pathway regulates cyclin E-CDK2; in addition, we performed all the experiments on a single cell line, and we have not yet validated our results based on animal models by using cell-derived xenografts, which are our research plans for the future. It must be acknowledged that the mechanism of SCC is complex and must therefore be taken into account holistically in the interpretation of the mechanism and future treatment. Moreover, a genomic approach is necessary for determining DNA mutations caused by arsenite treatment in future studies, which is important to understand whether arsenite is a carcinogenicity promoter or initiator.

5. Conclusions

In conclusion, we demonstrated that NRF2/NQO1 plays an important role in the arsenite-induced malignant transformation of cells. Increased regulation of NRF2/NQO1 increases cyclin E-cdk2 expression and promotes arsenite-induced disruption of the HaCaT cell cycle from the G0/G1 to S phase, which leads to cell proliferation and malignancy. In summary, NRF2 has been shown to regulate NQO1, and both contribute to the transformation of arsenite-induced cell malignancy (Figure 7). Our research provides important information about the molecular mechanisms of arsenic-induced SCC. We found that using siRNA to silence *NRF2* and *NQO1* led to a significant decrease in protein expression in genes downstream of NRF2 (NQO1) and Cyclin E-CDK2. Therefore, these results suggest that the silencing of *NRF2* and *NQO1* inhibits the growth of HaCaT cells and that arsenite-induced malignant phenotypes are alleviated. In summary, the findings from this study can assist in further understanding the health effect mechanism of arsenic pollution and facilitate future developments of new arsenic and other similar environmental factors (e.g., UV radiation)-induced SCC treatment and prevention.

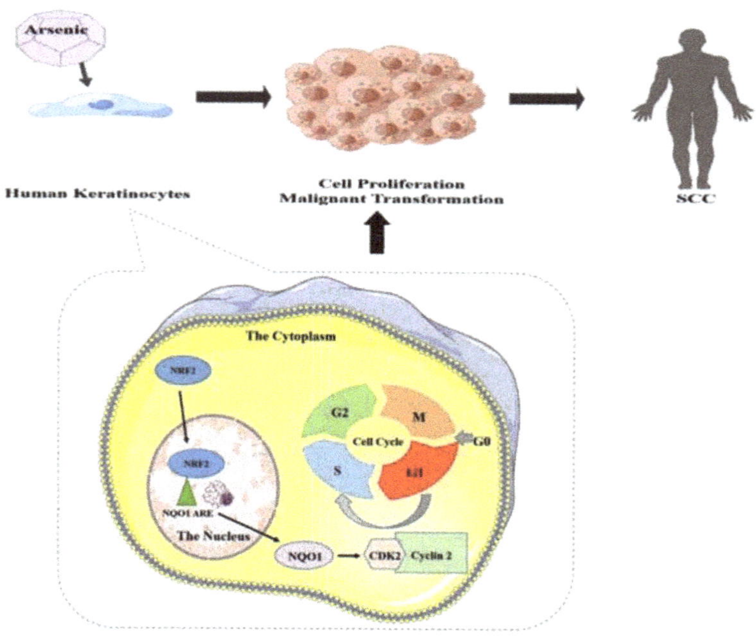

Figure 7. Proposed model by which NRF2 regulates NQO1 to promote proliferation in arsenite-induced malignant transformation.

Author Contributions: Conceptualization, Y.A.; methodology, Y.M.; software, H.X.; validation, R.Y. and Q.Y.; formal analysis, K.K.; investigation, H.D.; resources, H.Z.; data curation, X.W. and C.L.; writing—original draft preparation, Q.Y.; writing—review and editing, T.J.; visualization, Q.Y.; supervision, J.Z.; project administration, Y.A.; funding acquisition, T.J. and Y.A. All authors have read and agreed to the published version of the manuscript.

Funding: This work was supported in part by the National Natural Science Foundation of China (NSFC) (No. 81872646; 81811540034; 81573173); a project funded by the Priority Academic Program Development of Jiangsu Higher Education Institutions (PAPD); the open project of key laboratory of environmental pollution and disease surveillance, Ministry of Education (qian jiao he KY NO. [2018]483); and the Guizhou province science and technology planning project, Science and Technology Department of Guizhou Province (qian jiao he support NO. [2020]4Y235).

Institutional Review Board Statement: Not applicable.

Informed Consent Statement: Not applicable.

Data Availability Statement: Not applicable.

Acknowledgments: We would like to thank Yangming Gao from Dalian Medical University for his generously supplied HaCaT cells in this study.

Conflicts of Interest: The authors declare no conflict of interest.

Abbreviations

ANOVA	analysis of variance
ARE	antioxidant response element
CDK2	Cyclin-dependent kinase 2
DMEM	Dulbecco's modified Eagle's medium
FBS	Fetal bovine serum
HaCaT	immortalized human keratinocyte cells
Keap1	Kelch-like ECH-associated protein 1
$NaAsO_2$	sodium arsenite
NRF2	nuclear factor-erythroid-2 p45-related factor 2
NQO-1	NAD(P) H: Quinine oxidoreductase 1
NADPH	nicotinamide adenine dinucleotide phosphate
PBS	phosphate-buffered saline
PI	Propidium iodide
ROS	reactive oxygen species
T-HaCaT	arsenite-transformed HaCaT cells
HBE	human bronchial epithelial
MC	skin melanoma
NMSC	non-melanoma skin cancer
BCC	basal cell carcinoma
SCC	squamous cell carcinoma
UV	ultraviolet radiation
siRNA	small interfering RNA

References

1. Garelick, H.; Jones, H.; Dybowska, A.; Valsami-Jones, E. Arsenic pollution sources. *Rev. Environ. Contam. Toxicol.* **2008**, *197*, 17–60. [CrossRef] [PubMed]
2. Hong, Y.S.; Song, K.H.; Chung, J.Y. Health effects of chronic arsenic exposure. *J. Prev. Med. Public Health* **2014**, *47*, 245–252. [CrossRef] [PubMed]
3. Tchounwou, P.B.; Wilson, B.; Ishaque, A. Important considerations in the development of public health advisories for arsenic and arsenic-containing compounds in drinking water. *Rev. Environ. Health* **1999**, *14*, 211–229. [CrossRef] [PubMed]
4. Rousselot, P.; Labaume, S.; Marolleau, J.P.; Larghero, J.; Noguera, M.H.; Brouet, J.C.; Fermand, J.P. Arsenic trioxide and melarsoprol induce apoptosis in plasma cell lines and in plasma cells from myeloma patients. *Cancer Res.* **1999**, *59*, 1041–1048.
5. Missimer, T.M.; Teaf, C.M.; Beeson, W.T.; Maliva, R.G.; Woolschlager, J.; Covert, D.J. Natural Background and Anthropogenic Arsenic Enrichment in Florida Soils, Surface Water, and Groundwater: A Review with a Discussion on Public Health Risk. *Int. J. Environ. Res. Public Health* **2018**, *15*, 2278. [CrossRef]

6. Oremland, R.S.; Stolz, J.F. The ecology of arsenic. *Science* **2003**, *300*, 939–944. [CrossRef]
7. Hull, E.A.; Barajas, M.; Burkart, K.A.; Fung, S.R.; Jackson, B.P.; Barrett, P.M.; Neumann, R.B.; Olden, J.D.; Gawel, J.E. Human health risk from consumption of aquatic species in arsenic-contaminated shallow urban lakes. *Sci. Total Environ.* **2021**, *770*, 145318. [CrossRef]
8. Xue, L.; Zhao, Z.; Zhang, Y.; Liao, J.; Wu, M.; Wang, M.; Sun, J.; Gong, H.; Guo, M.; Li, S.; et al. Dietary exposure to arsenic and human health risks in western Tibet. *Sci. Total Environ.* **2020**, *731*, 138840. [CrossRef]
9. Monteiro, D.O.E.; Caixeta, E.S.; Santos, V.; Pereira, B.B. Arsenic exposure from groundwater: Environmental contamination, human health effects, and sustainable solutions. *J. Toxicol. Environ. Health B Crit. Rev.* **2021**, *24*, 119–135. [CrossRef]
10. Byeon, E.; Kang, H.M.; Yoon, C.; Lee, J.S. Toxicity mechanisms of arsenic compounds in aquatic organisms. *Aquat Toxicol.* **2021**, *237*, 105901. [CrossRef]
11. Rahman, M.A.; Hasegawa, H.; Lim, R.P. Bioaccumulation, biotransformation and trophic transfer of arsenic in the aquatic food chain. *Environ. Res.* **2012**, *116*, 118–135. [CrossRef]
12. Wei, Z.; Zhiqiang, G.; Dongdong, S.; Sen, D.; Li, Z. Arsenic speciation in wild marine organisms and a health risk assessment in a subtropical bay of China. *Sci. Total Environ.* **2018**, *626*, 621–629.
13. Vahter, M.; Skroder, H.; Rahman, S.M.; Levi, M.; Derakhshani, H.J.; Kippler, M. Prenatal and childhood arsenic exposure through drinking water and food and cognitive abilities at 10 years of age: A prospective cohort study. *Environ. Int.* **2020**, *139*, 105723. [CrossRef]
14. Mondal, D.; Rahman, M.M.; Suman, S.; Sharma, P.; Siddique, A.B.; Rahman, M.A.; Bari, A.; Kumar, R.; Bose, N.; Singh, S.K.; et al. Arsenic exposure from food exceeds that from drinking water in endemic area of Bihar, India. *Sci. Total Environ.* **2021**, *754*, 142082. [CrossRef]
15. Joardar, M.; Das, A.; Chowdhury, N.R.; Mridha, D.; De, A.; Majumdar, K.K.; Roychowdhury, T. Health effect and risk assessment of the populations exposed to different arsenic levels in drinking water and foodstuffs from four villages in arsenic endemic Gaighata block, West Bengal, India. *Environ. Geochem. Health* **2021**, *43*, 3027–3053. [CrossRef]
16. Baker, B.A.; Cassano, V.A.; Murray, C. Arsenic Exposure, Assessment, Toxicity, Diagnosis, and Management: Guidance for Occupational and Environmental Physicians. *J. Occup. Environ. Med.* **2018**, *60*, e634–e639. [CrossRef]
17. Chou, C.H.; De Rosa, C.T. Case studies—Arsenic. *Int. J. Hyg. Environ. Health* **2003**, *206*, 381–386. [CrossRef]
18. Jarup, L. Hazards of heavy metal contamination. *Br. Med. Bull.* **2003**, *68*, 167–182. [CrossRef]
19. Chakraborti, D.; Rahman, M.M.; Chatterjee, A.; Das, D.; Das, B.; Nayak, B.; Pal, A.; Chowdhury, U.K.; Ahmed, S.; Biswas, B.K.; et al. Fate of over 480 million inhabitants living in arsenic and fluoride endemic Indian districts: Magnitude, health, socio-economic effects and mitigation approaches. *J Trace Elem Med Biol* **2016**, *38*, 33–45. [CrossRef]
20. Tchounwou, P.B.; Yedjou, C.G.; Udensi, U.K.; Pacurari, M.; Stevens, J.J.; Patlolla, A.K.; Noubissi, F.; Kumar, S. State of the science review of the health effects of inorganic arsenic: Perspectives for future research. *Environ. Toxicol.* **2019**, *34*, 188–202. [CrossRef]
21. Pye-Smith, R.J. Arsenic Cancer, with Description of a Case. *Proc. R. Soc. Med.* **1913**, *6*, 229–236. [CrossRef]
22. Bungeler, W. Arsenic cancer. *Munch. Med. Wochenschr.* **1958**, *100*, 1117–1118.
23. Wei, S.; Zhang, H.; Tao, S. A review of arsenic exposure and lung cancer. *Toxicol. Res.* **2019**, *8*, 319–327. [CrossRef]
24. Mayer, J.E.; Goldman, R.H. Arsenic and skin cancer in the USA: The current evidence regarding arsenic-contaminated drinking water. *Int. J. Dermatol.* **2016**, *55*, e585–e591. [CrossRef]
25. Jackson, R.; Grainge, J.W. Arsenic and cancer. *Can. Med. Assoc. J.* **1975**, *113*, 396–401.
26. Hamilton, A. *Industrial Toxicology*, 2nd ed.; P. B. Hoeber: New York, NY, USA, 1949.
27. Gordon, R. Skin cancer: An overview of epidemiology and risk factors. *Semin. Oncol. Nurs.* **2013**, *29*, 160–169. [CrossRef]
28. Rosales-Castillo, J.A.; Acosta-Saavedra, L.C.; Torres, R.; Ochoa-Fierro, J.; Borja-Aburto, V.H.; Lopez-Carrillo, L.; Garcia-Vargas, G.G.; Gurrola, G.B.; Cebrian, M.E.; Calderon-Aranda, E.S. Arsenic exposure and human papillomavirus response in non-melanoma skin cancer Mexican patients: A pilot study. *Int. Arch. Occup. Environ. Health* **2004**, *77*, 418–423. [CrossRef]
29. Surdu, S. Non-melanoma skin cancer: Occupational risk from UV light and arsenic exposure. *Rev. Environ. Health* **2014**, *29*, 255–264. [CrossRef]
30. Surdu, S.; Fitzgerald, E.F.; Bloom, M.S.; Boscoe, F.P.; Carpenter, D.O.; Haase, R.F.; Gurzau, E.; Rudnai, P.; Koppova, K.; Vahter, M.; et al. Polymorphisms in DNA repair genes XRCC1 and XRCC3, occupational exposure to arsenic and sunlight, and the risk of non-melanoma skin cancer in a European case-control study. *Environ. Res.* **2014**, *134*, 382–389. [CrossRef]
31. Kim, T.H.; Seo, J.W.; Hong, Y.S.; Song, K.H. Case-control study of chronic low-level exposure of inorganic arsenic species and non-melanoma skin cancer. *J. Dermatol.* **2017**, *44*, 1374–1379. [CrossRef]
32. Hanahan, D. Hallmarks of Cancer: New Dimensions. *Cancer Discov.* **2022**, *12*, 31–46. [CrossRef] [PubMed]
33. Wu, S.; Lu, H.; Bai, Y. Nrf2 in cancers: A double-edged sword. *Cancer Med.* **2019**, *8*, 2252–2267. [CrossRef] [PubMed]
34. Fan, Z.; Wirth, A.K.; Chen, D.; Wruck, C.J.; Rauh, M.; Buchfelder, M.; Savaskan, N. Nrf2-Keap1 pathway promotes cell proliferation and diminishes ferroptosis. *Oncogenesis* **2017**, *6*, e371. [CrossRef] [PubMed]
35. Murakami, S.; Motohashi, H. Roles of Nrf2 in cell proliferation and differentiation. *Free Radic. Biol. Med.* **2015**, *88*, 168–178. [CrossRef] [PubMed]
36. Geismann, C.; Arlt, A.; Sebens, S.; Schafer, H. Cytoprotection "gone astray": Nrf2 and its role in cancer. *OncoTargets Ther.* **2014**, *7*, 1497–1518. [CrossRef]
37. Montserrat, R.D.L.V.; Eli, C.; Donna, D.Z. NRF2 and the Hallmarks of Cancer. *Cancer Cell* **2018**, *34*, 21–43.

38. Chorley, B.N.; Campbell, M.R.; Wang, X.; Karaca, M.; Sambandan, D.; Bangura, F.; Xue, P.; Pi, J.; Kleeberger, S.R.; Bell, D.A. Identification of novel NRF2-regulated genes by ChIP-Seq: Influence on retinoid X receptor alpha. *Nucleic Acids Res.* **2012**, *40*, 7416–7429. [CrossRef] [PubMed]
39. Malhotra, D.; Portales-Casamar, E.; Singh, A.; Srivastava, S.; Arenillas, D.; Happel, C.; Shyr, C.; Wakabayashi, N.; Kensler, T.W.; Wasserman, W.W.; et al. Global mapping of binding sites for Nrf2 identifies novel targets in cell survival response through ChIP-Seq profiling and network analysis. *Nucleic Acids Res.* **2010**, *38*, 5718–5734. [CrossRef]
40. Dinkova-Kostova, A.T.; Talalay, P. NAD(P)H:quinone acceptor oxidoreductase 1 (NQO1), a multifunctional antioxidant enzyme and exceptionally versatile cytoprotector. *Arch. Biochem. Biophys.* **2010**, *501*, 116–123. [CrossRef]
41. Ross, D.; Kepa, J.K.; Winski, S.L.; Beall, H.D.; Anwar, A.; Siegel, D. NAD(P)H:quinone oxidoreductase 1 (NQO1): Chemoprotection, bioactivation, gene regulation and genetic polymorphisms. *Chem. Biol. Interact.* **2000**, *129*, 77–97. [CrossRef]
42. Yang, Y.; Zhu, G.; Dong, B.; Piao, J.; Chen, L.; Lin, Z. The NQO1/PKLR axis promotes lymph node metastasis and breast cancer progression by modulating glycolytic reprogramming. *Cancer Lett.* **2019**, *453*, 170–183. [CrossRef]
43. Butsri, S.; Kukongviriyapan, V.; Senggunprai, L.; Kongpetch, S.; Zeekpudsa, P.; Prawan, A. Downregulation of NAD(P)H:quinone oxidoreductase 1 inhibits proliferation, cell cycle and migration of cholangiocarcinoma cells. *Oncol. Lett.* **2017**, *13*, 4540–4548. [CrossRef]
44. Cheng, Y.; Li, J.; Martinka, M.; Li, G. The expression of NAD(P)H:quinone oxidoreductase 1 is increased along with NF-kappaB p105/p50 in human cutaneous melanomas. *Oncol. Rep.* **2010**, *23*, 973–979. [CrossRef]
45. Wang, D.; Ma, Y.; Yang, X.; Xu, X.; Zhao, Y.; Zhu, Z.; Wang, X.; Deng, H.; Li, C.; Gao, F.; et al. Hypermethylation of the Keap1 gene inactivates its function, promotes Nrf2 nuclear accumulation, and is involved in arsenite-induced human keratinocyte transformation. *Free Radic. Biol. Med.* **2015**, *89*, 209–219. [CrossRef]
46. Luo, S.; Lei, K.; Xiang, D.; Ye, K. NQO1 Is Regulated by PTEN in Glioblastoma, Mediating Cell Proliferation and Oxidative Stress. *Oxid. Med. Cell. Longev.* **2018**, *2018*, 9146528. [CrossRef]
47. Yang, X.; Wang, D.; Ma, Y.; Xu, X.; Zhu, Z.; Wang, X.; Deng, H.; Li, C.; Chen, M.; Tong, J.; et al. Continuous activation of Nrf2 and its target antioxidant enzymes leads to arsenite-induced malignant transformation of human bronchial epithelial cells. *Toxicol. Appl. Pharmacol.* **2015**, *289*, 231–239. [CrossRef]
48. Tong, Y.H.; Zhang, B.; Yan, Y.Y.; Fan, Y.; Yu, J.W.; Kong, S.S.; Zhang, D.; Fang, L.; Su, D.; Lin, N.M. Dual-negative expression of Nrf2 and NQO1 predicts superior outcomes in patients with non-small cell lung cancer. *Oncotarget* **2017**, *8*, 45750–45758. [CrossRef]
49. Kong, Q.; Deng, H.; Li, C.; Wang, X.; Shimoda, Y.; Tao, S.; Kato, K.; Zhang, J.; Yamanaka, K.; An, Y. Sustained high expression of NRF2 and its target genes induces dysregulation of cellular proliferation and apoptosis is associated with arsenite-induced malignant transformation of human bronchial epithelial cells. *Sci. Total Environ.* **2021**, *756*, 143840. [CrossRef]
50. Boukamp, P.; Petrussevska, R.T.; Breitkreutz, D.; Hornung, J.; Markham, A.; Fusenig, N.E. Normal keratinization in a spontaneously immortalized aneuploid human keratinocyte cell line. *J. Cell Biol.* **1988**, *106*, 761–771. [CrossRef]
51. Schafer, K.A. The Cell Cycle: A Review. *Vet. Pathol.* **1998**, *35*, 461–478. [CrossRef]
52. Yu, S.; Liao, W.T.; Lee, C.H.; Chai, C.Y.; Yu, C.L.; Yu, H.S. Immunological dysfunction in chronic arsenic exposure: From subclinical condition to skin cancer. *J. Dermatol.* **2018**, *45*, 1271–1277. [CrossRef]
53. Wang, Z.; Zhao, Y.; Smith, E.; Goodall, G.J.; Drew, P.A.; Brabletz, T.; Yang, C. Reversal and prevention of arsenic-induced human bronchial epithelial cell malignant transformation by microRNA-200b. *Toxicol. Sci.* **2011**, *121*, 110–122. [CrossRef]
54. Ge, Y.; Zhu, J.; Wang, X.; Zheng, N.; Tu, C.; Qu, J.; Ren, X. Mapping dynamic histone modification patterns during arsenic-induced malignant transformation of human bladder cells. *Toxicol. Appl. Pharmacol.* **2018**, *355*, 164–173. [CrossRef]
55. Benbrahim-Tallaa, L.; Waterland, R.A.; Styblo, M.; Achanzar, W.E.; Webber, M.M.; Waalkes, M.P. Molecular events associated with arsenic-induced malignant transformation of human prostatic epithelial cells: Aberrant genomic DNA methylation and K-ras oncogene activation. *Toxicol. Appl. Pharmacol.* **2005**, *206*, 288–298. [CrossRef]
56. Hunt, K.M.; Srivastava, R.K.; Elmets, C.A.; Athar, M. The mechanistic basis of arsenicosis: Pathogenesis of skin cancer. *Cancer Lett.* **2014**, *354*, 211–219. [CrossRef]
57. Rusanov, A.L.; Romashin, D.D.; Zgoda, V.G.; Butkova, T.V.; Luzgina, N.G. Protein dataset of immortalized keratinocyte HaCaT cells and normal human keratinocytes. *Data Brief.* **2021**, *35*, 106871. [CrossRef]
58. Valerio, H.P.; Ravagnani, F.G.; Ronsein, G.E.; Di Mascio, P. A single dose of Ultraviolet-A induces proteome remodeling and senescence in primary human keratinocytes. *Sci. Rep.* **2021**, *11*, 23355. [CrossRef]
59. Lee, R.H.; Oh, J.D.; Hwang, J.S.; Lee, H.K.; Shin, D. Antitumorigenic effect of insect-derived peptide poecilocorisin-1 in human skin cancer cells through regulation of Sp1 transcription factor. *Sci. Rep.* **2021**, *11*, 18445. [CrossRef]
60. Bicheng, F.; Pradeep, D.; Fengjie, L.; Laura, G.; Martina, B.; Alfonso, M.; Deniz, T. Pyrenosetin D, a New Pentacyclic Decalinoyltetramic Acid Derivative from the Algicolous Fungus *Pyrenochaetopsis* sp. FVE-087. *Mar. Drugs* **2020**, *18*, 281.
61. Makoto, S.; Kazunori, H.; Masafumi, Y.; Mohammad, D.A.; Said, H.F.; Nobuyuki, H.; Lisa, K.; Kiyoshi, Y.; Masashi, K. Lithium promotes malignant transformation of nontumorigenic cells in vitro. *Sci. Total Environ.* **2020**, *744*, 140830.
62. Wright, C.; Iyer, A.K.V.; Wang, L.; Wu, N.; Yakisich, J.S.; Rojanasakul, Y.; Azad, N. Effects of titanium dioxide nanoparticles on human keratinocytes. *Drug Chem. Toxicol.* **2017**, *40*, 90–100. [CrossRef] [PubMed]

63. Zhang, H.S.; Zhang, Z.G.; Du, G.Y.; Sun, H.L.; Liu, H.Y.; Zhou, Z.; Gou, X.M.; Wu, X.H.; Yu, X.Y.; Huang, Y.H. Nrf2 promotes breast cancer cell migration via up-regulation of G6PD/HIF-1alpha/Notch1 axis. *J. Cell Mol. Med.* **2019**, *23*, 3451–3463. [CrossRef] [PubMed]
64. Xu, P.; Jiang, L.; Yang, Y.; Wu, M.; Liu, B.; Shi, Y.; Shen, Q.; Jiang, X.; He, Y.; Cheng, D.; et al. PAQR4 promotes chemoresistance in non-small cell lung cancer through inhibiting Nrf2 protein degradation. *Theranostics* **2020**, *10*, 3767–3778. [CrossRef] [PubMed]
65. Mitsuishi, Y.; Taguchi, K.; Kawatani, Y.; Shibata, T.; Nukiwa, T.; Aburatani, H.; Yamamoto, M.; Motohashi, H. Nrf2 redirects glucose and glutamine into anabolic pathways in metabolic reprogramming. *Cancer Cell* **2012**, *22*, 66–79. [CrossRef]
66. Homma, S.; Ishii, Y.; Morishima, Y.; Yamadori, T.; Matsuno, Y.; Haraguchi, N.; Kikuchi, N.; Satoh, H.; Sakamoto, T.; Hizawa, N.; et al. Nrf2 enhances cell proliferation and resistance to anticancer drugs in human lung cancer. *Clin. Cancer Res.* **2009**, *15*, 3423–3432. [CrossRef]
67. Jaramillo, M.C.; Zhang, D.D. The emerging role of the Nrf2-Keap1 signaling pathway in cancer. *Genes Dev.* **2013**, *27*, 2179–2191. [CrossRef]
68. Yamadori, T.; Ishii, Y.; Homma, S.; Morishima, Y.; Kurishima, K.; Itoh, K.; Yamamoto, M.; Minami, Y.; Noguchi, M.; Hizawa, N. Molecular mechanisms for the regulation of Nrf2-mediated cell proliferation in non-small-cell lung cancers. *Oncogene* **2012**, *31*, 4768–4777. [CrossRef]
69. Hosein, A.N.; Beg, M.S. Pancreatic Cancer Metabolism: Molecular Mechanisms and Clinical Applications. *Curr. Oncol. Rep.* **2018**, *20*, 56. [CrossRef] [PubMed]
70. Marco, G.; Aijaz, A.W.; Gang, L. The NAD(P)H:Quinone Oxidoreductase 1 induces cell cycle progression and proliferation of melanoma cells. *Free Radical. Bio. Med.* **2010**, *48*, 1601–1609.
71. Marrot, L.; Jones, C.; Perez, P.; Meunier, J.R. The significance of Nrf2 pathway in (photo)-oxidative stress response in melanocytes and keratinocytes of the human epidermis. *Pigment. Cell Melanoma Res.* **2008**, *21*, 79–88. [CrossRef]
72. Yang, Y.; Zheng, J.; Wang, M.; Zhang, J.; Tian, T.; Wang, Z.; Yuan, S.; Liu, L.; Zhu, P.; Gu, F.; et al. NQO1 promotes an aggressive phenotype in hepatocellular carcinoma via amplifying ERK-NRF2 signaling. *Cancer Sci.* **2021**, *112*, 641–654. [CrossRef]
73. Cedric, O.R.; Panos, G.Z.; Dionysios, V.C.; Dionysios, V.C.; Massimo, B.; Gerasimos, P.S. Keap1/Nrf2 Signaling: A New Player in Thyroid Pathophysiology and Thyroid Cancer. *Front. Endocrinol.* **2019**, *10*, 510.
74. Carpenter, E.L.; Becker, A.L.; Indra, A.K. NRF2 and Key Transcriptional Targets in Melanoma Redox Manipulation. *Cancers* **2022**, *14*, 1531. [CrossRef]
75. Chu, C.; Geng, Y.; Zhou, Y.; Sicinski, P. Cyclin E in normal physiology and disease states. *Trends Cell Biol.* **2021**, *31*, 732–746. [CrossRef]
76. Siu, K.T.; Rosner, M.R.; Minella, A.C. An integrated view of cyclin E function and regulation. *Cell Cycle* **2012**, *11*, 57–64. [CrossRef]
77. Hwang, H.C.; Clurman, B.E. Cyclin E in normal and neoplastic cell cycles. *Oncogene* **2005**, *24*, 2776–2786. [CrossRef]
78. Lee, G.; Lee, C.; An, H.; Kang, H.C.; Lee, H.S.; Lee, J.Y.; Oh, S.; Cho, S.; Kim, D.J.; Cho, Y. Fargesin Inhibits EGF-Induced Cell Transformation and Colon Cancer Cell Growth by Suppression of CDK2/Cyclin E Signaling Pathway. *Int. J. Mol. Sci.* **2021**, *22*, 2073. [CrossRef]
79. Liang, Y.; Gao, H.; Lin, S.Y.; Goss, J.A.; Brunicardi, F.C.; Li, K. siRNA-based targeting of cyclin E overexpression inhibits breast cancer cell growth and suppresses tumor development in breast cancer mouse model. *PLoS ONE* **2010**, *5*, e12860. [CrossRef]
80. Lastra, D.; Escoll, M.; Cuadrado, A. Transcription Factor NRF2 Participates in Cell Cycle Progression at the Level of G1/S and Mitotic Checkpoints. *Antioxidants* **2022**, *11*, 946. [CrossRef]
81. Reddy, N.M.; Kleeberger, S.R.; Bream, J.H.; Fallon, P.G.; Kensler, T.W.; Yamamoto, M.; Reddy, S.P. Genetic disruption of the Nrf2 compromises cell-cycle progression by impairing GSH-induced redox signaling. *Oncogene* **2008**, *27*, 5821–5832. [CrossRef]
82. Zhang, K.; Chen, D.; Ma, K.; Wu, X.; Hao, H.; Jiang, S. NAD(P)H:Quinone Oxidoreductase 1 (NQO1) as a Therapeutic and Diagnostic Target in Cancer. *J. Med. Chem.* **2018**, *61*, 6983–7003. [CrossRef]
83. Oh, E.T.; Park, H.J. Implications of NQO1 in cancer therapy. *BMB Rep.* **2015**, *48*, 609–617. [CrossRef]
84. Russo, M.; Spagnuolo, C.; Russo, G.L.; Skalicka-Wozniak, K.; Daglia, M.; Sobarzo-Sanchez, E.; Nabavi, S.F.; Nabavi, S.M. Nrf2 targeting by sulforaphane: A potential therapy for cancer treatment. *Crit. Rev. Food Sci. Nutr.* **2018**, *58*, 1391–1405. [CrossRef]
85. Hammad, A.; Namani, A.; Elshaer, M.; Wang, X.J.; Tang, X. "NRF2 addiction" in lung cancer cells and its impact on cancer therapy. *Cancer Lett.* **2019**, *467*, 40–49. [CrossRef]
86. Yang, Y.; Yang, I.; Cao, M.; Su, Z.Y.; Wu, R.; Guo, Y.; Fang, M.; Kong, A.N. Fucoxanthin Elicits Epigenetic Modifications, Nrf2 Activation and Blocking Transformation in Mouse Skin JB6 P+ Cells. *AAPS J.* **2018**, *20*, 32. [CrossRef]
87. Chun, K.S.; Kundu, J.; Kundu, J.K.; Surh, Y.J. Targeting Nrf2-Keap1 signaling for chemoprevention of skin carcinogenesis with bioactive phytochemicals. *Toxicol. Lett.* **2014**, *229*, 73–84. [CrossRef]
88. Yang, H.L.; Lee, C.L.; Korivi, M.; Liao, J.W.; Rajendran, P.; Wu, J.J.; Hseu, Y.C. Zerumbone protects human skin keratinocytes against UVA-irradiated damages through Nrf2 induction. *Biochem. Pharmacol.* **2018**, *148*, 130–146. [CrossRef]
89. Sivinski, J.; Zhang, D.D.; Chapman, E. Targeting NRF2 to treat cancer. *Semin. Cancer Biol.* **2021**, *76*, 61–73. [CrossRef]
90. Yang, Y.; Yin, R.; Wu, R.; Ramirez, C.N.; Sargsyan, D.; Li, S.; Wang, L.; Cheng, D.; Wang, C.; Hudlikar, R.; et al. DNA methylome and transcriptome alterations and cancer prevention by triterpenoid ursolic acid in UVB-induced skin tumor in mice. *Mol. Carcinog.* **2019**, *58*, 1738–1753. [CrossRef]

International Journal of
Environmental Research and Public Health

MDPI

Article

Identifying Algicides of *Enterobacter hormaechei* F2 for Control of the Harmful Alga *Microcystis aeruginosa*

Bin Zhang [1,†], Ying Yang [2,3,†], Wenjia Xie [1], Wei He [1], Jia Xie [1] and Wei Liu [1,4,*]

1 School of Environmental Science and Engineering, Sun Yat-sen University, Guangzhou 510006, China; zhangb268@mail2.sysu.edu.cn (B.Z.); xiewj3@mail2.sysu.edu.cn (W.X.); hewei67@mail2.sysu.edu.cn (W.H.); xiej79@mail2.sysu.edu.cn (J.X.)
2 School of Marine Sciences, Sun Yat-sen University, Zhuhai 519082, China; yangying6@mail.sysu.edu.cn
3 Southern Marine Science and Engineering Guangdong Laboratory (Zhuhai), Zhuhai 519080, China
4 Guangdong Provincial Key Laboratory of Environmental Pollution and Remediation Technology, Guangzhou 510006, China
* Correspondence: esslw@mail.sysu.edu.cn
† These authors contributed equally to this work.

Abstract: Eutrophication has become an increasingly serious environmental issue and has contributed towards an explosion in harmful algal blooms (HABs) affecting local development. HABs can cause serious threats to ecosystems and human health. A newly isolated algicidal strain, *Enterobacter hormaechei* F2, showed high algicidal activity against the typical HAB species *Microcystis aeruginosa*. Potential algicides were detected through liquid chromatograph–mass spectrometer analysis, revealing that prodigiosin is an algicide and PQS is a quorum sensing molecule. RNA-seq was used to understand the algicidal mechanisms and the related pathways. We concluded that the metabolism of prodigiosin and PQS are active at the transcriptional level. The findings indicate that *E. hormaechei* F2 can be used as a potential biological agent to control harmful algal blooms to prevent the deterioration of the ecological and economic value of water bodies.

Keywords: algicidal bacteria; prodigiosin; quorum sensing molecular; transcriptome

Citation: Zhang, B.; Yang, Y.; Xie, W.; He, W.; Xie, J.; Liu, W. Identifying Algicides of *Enterobacter hormaechei* F2 for Control of the Harmful Alga *Microcystis aeruginosa*. *Int. J. Environ. Res. Public Health* **2022**, *19*, 7556. https://doi.org/10.3390/ijerph19137556

Academic Editor: Cheng Yan

Received: 15 April 2022
Accepted: 16 June 2022
Published: 21 June 2022

1. Introduction

Harmful algal blooms (HABs) can cause serious damage to the culture and quality of aquatic environments [1]. The cyanobacterium *Microcystis aeruginosa* is a typical HAB species, which can produce toxic microcystins (MCs). Concentrations of MCs at 30 μg/L were found to cause damage to mice and zebrafish [2,3]; however, the concentrations of these compounds in numerous surface water tests have exceeded this level [4–6].

HABs can be treated by physical, chemical, and biological methods, among which biological treatment utilizing microbial agents is effective at inhibiting their growth and reducing eutrophic substances in natural water bodies. Algicidal bacteria act on algal cells via direct contact and indirect attack. The former involves directly attacking algal cells through the direct contact of algal surfaces, whereas the latter involves killing algal cells by secreting algicides [7,8]. Due to the high bacterial diversity and difficulty in separating algicidal substances, many algicides are still unknown, and only a few algicides have been identified, which include peptides [9,10], alkaloids [11,12], and lipids [13,14]. One common algicide is prodigiosin, a red pigment belonging to the antibiotic family, and is an efficient algicide [15,16] against *Phaeocystis globosa* [17], red tide phytoplankton [18], and *M. aeruginosa* [19–21]. Several extracellular enzymes have also been identified as potent algicides. The activity of extracellular enzymes of white rot fungi *Phanerochaete chrysosporium* varied with the algicidal process, and different enzymes functioned in different algicidal phases, for example, laccase and MnP mainly function in the early and late stages, respectively [22].

Algicidal bacteria degrade algal cells by destroying their cell morphology through cell membrane and cell wall damage, as well as affecting their physiological activities [23].

Some studies have revealed that more algal cells are propidium iodide-stained after co-culture with algicides, indicating the effect of algicides on cell membrane permeability, thereby allowing the entry of foreign substances into the cells, and consequently damaging cellular organelles [24]. Algicidal bacteria can also reduce the photosynthesis efficiency and antioxidant activity of algal cells by destroying the photosynthesis system that absorbs light and suppressing electron transfer in photosynthesis reactions [17,19,25]. Furthermore, algicidal bacteria can increase superfluous reactive oxygen species content in algal cells and ultimately cause algal cell death [19,26].

Previous studies on algicidal bacteria have mainly focused on purifying and identifying algicides or methods to promote algicidal efficiency. Studies on algicidal mechanisms have primarily focused on algal cell damage, whereas studies based on molecular biology perspectives are limited. Kwon et al. [27] found a gene cluster related to prodigiosin secretion with an algicidal effect in *Hahella chejuensis* and *Streptomyces coeclicolor A3*. By using transcriptional analysis, Zhang et al. [28] indicated that the algicidal activity of *Brevibacillus laterosporus* was performed by suppressed photosynthesis and oxidative phosphorylation of *M. aeruginosa*, which could block electron transport and affect energy absorption of the cyanobacterium.

The aim of this study was to identify the algicides secreted by *Enterobacter hormaechei* F2, an algicidal bacterium isolated in a previous study, and determine the relevant mechanisms of bacterial gene expression during the algicidal process using transcriptomic analysis. Finally, transmission electron microscopy (TEM) and scanning electron microscopy (SEM) were used to observe morphological changes in algal cells during the algicidal process. The results could provide useful information for the bio-control of HABs, which pose a severe environmental threat.

2. Materials and Methods

2.1. Growth Conditions of Algicidal Bacteria and Cyanobacterial Cells

The algicidal bacteria *E. hormaechei* F2 was isolated from the Pearl River, Guangzhou, China. The strain was cultured in a Luria–Bertani medium (LB medium: 5 g yeast extract, 10 g tryptone in L distilled water, pH 7.0) in an incubator shaker at 150 rpm and 37 °C for 48 h. The experimental cyanobacterium, *M. aeruginosa* (FACHB-315), was provided by the Freshwater Algae Culture collection at the Institute of Hydrobiology of Chinese Academy of Sciences (FACHB, Wuhan, China), and was cultured in BG11 medium (BG11 medium: 1.500 g $NaNO_3$, 0.040 g K_2HPO_4, 0.001 g EDTA·2Na, 0.075 g $MgSO_4 \cdot 7H_2O$, 0.036 g $CaCl_2 \cdot 2H_2O$, 0.006 g ferric ammonium citrate, 0.006 g citric acid, 0.020 g Na_2CO_3, 1 mL A_5+Co solution in 1 L distilled water, pH 7.1). The A_5+Co solution was prepared by dissolving the following in 1 L of distilled water: 2.860 g H_3BO_3, 1.810 g $MnCl_2 \cdot 4H_2O$, 0.049 g $Co(NO_3)_2 \cdot 6H_2O$, 0.222 g $ZnSO_4 \cdot 7H_2O$, 0.079 g $Na_2MoO4 \cdot 2H_2O$, 0.390 g $CuSO_4 \cdot 5H_2O$. The cyanobacterial culture was kept at 25 °C under a 15:9 h light-dark cycle, with a light intensity of 2500 Lux.

2.2. Algicidal Activity

To determine the algicidal activity of *E. hormaechei* F2 on *M. aeruginosa*, *E. hormaechei* F2 was inoculated in 100 mL LB medium and grew to a stationary phase for 48 h in a shaker at 150 rpm. Then, the bacterial culture was added to the cyanobacterial cultures at a concentration of 2.0% (final concentration at 2×10^5 CFU/mL), whereas the cyanobacterial cultures with the same concentration of LB medium were established as the control group. The cell density of *M. aeruginosa* was determined by measuring the chlorophyll *a* (Chl *a*) concentration using the acetone method [29].

Microcystis aeruginosa cells were counted by a hemocytometer plate according to Equation (1).

$$\text{Cells/mL} = \frac{N}{100} \times 400 \times 10^4 \times \text{Dilution factor} \quad (1)$$

where N represents the cell numbers in 100 small squares.

The algicidal ratio was calculated according to Equation (2).

$$\text{Algicidal ratio} = \frac{Nc - Ne}{Nc} \times 100 \quad (2)$$

where *Nc* and *Ne* represent the algal chlorophyll *a* concentration in the control and experimental groups, respectively. The algicidal ratio was determined after inoculation for 4 days.

Prodigiosin (Tokyo Chemical Industry, Tokyo, Japan), PQS (Bidepharm, Shanghai, China) and phenazine (Bidepharm, Shanghai, China) were dissolved in dimethyl sulfoxide for algicidal activity. The mother liquor concentration was 2 mol/L, with filter sterilization. Equal amounts of dimethyl sulfoxide were added to the control group for these experiments.

Laccase activity was performed by the ABTS method [22].

2.3. Nucleic Acid-Related Manipulation

Genomic DNA was extracted with a DNA Kit (Omega, Hampton, NH, USA, D3195-01). PCR amplification was performed using Phusion High-Fidelity DNA Polymerase (Thermo Scientific, Waltham, MA, USA, lot:00528748). DNA fragment elution was performed using a Gel Extraction Kit (Omega, D2500-02) and/or a Cycle Pure Kit (Omega, D6492-02). For the Southern blot assay, restriction enzymes used for the digestion of genomic DNA were from NEB (New England Biolabs (Beijing) Ltd., Beijing, China). The Labeling and Detection Starter Kit I (Roche, Basel, Switzerland, 11745832910) was used to label PCR amplified fragments as a probe. Amersham Hybond TM-N+ (GE Healthcare, Chicago, IL, USA, RFN303B) membranes were used for blotting. NBT/BCIP Stock Solution (Roche, 11681451001) was used for probed band detection.

For the transcriptome assay, the co-cultivation of *E. hormaechei* F2 and *M. aeruginosa* on 0 d were used as controls, and 4 d and 7 d as the experimental groups. The cells were collected by centrifugation (5000× *g*, 10 min, 4 °C) and total RNA extraction was performed by a Qiagen RNeasy Mini kit (74104). The Ambion® TURBO DNA-free™ kit (Invitrogen, Waltham, MA, USA, AM1907) was used to remove the contaminating DNA from the RNA preparations. The TransScript® First-Strand cDNA Synthesis Super Mix (Transgen, Beijing, China, AT301-02) was used for cDNA synthesis. The differentially expressed genes (DEGs) screening conditions were as follows: $\log2 | \text{fold change} | > 1$, p-value < 0.05.

The identified genes were annotated using the Kyoto Encyclopedia of Genes and Genomes (KEGG) Compound database (http://www.kegg.jp/kegg/compound/ (accessed on 10 February 2020)) and mapped to the KEGG Pathway database (http://www.kegg.jp/kegg/pathway.html (accessed on 10 February 2020)), as well as the Gene Ontology (GO) database (http://geneontology.org/ (accessed on 10 February 2020)). Pathways with significantly regulated genes and metabolites mapped were then fed into metabolite set enrichment analysis (MSEA), and their significance was determined using hypergeometric test *p*-values.

For real-time qPCR, the sample collection and RNA extraction were the same as the transcriptome assay. We used PowerUp™ SYBR® Green Master Mix (Applied Biosystems, Waltham, MA, USA, A25742), and the reaction (95 °C 5 min, (95 °C 10 s, 60 °C 30 s) 30 cycles, 95 °C 15 s, 60 °C 60 s, 95 °C 15 s) ran on the QuantStudio 6 Flex Real-Time PCR System (Thermo Fisher Scientific, Waltham, MA, USA). Next, $2^{-\Delta\Delta Ct}$ was used to calculate the relative fold change. The primers used for qRT-PCR analysis are listed in Table S2.

2.4. Extraction and Purification of the Algicidal Compound

The algicidal bacteria was grown in a 1 L flask with LB medium in a shaker (150 rpm, 37 °C) and then centrifuged (5000× *g*, 10 min, 4 °C) to obtain the cell-free medium. Then, the solutions were concentrated to 100 mL using a rotary evaporator (40 °C), supplemented with 65 mL of 95% ethanol, and placed into the fridge (4 °C) overnight. Then, the culture was centrifuged (5000× *g*, 10 min, 4 °C) to obtain the supernatant. Petroleum ether, ethyl acetate, n-butyl alcohol, and chloroform were utilized as extractants and were added with the extract to the cyanobacterial culture at a concentration of 1 g/L to test the algicidal effect. The n-butyl alcohol extract of the cell-free medium showed the strongest algicidal activity. Then, the n-butyl alcohol extract was separated with a silica gel chromatographic column via a degraded chloroform and methanol series (*v/v* = 1:0, 5:1, 2:1, 1.5:1, 1:1, 1:1.5 1:2, 1:5, 0:1) at a rate of 0.75 mL/min, and each fraction was evaporated and added to the cyanobacterial culture at a concentration of 1 g/L to test the algicidal activity. The fractions were defined as follows: extract 1 (*v/v* = 1:0), extract 2 (*v/v* = 5:1), extract 3 (*v/v* = 2:1), extract 4 (*v/v* = 1.5:1), extract 5 (*v/v* = 1:1), extract 6 (*v/v* = 1:1.5), extract 7 (*v/v* = 1:2), extract 8 (*v/v* = 1:5), extract 9 (*v/v* = 0:1). Among these fractions, extract 2 (*v/v* = 5:1) and extract 7 (*v/v* = 1:2) exhibited strong algicidal effects.

2.5. Liquid Chromatograph–Mass Spectrometer (LC–MS) Analysis

Extract 2 and 7, prodigiosin (Tokyo Chemical Industry, Tokyo, Japan) and PQS (Bidepharm, Shanghai, China) were dissolved in methanol, and the solutions were filtered with a 0.22 μm filter membrane. Compound separation was performed with a reverse-phase chromatographic column (Agilent XB-C18, 4.6 mm × 50 mm, 2.6 μm) at 40 °C, with mobile phase A consisting of 0.1% HCOOH, mobile phase B comprising 0,1% HCOOH-CAN, and a flow rate of 1.2 mL/min. The effluent was transferred to the mass spectrometer with an electro-spray ionization source. The mass range was between 105 and 1500 m/z. The calculated method referred to Liu et al. [30].

2.6. Transmission Electron Microscopy

Cyanobacterial cells were incubated with the bacterial culture for 0, 4, and 7 d, and then prepared for TEM. Next, 30 mL of cyanobacterial culture was collected (5000× *g*, 10 min, 4 °C) and fixed with 2.5% glutaraldehyde (*v/v*) for 24 h. The sample was then rinsed three times with PBS buffer and fixed with 1% OsO_4 for 2 h. The samples were then rinsed with the PBS buffer three times, dehydrated in a degraded ethanol series (30%, 50%, 70%, 80%, 90%, 95% and 100%), and then stored in tertiary butanol for 20 min. The samples were embedded with epoxy resin and acetone (*v:v* = 3:1) for 3 h and then embedded for one night at 70 °C. The sections (70–90 nm) were obtained with an ultramicrotome (LEICA-EM-UC7) and stained with lead citrate and uranium-citric for 5 min. The samples were viewed by TEM (H7650).

2.7. Scanning Electron Microscopy

Cyanobacterial cells were incubated with bacterial culture for 0, 4, and 7 d and then prepared for TEM. Here, 30 mL of cyanobacterial culture was collected (5000× *g*, 10 min, 4 °C) and fixed with 2.5% glutaraldehyde (*v/v*) for 24 h. The sample was then rinsed with PBS buffer and fixed with 1% OsO_4 for 2 h. The samples were rinsed with PBS buffer three times, dehydrated in a degraded ethanol series (30%, 50%,70%, 80%, 90%, 95% and 100%), and then stored in tertiary butanol for 20 min. Next, the samples were treated with ethanol and isoamyl acetate solution (*v/v* = 1/1) for 30 min and stored in isoamyl acetate overnight. The samples were critical-point-dried and coated with gold-palladium and finally imaged by SEM (SU8010).

2.8. Statistical Analysis

In the transcriptome analysis, unsupervised principal component analysis (PCA) was performed using the statistical function prcomp in R (www.r-project.org (accessed on 3 March 2020)). The data were unit variance-scaled before unsupervised PCA. The hierarchical cluster analysis (HCA) results of the samples and metabolites were presented as heatmaps with dendrograms, whereas Pearson correlation coefficients (PCC) between the samples were calculated using the cor function in R and presented as heatmaps. Both HCA and PCC were performed using the R package ComplexHeatmap. For HCA, the normalized signal intensities of the genes and metabolites (unit variance scaling) were visualized as a color spectrum.

In the algicidal activity analysis, the figures and statistics are based on mean value ± standard error (S.E.). Significance analysis of difference was performed by paired Student's *t*-test.

3. Results and Discussion

3.1. Algicidal Effects of the Strain

An algicidal bacterium, which was identified and named as *E. hormaechei* F2 in a previous study, showed effects on the damage or death of cyanobacterium cells (Figure 1A). The cyanobacterial solution appeared green at the beginning of bacteria-cyanobacterium co-cultivation, and the color of the cyanobacterial solution faded slightly at 4 d and became nearly colorless at 7 d of co-cultivation. The color change during the co-cultivation period suggested the algicidal effect of *E. hormaechei* F2 on *M. aeruginosa*.

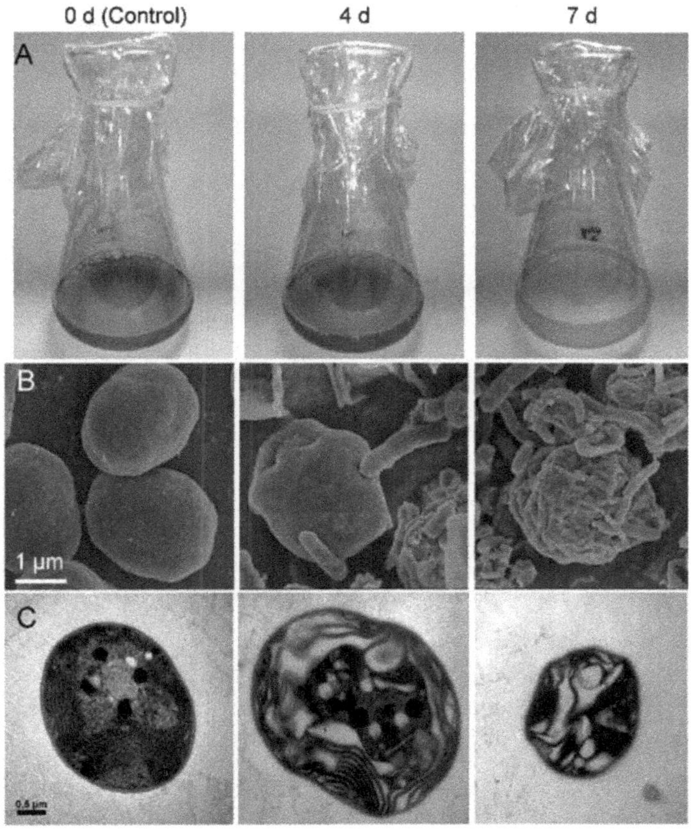

Figure 1. Electron microscopic images of *Microcystis aeruginosa* co-incubated with *Enterobacter hormaechei* F2. (**A**) Phenotype-based images of algicidal effects. (**B**) Scanning electron microscopy images, bar = 1 μm; to enable easier distinguishing of bacteria from algae, the photo was processed in color, with cyanobacterial cells in green and bacteria in yellow. (**C**) Transmission electron microscopy images, bar = 0.5 μm.

Electron microscopy was used to observe the morphological changes in the cyanobacterium under the influence of algicidal *E. hormaechei* F2 (Figure 1B). Before co-cultivation (0 d of co-cultivation), cyanobacterial cells exhibited a spherical shape with a smooth surface; however, after 4 d of co-cultivation, cyanobacterial cells shrunk, and the SEM images showed the adherence of bacterial cells to the cyanobacterial cell surface. After 7 d of co-cultivation, the cyanobacterial surface became significantly rugose and the bacteria adhered to the cyanobacterial cells, which demonstrated direct contact algicidal activity [31,32].

Changes inside of the cyanobacterial cells are shown by the TEM images (Figure 1C). Before co-cultivation, the cells had a complete membrane structure, the cytoplasm was evenly distributed, and the thylakoid structure was closely arranged. With an increase in time, the structure of the cell membrane became irregular, vacuoles appeared in the cells, and thylakoids were loosely and irregularly arranged. By the end of co-cultivation, the cyanobacterial cells became smaller and the cytoplasm and thylakoids were reduced, indicating the death of the cyanobacterial cells.

3.2. Detection of Algicides

3.2.1. Separation and Identification of Algicides

Algicidal bacteria can secrete various algicidal compounds [7,8,19–21]. To investigate the algicides secreted by *E. hormaechei* F2, components from the cultivation medium of the strain were extracted using nine different volume ratios of chloroform and methanol, and the algicidal efficiency of each extract were further tested. As shown in Figure 2A, Extracts 2, 5, 6, and 7 demonstrated higher inhibition efficiencies than other extracts, which were 59.54%, 54.60%, 57.49%, and 59.54%, respectively. However, the inhibition efficiencies of all the extracts were lower than those of the co-cultivation system, suggesting that the effective algicidal performance of *E. hormaechei* F2 (84.2%) was a combined result of different algicides and mechanisms.

To further explore the potential algicides in extracellular substances, liquid chromatograph–mass spectrometer analysis (LC–MS) was employed to detect the chromatograms of Extracts 2, 5, 6, and 7, using the widely reported algicide prodigiosin and AQs family (2-alkyl-4-quinolone family) [17,25,27,33] as standard substances. A characteristic peak of PQS-like (2-heptyl-3-hydroxy-quinolone, *Pseudomonas* quinolone signal) substances, which belong to the AQs family, was found in Extract 2, indicating the existence of PQS. The presence of prodigiosin was also shown in the chromatogram of Extract 7 (Figure 2C). Neither a PQS-like substance nor prodigiosin was found in Extract 5 and 6, suggesting the presence of an unknown algicide.

3.2.2. Algicidal Effects of Identified Substances

Prodigiosin as a Potential Algicide

The chromatogram of the extracellular substances demonstrated that PQS and prodigiosin are potential algicidal compounds of *E. hormaechei* F2, and they were constitutively expressed and produced. Further studies were conducted to determine the algicidal effects of these compounds. However, unlike other reported AQs families, such as HHQ (2-heptyl-4-quinolone) [34,35], the growth of *M. aeruginosa* was not affected after a 4 d treatment with PQS (Figure 3A). Prodigiosin, the algicide identified in Extract 6, belonged to a family of red pigments [17,25,27,33]. It has also been reported that phenazine-like components can control red tide organisms [36]. Prodigiosin showed algicidal activity at a concentration of 0.5 μM (Figure 3), as indicated by the fading of the cyanobacterial color and the significant decrease in cyanobacterial cell concentrations from 1×10^7 to 5×10^5 cells/mL.

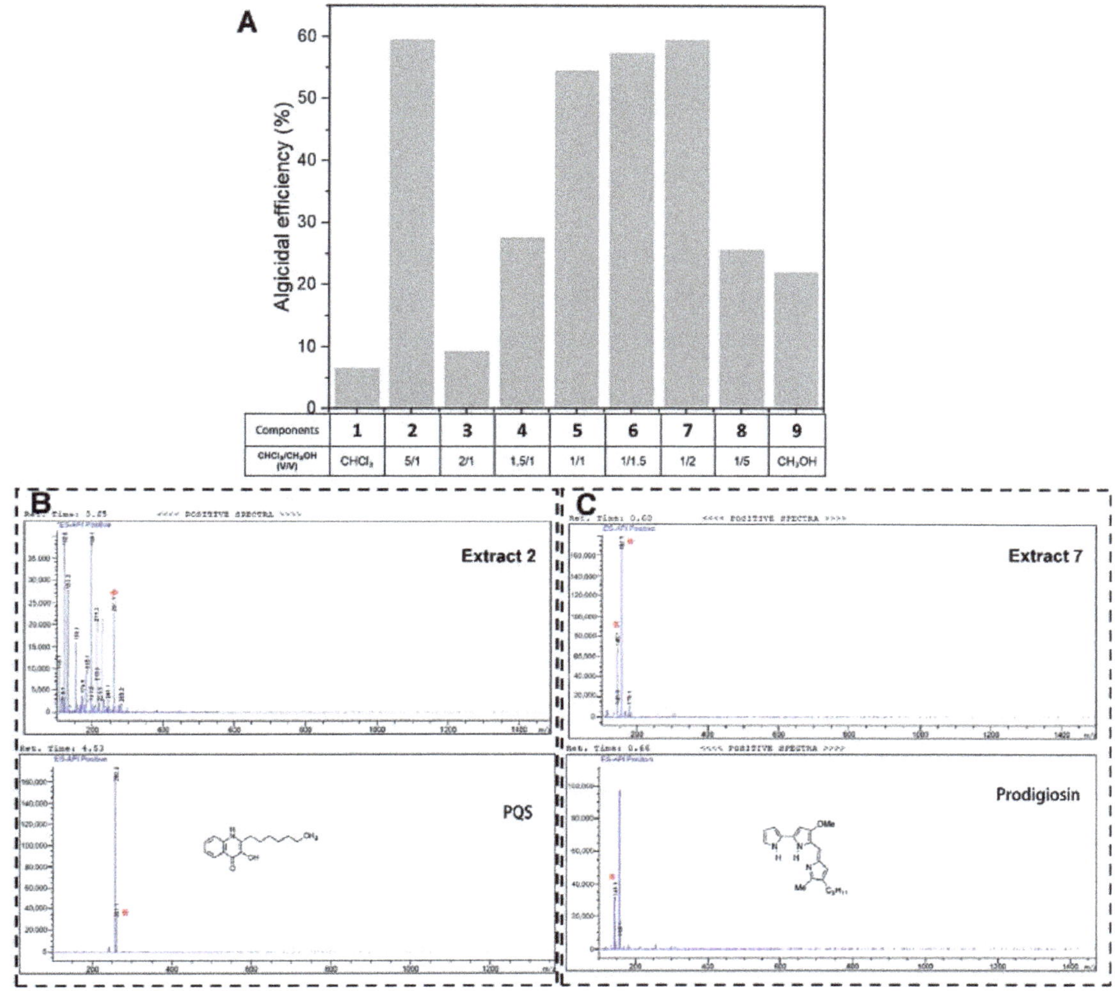

Figure 2. Algicidal effect of extracellular substances and potential algicides. (**A**) Algicidal efficiency of *Enterobacter hormaechei* F2 extracellular substances from different extracts. (**B**) LC–MS analysis to detect the presence of 2-heptyl-3-hydroxy-quinolone (PQS). (**C**) LC–MS analysis to detect the presence of prodigiosin; "*" indicates characteristic peaks of similar retention times.

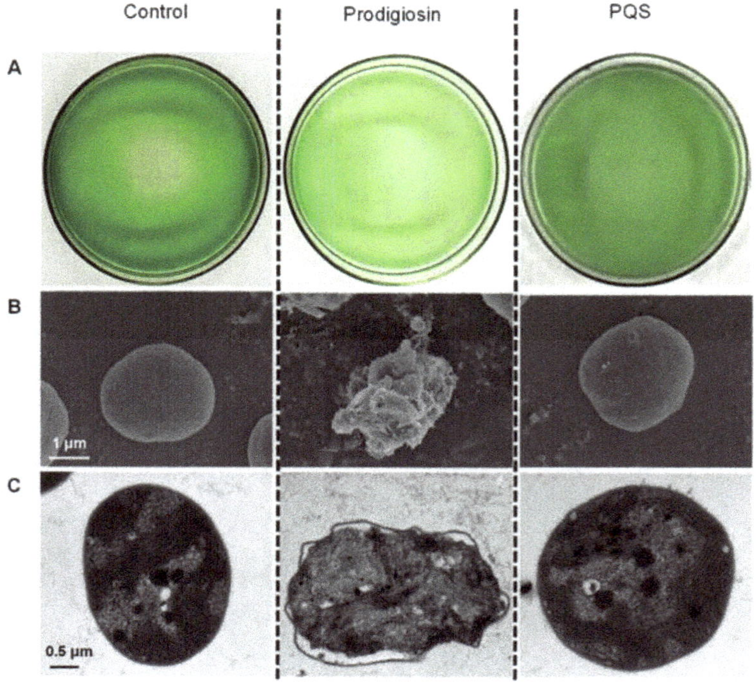

Figure 3. Algicidal activity of prodigiosin and 2-heptyl-3-hydroxy-quinolone (PQS). (**A**) Phenotype images that were taken from the bottom of the flask; (**B**) scanning electron microscopy images, bar = 1 μm; (**C**) transmission electron microscopy images, bar = 0.5 μm.

SEM images revealed that with prodigiosin treatment, the cyanobacterial cells were severely deformed when compared with the control cells (Figure 3B). TEM images showed the inner damage of the cyanobacterium; when the cyanobacterial cells were treated with prodigiosin, the cell wall remained intact, the peptidoglycan layer remained intact (the cell wall was maintained as a double-layer structure), chromatin contracted, and the carboxysome structure was maintained (Figure 3C). These phenomena are similar to the previously observed algicidal effects of the strain, which suggests that prodigiosin is a potential algicide of *E. hormaechei* F2.

PQS Might Act as a QS Molecule That Regulates Algicidal Genes and Enzymes

The direct algicidal effect of PQS from *E. hormaechei* F2 was not observed in this study; however, a previous study proposed that PQS might act as QS signaling molecule that can regulate strain algicidal activity [35]. Therefore, we tested this hypothesis by adding pure PQS into the bacteria-cyanobacterium co-cultivated system. The results of the cyanobacterial cell concentration and chlorophyll *a* content analyses both showed that the addition of the PQS-like substance could effectively enhance the algicidal effect of *E. hormaechei* F2 (Figure 4A,B). The fading of the green color was accelerated by an increase in the dosage of the PQS-like substance. As shown in Figure 4C, treatment with 2 μM of the PQS-like substance from *E. hormaechei* F2 for 4 d resulted in the cyanobacterial solution becoming almost transparent, compared with the control. Furthermore, there was a significant reduction in both the cyanobacterial cell concentration and chlorophyll *a* content, suggesting a better algicidal effect than in other groups at 4 d, and a more significant effect at 7 d. In the subsequent experiment, 2 μM PQS was used to study the effects of this substance and mechanism.

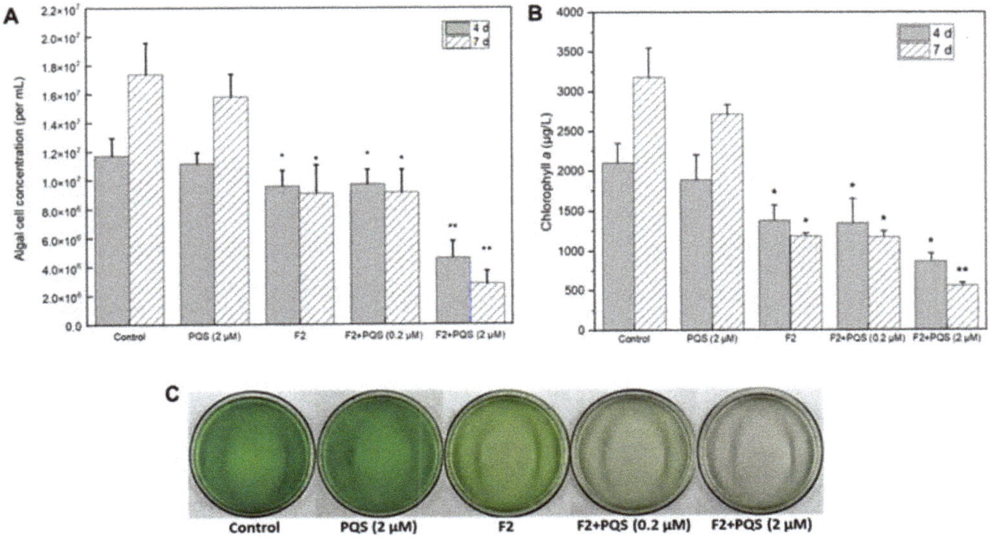

Figure 4. Algicidal activity of PQS from *Enterobacter hormaechei* F2. (**A**) Cyanobacterial cell concentration; (**B**) chlorophyll *a* content; (**C**) phenotype images that were taken from the bottom of the flask. "*" indicates a significant difference ($0.01 \leq p$ value ≤ 0.05), "**" indicates a very significant difference (p value ≤ 0.01).

As QS molecules generally promote bacterial growth [37], the effect of PQS on *E. hormaechei* F2 growth was assessed. Notably, the growth of *E. hormaechei* F2 was slightly suppressed after 16 h of incubation with PQS treatment (Figure 5A), implying that PQS promotes cyanobacterial inhibition by regulating gene expression rather than by increasing bacteria biomass. To verify this assumption, enzymatic activity and qRT-PCR analyses were performed. Laccase, which was reported to be an important algicidal enzyme [22], showed significantly higher activity when co-incubated with *M. aeruginosa*, suggesting the involvement of laccase in the algicidal processes of *E. hormaechei* F2. With the presence of PQS at a concentration of 2 μM, the activity of laccase became significantly higher, indicating that PQS can increase the activity of the enzyme, which promoted the algicidal activity of *E. hormaechei* F2 (Figure 5B).

Three algicide-related genes, including K00059, K00652, and K01657, were chosen for an expression test [17–19,27]. The results showed that in the absence of PQS, the expression level of K00059 did not change from 4 to 7 d. However, K00652 and K01657 were up-regulated at 7 d relative to levels at 4 d. When PQS was added, the expression level of these genes remained similar to that in the absence of PQS on day 4 but were highly up-regulated at 7 d, indicating that the regulatory effect of PQS was most pronounced after 4 d. In summary, PQS participates in the algicidal activity as a QS molecule that regulates algicide-related enzyme activity and gene expression.

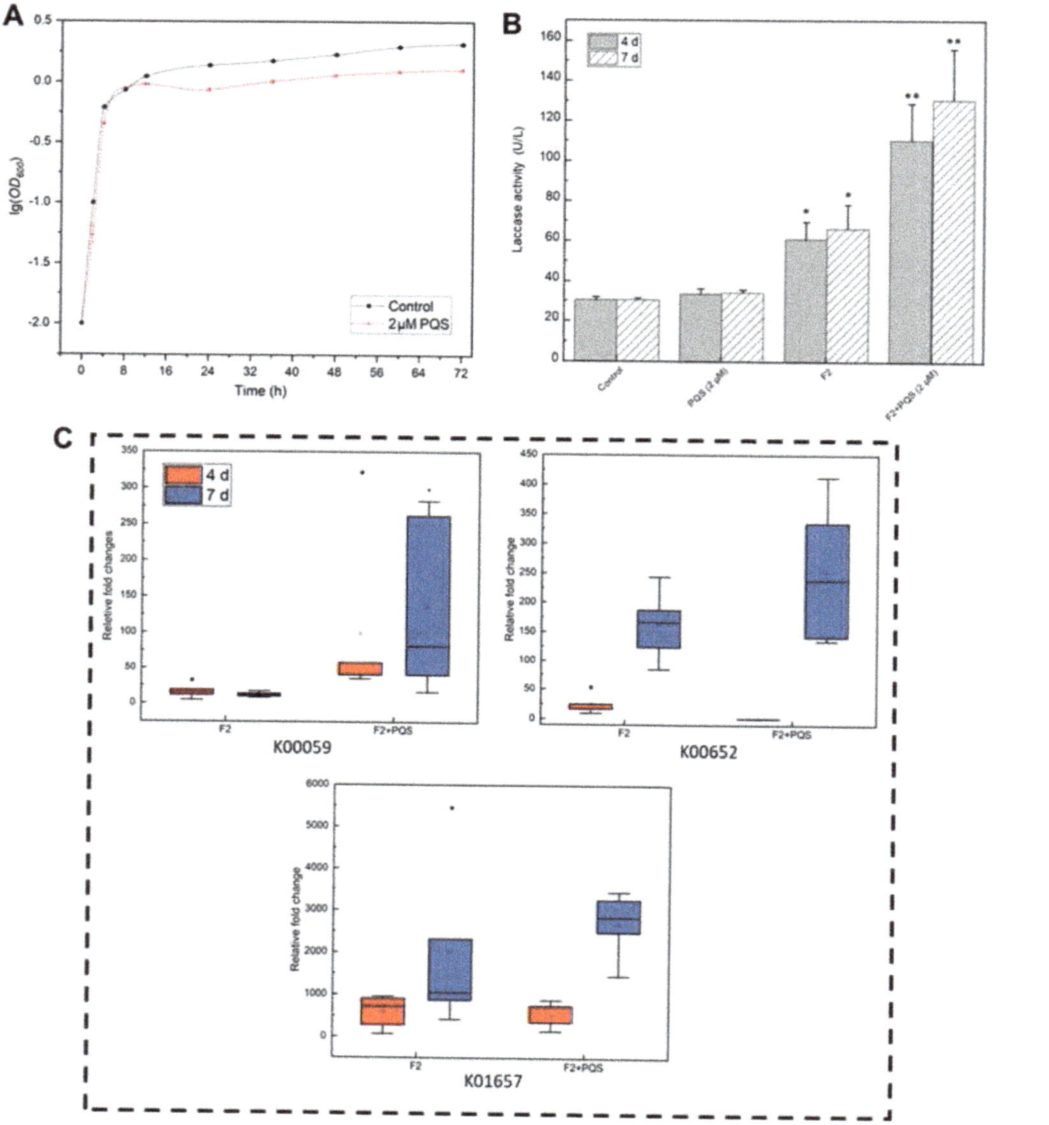

Figure 5. Effects of PQS from *Enterobacter hormaechei* F2. (**A**) Growth curve of *E. hormaechei* F2; the curve was prepared by plotting the logarithmic values of OD_{600} (optical density at 600 nm) vs. incubation time. Mean ± S.E. (**B**) Laccase activity in *E. hormaechei* F2; "*" indicates a significant difference ($0.01 \leq p$ value ≤ 0.05), "**" indicates a very significant difference (p value ≤ 0.01). (**C**) Relative gene expression fold-change, as calculated based on 16S rRNA as a housekeeping control.

3.3. *Transcriptome Analysis Reveals the Algicidal Mechanism of the Strain*

3.3.1. Enrichment of Differentially Expressed Genes (DEGs) during the Algicidal Process

To better understand the algicidal mechanism and the algicide-related pathways, a transcriptome analysis of gene expression in the algicidal bacterium *E. hormaechei* F2 in the co-cultivation system was performed. DEGs were assessed among 0 d, 4 d, and 7 d timepoints.

The overall distribution trend among all the samples can be observed using PCA, and the possible discrete points can be determined. PCA score plots are presented in

Figure S1, the first principal component accounted for 61.49% and the second principal component for 12.14% of the variance; these two principal components contributed over 70%, which represent the characteristics of the data. The PCA score plots showed samples at 4 d (blue dots) and were distinguished from samples at 0 d (red dots) and 7 d (yellow dots), which indicated a specificity of samples at 4 d (Figure S1A). Between 0 d and 4 d, 1023 genes were differentially expressed, with 614 up-regulated and 409 down-regulated. In contrast, 74 up-regulated and 63 down-regulated genes were detected between 4 d and 7 d, indicating that *E. hormaechei F2* was more active in the early stage, based on a higher number of DEGs identified between 0 d and 4 d (Figure S1B).

Hierarchical clustering analysis is a common method for data mining and statistics. As shown in the heatmaps and dendrogram (Figure S1C), genes or samples with similar variation trends in the absolute values of expression fold change extent were positioned closer together. As for the samples, 4 d and 7 d samples were clustered into one branch, indicating that the variation trends of DEGs were similar.

To further investigate the algicidal functions of *E. hormaechei* F2, a gene set enrichment analysis (GSEA) was performed based on the GO database. The GO terms of the identified DEGs at 0 and 4 d were enriched in the nitrate reductase complex of the cellular component category, and ATP synthesis coupled proton transport and phenylacetate catabolic process were enriched in the biological process category (Figure 6A). However, the GO terms of the identified DEGs between 0 and 7 d were enriched in catalytic activity, nitrate reductase activity, and oxidoreductase activity in the molecular function category; nitrate reductase complex in the cellular component category; and phenylacetate catabolic process, oxidation-reduction process, and nitrate metabolic process in the biological process category (Figure 2B). Furthermore, the GO terms of the identified DEGs at 4 and 7 d were enriched in proton-transporting ATP synthase activity (rotational mechanism) in the molecular function category; proton-transporting ATP synthase complex, catalytic core F (1) in the cellular component category; and ATP synthesis coupled proton transport and peptide transport in the biological process category (Figure 2C). In summary, the results of the GO analysis suggest that energy metabolism and aromatic compound metabolism are involved in the algicidal process. Similarly, the KEGG analysis showed that signaling system, energy metabolism, and aromatic amino acid metabolism pathways were crucial to the algicidal activity (Figure S2).

3.3.2. Comparative Genome and RNA Seq Analysis to Decipher the Associated Pathway

The enrichment analysis demonstrated that the signaling system, energy metabolism, and aromatic amino acid metabolism participated in the algicidal process. The identified algicide-related substance (i.e., PQS and prodigiosin) were the final products of the pathway of prodigiosin biosynthesis (map00333) and phenazine biosynthesis (map00405) pathways. To reveal the changes in these pathways at the transcriptional level, transcript profiling of DEGs was conducted (Figure 7). Similar to the analysis of DEGs, most were upregulated at 4 d of co-cultivation and slightly down-regulated at 7 d of co-cultivation, further suggesting that the fourth day is the most active in the algicidal process in terms of the transcriptional level.

Energy metabolism-related genes were upstream of the pathway, indicating an increasing extracellular energy metabolism, such as that related to sucrose, cellobiose, and D-glucose, which were most likely from the breakdown of cyanobacterium. With the enhanced downstream transcriptional activity, phenylalanine, tyrosine, and tryptophan metabolism showed upregulation, which suggested the potential algicide-related compounds in the pathways.

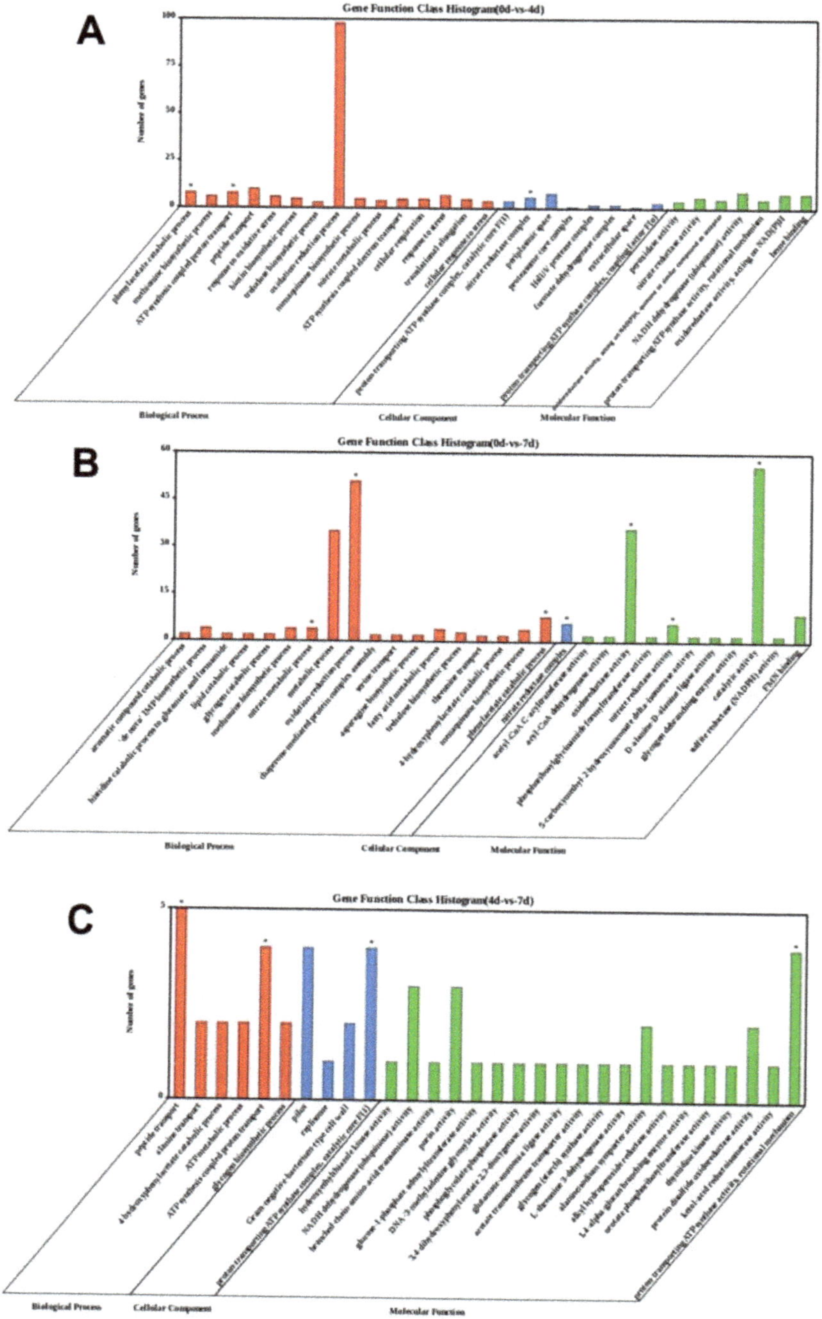

Figure 6. Gene Ontology (GO) functional analysis of differentially expressed genes (DEGs) for (**A**) 0 vs. 4 d (**B**) 0 vs. 7 d, and (**C**) 4 vs. 7 d. "*" indicates a significant enrichment of the GO term or KEGG pathway (p-value ≤ 0.05).

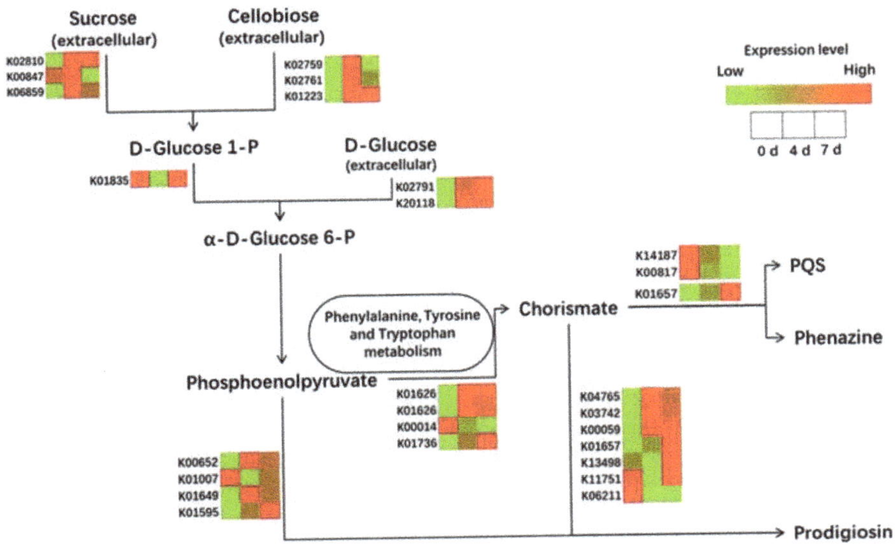

Figure 7. Transcript profiling of differentially expressed genes (DEGs). The color of the grids within a row indicates expression value of a DEG at 0, 4, and 7 d.

Comparative genomic analysis between *E. hormaechei* F2 and other algicidal bacteria (*H. chejuensis* KCTC2396, *Zobellia galactanivorans* DsijT, *Mangrovimonas yunxiaonensis* LY01, and Rhodobacteraceae bacterium PD-2) was employed to simplify the data (Table S1A,B). The results showed that the ortholog genes K00652 and K00059, reported as algicide-related genes that are related to the biosynthesis of prodigiosin and belong to the family of red pigments, were produced as algicides by bacteria [17–19,27]. Furthermore, phenazine-like pigments could control red tide organisms [36], which were observed in the transcript profiling (Figure 7), indicating that this could be an algicide of *E. hormaechei* F2.

Interestingly, phenazine-like pigments have been reported to control red tide organisms [36], and this is a product of the same pathway as PQS (Figure 6). The algicidal effect of phenazine was also assessed, which showed marked algicidal effects, as the color of the cyanobacterium solution almost completely faded after treatment (Figure S3A). As indicated by red arrows in Figure S3C, the cyanobacterial cell wall ruptured and the cytoplasm leaked, which are typical features of cell necrosis [10]. This result suggested that phenazine might be an important algicide for the strain.

The algicidal process of *E. hormaechei* F2 can be summarized as follows: (1) prodigiosin and phenazine may act as algicides of *E. hormaechei* F2. (2) PQS, which might act as a QS molecule, improved algicidal activity by enhancing gene expression and enzyme activity of *E. hormaechei* F2. (3) Simultaneously, *E. hormaechei* F2 adhered to cyanobacterial cells, resulting in the disruption of the cyanobacterial cell membrane in the next 4 d, thereby causing a loss of selective permeability. (4) Chlorophyll *a* content decreased, which could affect the photosynthetic efficiency. (5) Finally, the algal cells were fragmented into small molecules that can be used as nutrient sources.

4. Conclusions

In this study, *E. hormaechei* F2 showed significant algicidal activity against *M. aeruginosa* by secreting algicides. Subsequent analyses demonstrated that the potential algicides were PQS and prodigiosin, which could be detected in the extracellular substances of the strain. Prodigiosin showed high algicidal activity, and this phenomenon was also similar to the algicidal effect of the strain. Meanwhile, PQS did not show direct algicidal activity but could act as a QS molecule that regulates enzyme activity and gene expression. The transcriptome data suggested that throughout the entire algicidal process, amino acids and energy metabolism, a two-component system, the biosynthesis of secondary metabolites and quorum sensing participate in algicidal activity. PQS and prodigiosin are, thus, algicide-related compounds of the strain. The identification of algicides from *E. hormaechei* F2 and an understanding of the potential mechanism underlying their effects could promote our understanding of the performance of this algicidal bacteria, providing an effective way to control HABs, which pose a severe threat to inland aquatic ecosystems and human health.

Supplementary Materials: The following supporting information can be downloaded at: https://www.mdpi.com/article/10.3390/ijerph19137556/s1. Figure S1: Statistical analysis of differentially expressed genes (DEGs), Figure S2: Statistical analysis of differentially expressed genes (DEGs) and Kyoto Encyclopedia of Genes and Genomes (KEGG) functional analysis of DEGs, Figure S3: Algicidal activity of phenazine, Table S1: A Differentially expressed genes between 0 and 4 d, Table S2: Primers used in the study.

Author Contributions: B.Z., Y.Y. and W.X. carried out the design and operation of the experiment and data processing and completed paper writing. W.H. and J.X. assisted with the experiment. W.L. conceived and supervised the project. All authors have read and agreed to the published version of the manuscript.

Funding: This study was financially supported by the National Natural Science Foundation of China (51879290), Guangdong Provincial Special Science and Technology Fund for Practical Research and Development (NO. 2015B020235006) and the Guangdong Provincial Water Conservancy Scientific and Technological Innovation Project (NO. 2017-20).

Institutional Review Board Statement: Not applicable.

Informed Consent Statement: Not applicable.

Data Availability Statement: The genome sequence of the bacteria was uploaded to the NCBI database (http://www.ncbi.nlm.nih.gov/nuccore/CP047570.1 (accessed on 15 October 2021), GenBank accession number CP047570.1).

Acknowledgments: We would like to thank Editage (app.editage.cn/ (accessed on 15 October 2021)) for its linguistic assistance during the preparation of this manuscript. We also thank Wuhan Metware Biotechnology Co., Ltd., Wuhan, China, for their contributions and assistance in this study.

Conflicts of Interest: The funders had no role in the design of the study; in the collection, analyses, or interpretation of data; in the writing of the manuscript, or in the decision to publish the results.

References

1. Janssen, E.M.-L. Cyanobacterial peptides beyond microcystins—A review on co-occurrence, toxicity, and challenges for risk assessment. *Water Res.* **2019**, *151*, 488–499. [CrossRef] [PubMed]
2. Lin, W.; Hou, J.; Guo, H.H.; Li, L.; Wang, L.K.; Zhang, D.D.; Li, D.P.; Tang, R. The synergistic effects of waterborne microcystin-LR and nitrite on hepatic pathological damage, lipid peroxidation and antioxidant responses of male zebrafish. *Environ. Pollut.* **2018**, *235*, 197–206. [CrossRef] [PubMed]
3. Wang, J.; Chen, Y.B.; Zhang, C.L.; Xiang, Z.; Ding, J.; Han, X.D. Learning and memory deficits and alzheimer's disease-like changes in mice after chronic exposure to microcystin-LR. *J. Hazard. Mater.* **2019**, *373*, 504–518. [CrossRef] [PubMed]
4. Gurbuz, F.; Uzunmehmetoglu, O.Y.; Diler, O.; Metcalf, J.S.; Codd, G.A. Occurrence of microcystins in water, bloom, sediment and fish from a public water supply. *Sci. Total Environ.* **2016**, *562*, 860–868. [CrossRef]
5. Miller, T.R.; Beversdorf, L.J.; Weirich, C.A.; Bartlett, S.L. Cyanobacterial Toxins of the Laurentian Great Lakes, Their Toxicological Effects, and Numerical Limits in Drinking Water. *Mar. Drugs* **2017**, *15*, 160. [CrossRef]
6. Wan, X.; Steinman, A.D.; Gu, Y.R.; Zhu, G.W.; Shu, X.B.; Xue, Q.J.; Zou, W.; Xie, L.Q. Occurrence and risk assessment of microcystin and its relationship with environmental factors in lakes of the eastern plain ecoregion, China. *Environ. Sci. Pollut. Res.* **2020**, *27*, 45095–45107. [CrossRef] [PubMed]
7. Kodani, S.; Imoto, A.; Mitsutani, A.; Murakami, M. Isolation and identification of the antialgal compound, harmane (1-methyl-beta-carboline), produced by the algicidal bacterium, *Pseudomonas* sp. K44-1. *J. Appl. Phycol.* **2002**, *14*, 109–114. [CrossRef]
8. Sukenik, A.; Kaplan, A. Cyanobacterial Harmful Algal Blooms in Aquatic Ecosystems: A Comprehensive Outlook on Current and Emerging Mitigation and Control Approaches. *Microorganisms* **2021**, *9*, 1472. [CrossRef]
9. Park, S.C.; Lee, J.K.; Kim, S.W.; Park, Y. Selective Algicidal Action of Peptides against Harmful Algal Bloom Species. *PLoS ONE* **2011**, *6*, e26733. [CrossRef]

10. Kataev, A.A.; Andreeva-Kovalevskaya, Z.I.; Solonin, A.S.; Ternovsky, V.I. *Bacillus cereus* can attack the cell membranes of the alga *Chara corallina* by means of HlyII. *Biochim. Biophys. Acta Biomembr.* **2012**, *1818*, 1235–1241. [CrossRef] [PubMed]
11. Dai, W.; Zhang, S.L.; Lin, Y.J.; Bi, X.D.; Yan, R.J.; Xing, K.Z. Allelopathic Effects of Berberine, a Plant Alkaloid, on the Algae, *Microcystis aeruginosa* (FACHB-905), at Different Initial Densities. *Isr. J. Aquac. Bamidgeh* **2013**, *65*, 1–5.
12. Wang, J.; Kong, L.J.; Li, Y.H.; Zhang, J.Z.; Shi, Y.; Xie, S.L.; Li, B. Effect of protopine exposure on the physiology and gene expression in the bloom-forming cyanobacterium *Microcystis aeruginosa*. *Environ. Sci. Pollut. Res.* **2021**, *28*, 64666–64673. [CrossRef] [PubMed]
13. Lenneman, E.M.; Wang, P.; Barney, B.M. Potential application of algicidal bacteria for improved lipid recovery with specific algae. *FEMS Microbiol. Lett.* **2014**, *354*, 102–110. [CrossRef]
14. Wu, J.-T.; Chiang, Y.-R.; Huang, W.-Y.; Jane, W.-N. Cytotoxic effects of free fatty acids on phytoplankton algae and cyanobacteria. *Aquat. Toxicol.* **2006**, *80*, 338–345. [CrossRef]
15. Zhang, H.J.; Xie, W.J.; Hou, F.R.; Hu, J.; Yao, Z.Y.; Zhao, Q.F.; Zhang, D.M. Response of microbial community to the lysis of *Phaeocystis globosa* induced by a biological algicide, prodigiosin. *Environ. Pollut.* **2020**, *265*, 115047. [CrossRef] [PubMed]
16. Zhang, S.; Zheng, W.; Wang, H. Physiological response and morphological changes of *Heterosigma akashiwo* to an algicidal compound prodigiosin. *J. Hazard. Mater.* **2020**, *385*, 121530. [CrossRef] [PubMed]
17. Zhang, H.J.; Wang, H.; Zheng, W.; Yao, Z.Y.; Peng, Y.; Zhang, S.; Hu, Z.; Tao, Z.; Zheng, T.L. Toxic Effects of Prodigiosin Secreted by *Hahella* sp. KA22 on Harmful Alga *Phaeocystis globosa*. *Front. Microbiol.* **2017**, *8*, 999. [CrossRef]
18. Nakashima, T.; Miyazaki, Y.; Matsuyama, Y.; Muraoka, W.; Yamaguchi, K.; Oda, T. Producing mechanism of an algicidal compound against red tide phytoplankton in a marine bacterium γ-proteobacterium. *Appl. Microbiol. Biotechnol.* **2006**, *73*, 684–690. [CrossRef] [PubMed]
19. Yang, K.; Chen, Q.L.; Zhang, D.Y.; Zhang, H.J.; Lei, X.Q.; Chen, Z.R.; Li, Y.; Hong, Y.L.; Ma, X.H.; Zheng, W.; et al. The algicidal mechanism of prodigiosin from *Hahella* sp. KA22 against *Microcystis aeruginosa*. *Sci. Rep.* **2017**, *7*, 7750. [CrossRef]
20. Liu, W.J.; Yang, J.; Tian, Y.N.; Zhou, X.G.; Wang, S.W.; Zhu, J.R.; Sun, D.; Liu, C. An in situ extractive fermentation strategy for enhancing prodigiosin production from *Serratia marcescens* BWL1001 and its application to inhibiting the growth of *Microcystis aeruginosa*. *Biochem. Eng. J.* **2021**, *166*, 107836. [CrossRef]
21. Wei, J.; Xie, X.; Huang, F.Y.; Xiang, L.; Wang, Y.; Han, T.R.; Massey, I.Y.; Liang, G.Y.; Pu, Y.P.; Yang, F. Simultaneous *Microcystis* algicidal and microcystin synthesis inhibition by a red pigment prodigiosin. *Environ. Pollut.* **2020**, *256*, 113444. [CrossRef] [PubMed]
22. Du, J.J.; Pu, G.Z.; Shao, C.; Cheng, S.J.; Cai, J.; Zhou, L.; Jia, Y.; Tian, X.J. Potential of extracellular enzymes from *Trametes versicolor* F21 a in *Microcystis* spp. degradation. *Mater. Sci. Eng. C* **2015**, *48*, 138–144. [CrossRef] [PubMed]
23. Wu, L.S.; Guo, X.L.; Liu, X.L.; Yang, H. NprR-NprX Quorum-Sensing System Regulates the Algicidal Activity of *Bacillus* sp. Strain S51107 against Bloom-Forming Cyanobacterium *Microcystis aeruginosa*. *Front. Microbiol.* **2017**, *8*, 1968. [CrossRef] [PubMed]
24. Cai, G.J.; Yang, X.J.; Lai, Q.L.; Yu, X.Q.; Zhang, H.J.; Li, Y.; Chen, Z.R.; Lei, X.Q.; Zheng, W.; Xu, H.; et al. Lysing bloom-causing alga *Phaeocystis globosa* with microbial algicide: An efficient process that decreases the toxicity of algal exudates. *Sci. Rep.* **2016**, *6*, 20081. [CrossRef] [PubMed]
25. Durrant, J.R.; Hastings, G.; Joseph, D.M.; Barber, J.; Porter, G.; Klug, D.R. Subpicosecond equilibration of excitation energy in isolated photosystem II reaction centers. *Proc. Natl. Acad. Sci. USA* **1992**, *89*, 11632–11636. [CrossRef]
26. Guo, X.; Liu, X.; Pan, J.; Yang, H. Synergistic algicidal effect and mechanism of two diketopiperazines produced by *Chryseobacterium* sp. strain GLY-1106 on the harmful bloom-forming *Microcystis aeruginosa*. *Sci. Rep.* **2015**, *5*, 14720. [CrossRef]
27. Kwon, S.K.; Park, Y.K.; Kim, J.F. Genome-Wide Screening and Identification of Factors Affecting the Biosynthesis of Prodigiosin by *Hahella chejuensis*, Using *Escherichia coli* as a Surrogate Host. *Appl. Environ. Microbiol.* **2010**, *76*, 1661–1668. [CrossRef]
28. Zhang, Y.; Chen, D.; Zhang, N.; Li, F.; Luo, X.; Li, Q.; Li, C.; Huang, X. Transcriptional Analysis of *Microcystis aeruginosa* Co-Cultured with Algicidal Bacteria *Brevibacillus laterosporus*. *Int. J. Environ. Res. Public Health* **2021**, *18*, 8615. [CrossRef]
29. Welschmeyer, N.A. Fluorometric analysis of chlorophyll a in the presence of chlorophyll b and pheopigments. *Limnol. Oceanogr.* **1994**, *39*, 1985–1992. [CrossRef]
30. Liu, Z.-Y.; Zhou, X.-N.; Zhang, H.-H.; Wan, L.; Sun, Z.-L. An integrated method for degradation products detection and characterization using hybrid ion trap/time-of-flight mass spectrometry and data processing techniques: Application to study of the degradation products of danofloxacin under stressed conditions. *Anal. Bioanal. Chem.* **2011**, *399*, 2475–2486. [CrossRef]
31. Oh, J.I.; Kim, M.J.; Lee, J.Y.; Ko, I.J.; Kim, W.; Kim, S.W. Isolation and characterization of algicidal bacteria from *Cochlodinium polykrikoides* culture. *Biotechnol. Bioprocess Eng.* **2011**, *16*, 1124–1133. [CrossRef]
32. Cao, X.Y.; Li, H.; Zhou, Y.Y.; Song, C.L. The shift of mutualistic relationships among algae, free-living and attached bacteria through different nutrient addition mode: A mesocosm study. *J. Freshw. Ecol.* **2020**, *35*, 535–548. [CrossRef]
33. Long, R.A.; Qureshi, A.; Faulkner, D.J.; Azam, F. 2-n-pentyl-4-quinolinol produced by a marine Alteromonas sp and its potential ecological and biogeochemical roles. *Appl. Environ. Microbiol.* **2003**, *69*, 568–576. [CrossRef]
34. Pollara, S.B.; Becker, J.W.; Nunn, B.L.; Boiteau, R.; Repeta, D.; Mudge, M.C.; Downing, G.; Chase, D.; Harvey, E.L.; Whalen, K.E. Bacterial Quorum-Sensing Signal Arrests Phytoplankton Cell Division and Impacts Virus-Induced Mortality. *mSphere* **2021**, *6*, e00009-21. [CrossRef] [PubMed]
35. Zender, M.; Klein, T.; Henn, C.; Kirsch, B.; Maurer, C.K.; Kail, D.; Ritter, C.; Dolezal, O.; Steinbach, A.; Hartmann, R.W. Discovery and Biophysical Characterization of 2-Amino-oxadiazoles as Novel Antagonists of PqsR, an Important Regulator of *Pseudomonas aeruginosa* Virulence. *J. Med. Chem.* **2013**, *56*, 6761–6774. [CrossRef] [PubMed]

36. Gong, L.Y.; Wang, X.L.; Li, Y.B.; Zhang, C.S.; Li, A.S.K.; Zhu, C.J. Isolation and Purification of Phenazine Pigments Produced by *Pseudomonas aeruginosa* and its Effects on the Growth of Red Tide Organisms. *Fudan Xuebao Zirankexueban* **2004**, *43*, 494–499, 506.
37. Hoover, S.E.; Perez, A.J.; Tsui, H.-C.T.; Sinha, D.; Smiley, D.L.; DiMarchi, R.D.; Winkler, M.E.; Lazazzera, B.A. A new quorum-sensing system (TprA/PhrA) for *Streptococcus pneumoniae* D39 that regulates a lantibiotic biosynthesis gene cluster. *Mol. Microbiol.* **2015**, *97*, 229–243. [CrossRef]

International Journal of
Environmental Research and Public Health

MDPI

Article

Estimation of Children's Soil and Dust Ingestion Rates and Health Risk at E-Waste Dismantling Area

Yan Yang [1,2,3,*], Mengdi Zhang [1], Haojia Chen [1,2,3], Zenghua Qi [1], Chengcheng Liu [3], Qiang Chen [4] and Tao Long [4]

1 School of Environmental Science and Engineering, Institute of Environmental Health and Pollution Control, Guangdong University of Technology, Guangzhou 510006, China; mengdizhang0120@163.com (M.Z.); chenhaojia_gdut@163.com (H.C.); qizenghua@126.com (Z.Q.)
2 Chemistry and Chemical Engineering Guangdong Laboratory, Shantou 515041, China
3 Synergy Innovation Institute of GDUT, Shantou 515041, China; 13610227100@163.com
4 State Environmental Protection Key Laboratory of Soil Environmental Management and Pollution Control, Nanjing Institute of Environmental Sciences, Ministry of Ecology and Environment of China, Nanjing 210042, China; chenqiang@nies.org (Q.C.); longtao@nies.org (T.L.)
* Correspondence: yangyan1209@gdut.edu.cn; Tel.: +86-020-39322564

Abstract: Due to environmental health concerns, exposure to heavy metals and related adverse effects in electronic waste (e-waste) dismantling areas have attracted considerable interest in the recent years. However, little information is available about the Soil/Dust Ingestion Rates (SIR) of heavy metals for children living in such sites. This study estimated the soil ingestion of 66 children from e-waste disassembly areas by collecting and analyzing selected tracer elements in matched samples of their consumed food, feces, and urine, as well as soil samples from their play areas. The concentrations of tracer elements (including Al, Ba, Ce, Mn, Sc, Ti, Y, and V) in these samples were analyzed. The SIR was estimated to be 148.3 mg/day (median) and 383.3 mg/day (95th percentile) based on the Best Tracer Method (BTM). These values are somewhat higher than those observed in America, Canada, and other parts of China. Health risk assessments showed that Cr presented the greatest carcinogenic risk, at more than 10^{-6} in this typical polluted area, while As was second. These findings provide important insights into the exposure risks of heavy metals in e-waste dismantling sites and emphasize the health risk caused by Cr and As.

Keywords: e-waste; heavy metal pollution; children; soil and dust ingestion rates; health risk assessment

Citation: Yang, Y.; Zhang, M.; Chen, H.; Qi, Z.; Liu, C.; Chen, Q.; Long, T. Estimation of Children's Soil and Dust Ingestion Rates and Health Risk at E-Waste Dismantling Area. *Int. J. Environ. Res. Public Health* **2022**, *19*, 7332. https://doi.org/10.3390/ijerph19127332

Academic Editor: Paul B. Tchounwou

Received: 11 April 2022
Accepted: 9 June 2022
Published: 15 June 2022

1. Introduction

The potentially harmful environmental and human health effects of primitive electronic-waste (e-waste) recycling processes, including manual disassembly, roasting, acid leaching, and open burning, have caused concern around the world, particularly in rapidly industrializing and urbanizing developing countries such as China, India, and Vietnam [1,2]. Notably, heavy metal pollution is ubiquitous in the environment and bodies of people living near e-waste disposal sites [3,4]. Heavy metal elements accumulate in the human body and interfere with the human endocrine system [5], damage the body's cardiovascular and nervous systems [6,7], and can even lead to cancer [8]. It has been reported that the concentration of Cu was about three times higher than the Grade II guideline level (Soil Environmental Quality Standard, GB 15618–1995) in an e-waste disassembly area in China, and the Cu concentrations of soils from dumping, burning, and acid leaching sites were found to be 10, 40, and 60 higher, respectively, than the Grade II level. It was also reported that concentrations of blood lead, cadmium, and lead in meconium were higher in children and newborns living in e-waste disassembly areas than in neighboring areas [9,10]. In addition, children also have higher frequencies of mouthing behaviors, higher ingestion rates, lower body weights, and are more vulnerable to toxic substances, as compared to adults [11]. Therefore, it is necessary to assess the exposure levels and health risk caused by hand-to-mouth/object-to-mouth transfer that children face due to heavy metals in e-waste disassembly areas.

To more accurately assess the health risks that heavy metals post to children resulting from hand-to-mouth/object-to-mouth exposure, the Soil/Dust Ingestion Rate (SIR) is an important factor when estimating the risks caused when children are exposed to pollutants that are prone to binding to soils, such as heavy metals [12]. An estimation of the daily SIR of children from Gansu Province via hand-to-mouth contact showed that kindergarten and primary school children ingested 7.73 and 6.61 mg/day, respectively [13]. Lin et al. presented the first large-scale study of SIR for 177 Chinese children and recommended SIR values for the general population of Chinese children (from 2.5 to 12 years old): 52 mg/day was the central tendency and 217 mg/day was the upper percentile [14]. The SIR of children from 6 to 71 months old in the United States was found to be 85 mg/day [15]. This value was subsequently set as the recommended SIR value for children under one year of age by the USEPA; Chinese guidelines reference this. Due to various factors, including different living behaviors, different SIR values are observed in different regions. However, the coefficients used to compute SIR values are often based on the results of studies conducted in the United States, and directly applying these coefficients may lead to errors in studies conducted elsewhere. Therefore, independent SIR determinations should be conducted in individual districts. The need for accurate regional SIR data is enhanced by the fact that some areas, such as e-waste dismantling sites, have very high levels of heavy metal contamination in soil and dust. Consequently, reliable information on exposure factors and SIR in such regions is urgently needed to enable an accurate assessment of children's health risks.

At present, three experimental methods are available to estimate the SIR of children: the activity pattern-based methodology [16–18], the biokinetic modeling methodology [19,20], and the tracer element methodology [21–23]. The tracer element methodology, which is suitable for all situations and based on accurate experimental data, has been widely used to determine SIR since 1980 [14,24,25]. To calculate SIR (US guidelines), this method analyzes the concentration of tracer elements in the soil to which children are exposed, the children's intake of food, their excreted feces and urine, and the tracer element contents in the children's food, feces, and urine. However, because the lowest estimated SIR determined for a given tracer element will always be greater than the actual SIR of the human body, researchers developed the Limitation Tracer Method (LTM), which defines the soil intake as the lowest of the individual estimated values for a set of tracer elements [13,26]. Then, to improve the accuracy of the LTM which does not look at ingestion traces in food or medicine, the Best Tracer Method (BTM) was put forward to obtain a child's SIR [25]. At present the BTM is the most suitable method of accurately quantifying SIR [14,27].

Therefore, this study focuses on estimating the children's SIRs in e-waste disassembly areas by collecting and analyzing selected tracer elements in matched samples of their food, feces, and urine, as well as soil samples from their play areas, to further assess the health risk to children in e-waste dismantling areas.

2. Materials and Methods

2.1. Study Site and Sampling

The study examined a population sample of 66 children, most of whom lived in an e-waste dismantling area in South China. The e-waste dismantling area we studied is an E-Waste Recycling Town located in South China, where a possible human body burden and health consequences due to heavy metals exposure have been reported [10]. The ages, heights, and weights of the participating children and some associated descriptive statistics are presented in Figure S1. Ages ranging from 3 to 17 years old, with a median of 9 years old, were chosen. The median (maximum to minimum) weight and height were 19.0 kg (7.0–72.0 kg) and 113.0 cm (74.0–160.0 cm), respectively.

The children mainly came from a full-day school and kindergarten. Samples of their food, urine, and feces were collected daily by their parents, guardians, and teachers. Each participant was forbidden to take any drugs during the sampling period to reduce errors. It is generally assumed that there is a lag time of a 28-h from the ingestion of food and soil to the resulting fecal and urinary output [28]. Sample collections followed the USEPA recommendation of a 28-h period from food to feces and urine (for example, food collection from day-1 morning (approximately at 07:00 h, including breakfast) to day-2 morning (but not day-2 breakfast) and feces collection from day-2 morning (approximately at 07:00) to day-3 noon (approximately 11:00)). Therefore, the sample collection corresponding to one day lasted 52 h (collecting food in the first 24 h and collecting feces and urine after 28 h).

A "duplicate plate" method (two identical meals were prepared: one was for the subjects to eat and the other was mixed into a food sample and weighed for laboratory determination) was used to collect food samples, including breakfast, lunch, and children's dinner. Food samples (n = 66) were weighed before being lyophilized and then crushed. All feces (n = 62) and urine (n = 64) for each subject were collected daily using pre-labeled and pre-weighed portable sample containers, respectively. When the collection was completed, the samples were taken back to the laboratory and stored first in the refrigerator (−20 °C). Then, the feces were freeze-dried (under vacuum conditions, the vacuum freeze drier temperature ranges from −40 °C to −50 °C for 48 h) after measuring the weight with a vacuum freeze-dryer. The urine was stored after measuring the volume. Topsoil (n = 5) and dust (n = 3) samples were collected from campuses and green spaces where children generally play, respectively. Dust was collected indoors or outdoors by cleaning dust from areas such as tables and windowsills. A total of 20 g of soil and as much dust (5–20 g) as possible were collected during the sampling process. After air-drying, all samples were crushed with a ceramic mortar and pestle and then passed through a 0.25-mm sieve. This sampling method was also used to collect the 46 soil samples that were used to measure the heavy metals, including 24 residential areas and 20 park green areas. A combined assessment of local heavy metal pollution levels and children's SIR enabled accurate assessment of the health risks facing children due to local soil contamination and the associated heavy metal intake. All samples were collected in 2019 and stored at −20 °C for later processing and analysis.

2.2. Sample Preparation and Instrumental Analysis

Feces samples and food samples were pretreated in the same way. Dried samples (1 g) were digested to evaporate at low temperatures of 55 °C on a heating plate with 3 mL concentrated nitric acid, 3 mL hydrogen fluoride, and 1 mL perchloric acid (HNO_3–HF–$HClO_4$). The digestion process was repeated until the sample became sticky. Crushed evaporated samples were microwaved with 2 mL concentrated nitric acid and 3 mL hydrogen peroxide (HNO_3–H_2O_2). Digestion was performed at 120 °C for 5 min, then 160 °C for 5 min, and finally 180 °C for 15 min. The digested product was diluted to 30 mL with ultrapure water and then stored at 4 °C. The supernatant was extracted and analyzed by High-Resolution Inductively Coupled Plasma Mass Spectrometry (HR–ICP–MS, Nu Attom, North Wales, UK) to determine Mn, Al, Ba, Ti, Ce, V, Sc, and Y.

Urine samples (15 mL) were placed in a digestion tube and digested by microwaving with 2 mL of H_2O_2 and 3 mL of concentrated HNO_3. The conditions of microwave digestion are the same as mentioned above. Finally, the sample after digestion was analyzed by HR–ICP–MS.

Dried soil and dust samples (0.5 g) were digested the same as feces and food samples on a heating plate by HNO_3–HF–$HClO_4$. Then, samples were microwaved with HNO_3–H_2O_2. The supernatant was analyzed for Al, Ba, Mn, Ti, and V by Inductively Coupled Plasma Optical Emission Spectroscopy (ICP–OES, Spectra Arcos SOP, Kleve, Germany), which is a method of atomic emission spectroscopy analysis using a light source that generates plasma discharge through high-frequency inductive coupling and ICP–MS for Ce, Sc, and Y. In addition, dried soil samples from living spaces (0.5 g) were digested with the same pre-treatment and analyzed by ICP–MS for Pb, As, Cr, Cu, Ni, Cd, and Zn. The standard concentration curve was used to determine the sample concentration established by the heavy metal standard (all standards were from The Nonferrous Metals Society of China).

2.3. QA/QC Method for Analytical Data

The accuracy of the method used to analyze tracer elements in the collected food, feces, urine and soil samples was tested by using the same method to detect the same elements in substrate mixed standards, while certified reference materials replaced dust and soil substrates. Analysis of the certified reference materials with the same elements was used to show the accuracy of the method used for the analysis of heavy metals in soil samples. The tracer elements recoveries from soil and dust, food, and fecal and urine samples ranged from 74.2% to 102.3%, 83.4% to 108%, and 84.6% to 110%, respectively, for Al, Ba, Mn, Ti, and V analyzed by ICP–OES. The recovery of tracer elements from soil and dust, food, and fecal and urine samples ranged from 79.3% to 97.3%, 82.1% to 110%, and 81.1% to 101%, respectively, for Ce, Sc, and Y analyzed by ICP-MS. Furthermore, the recovery of Pb, As, Cr, Cu, Ni, Cd, and Zn analyzed by ICP-MS from soil and dust ranged from 71.3% to 93.9%. All reagents used in the analysis were of high purity. The experimental water was ultra-pure, and all the glassware was soaked for more than 36 h in 10% nitric acid before use. The results showed that the recovery of various heavy metal elements ranged from 95% to 105%. The Ce, Sc, and Y concentrations in the supernatant were defined as half of their detection limits, because all of them were under their detection limits, which were 0.001 ng/mL, 0.001 ng/mL and 0.002 ng/mL, respectively).

2.4. SIR Estimates and Best Mass-Tracer Method

Based on the experimental data, the daily SIR of each participating child was calculated using the following expression. After caluclating SIR values for each element, the best mass-tracer method was used to obtain a more representative SIR value.

$$SIR = \frac{\left[\left(W_{feces} \times C_{feces} + V_{urine} \times C_{urine}\right) - \left(W_{food} \times C_{food}\right)\right]}{C_{soil/dust}} \quad (1)$$

where SIR is the SIR (mg/day) each day over the study period for each kid. The dry weight of feces is W_{feces} (kg/day), the concentration of tracer elements in feces is C_{feces} (mg/kg), the urine volume is V_{urine} (mL/day), the concentration of tracer elements in urine is Curine (μg/mL), the weight of food consumed is W_{food} (kg/day), the concentration of tracer elements in food is C_{food} (mg/kg), and the concentration of tracer elements in soil and dust is $C_{soil/dust}$ (mg/kg) [28]. Data for W_{feces}, V_{urine}, and W_{food} are shown in Figure S2, while data on C_{feces}, C_{urine}, and C_{food} are shown in Tables S1–S3.

2.5. Human Health Risk Assessment

The Average Daily Dose (ADD, kg kg^{-1} bodyweight day^{-1}), which is used to assess the health risk due to intake of toxic materials [28], can be calculated with the following equation:

$$ADD = \left(C \times SIR \times EF \times \frac{ED}{BW} \times AT\right) \times CF \quad (2)$$

where the heavy metal concentration in soil/indoor dust (kg/kg) is C, the soil and dust ingestion rate is SIR (mg/day); the exposure frequency is EF (days/years) and is taken as 350 d/y; the exposure duration is ED (year) and is taken as 6 years for per child; the body weight is BW (kg); the average time is AT (day), for non-carcinogenic effect, which was taken to be 2190 days (for carcinogenic effect, AT is 27,740 days); the unit conversion factor of 10^{-6} is CF [28].

The Hazard Quotients (HQ) are the ratio of daily intake dose of pollutants to reference dose, which is used to characterize the levels of human exposure to non-carcinogenic contaminants through a single pathway which represents the level of non-carcinogenic risk. This is calculated as follows:

$$HQ = ADD/RfD \tag{3}$$

where the estimated maximum permissible dose to humans via oral ingestion exposure is *RfD* (mg/(kg·d)); in this study, the *RfD* is 0.003 for Cr, 0.02 for Ni, 0.04 for Cu, 0.3 for Zn, 0.0003 for As, 0.001 for Cd, and is 0.0035 for Pb [28].

To estimate carcinogenic risk, the Cancer Risk (CR) was computed. The CR is the incremental probability of an individual developing cancer over a lifetime due to exposure to a carcinogenic hazard, and is defined as follows:

$$\text{Risk} = ADD \times SF \tag{4}$$

where the cancer slope factor is *SF* ((kg·d)/mg). Of the tracer elements considered in this work, *SF* values for the hand-/object-to-mouth pathway have only been established for As and Cr; these values are 1.0 and 0.5, respectively [28].

3. Results and Discussion

3.1. The SIR Results

The SIR values, based on the measured tracer element concentrations of food, feces, urine, soil and dust are presented in Table 1 and Figure 1. The median (minimum and maximum) of SIR values were −124.3 (−278.0 to 228.2), −210.2 (−490.1 to 273.8), 27.1 (0.4 to 106.0), −22,532.8 (−29,443.8 to −6215.9), 23.9 (−45.3 to 268.0), 175.3 (−56.4 to 1040.7), 39.2 (−36.4 to 284.0), and −263.2 (−491.4 to 132.3) mg/day for Al, Ba, Ce, Mn, Sc, Ti, Y, and V. When comparing the ratios of the median SIR values obtained for individual elements in this work to the median SIR values reported previously for the same elements, it can be seen that the highest ratio is 1.6 (for Ce) and the lowest is 0 (for Sc). This relatively narrow range indicates that the SIR values obtained in this work are comparable to those reported previously. The coefficients of variation in the SIR values for the tracers examined in this work were not high, with the exception of Al, which was 596.3%. The frequency distribution histograms show few outliers, with most being high values (Figures S3a and S4). Experimental factors such as measurement error, source error, and transit time misalignment may lead some of these outliers [29]. Other outliers may be due to the behavior of certain children, such as pica behavior [28] or spending unusually large amounts of time playing in grassland. The medians of the SIR values after removing the high values (see Figures S3b and 1) were taken as the final SIR values for children living in the studied e-waste dismantling site in South China.

Table 1. The result of each day soil ingestion rate (SIR, mg/day) for 62 children based on tracer elements including Al, Ba, Ce, Mn, Sc, Ti, Y, and V.

	Al	Ba	Ce	Mn	Sc	Ti	Y	V
Max	228.2	273.8	106.0	−6215.9	268.0	1040.7	284.0	132.3
99.50%	224.8	267.3	101.0	−6438.2	261.5	988.2	269.5	131.6
97.50%	211.4	241.2	80.8	−7327.3	235.3	778.2	211.2	128.7
95.00%	183.9	213.5	61.4	−9089.1	201.0	579.2	150.1	125.0
90.00%	96.5	156.8	56.4	−14,893.6	128.9	537.3	104.7	99.0
75.00%	−12.8	1.5	48.5	−20,144.1	79.4	439.6	92.9	−32.6
Median	−124.3	−210.2	27.1	−22,532.8	23.9	175.5	39.2	−263.2
25.00%	−183.8	−363.4	13.5	−24,985.0	−15.8	13.7	−1.0	−333.3
10.00%	−233.3	−427.7	6.3	−27,811.6	−25.5	−24.0	−17.4	−414.1
2.50%	−243.7	−437.9	4.4	−28,870.4	−30.4	−30.2	−28.6	−443.8
0.50%	−274.2	−484.3	0.9	−29,391.0	−43.6	−53.5	−35.7	−486.3
Min	−278.0	−490.1	0.4	−29,443.8	−45.3	−56.4	−36.4	−491.4
Mean	112.5	192.6	18.8	4282.5	57.5	227.9	53.0	166.4

Table 1. *Cont.*

	Al	Ba	Ce	Mn	Sc	Ti	Y	V
SD	138.0	225.9	24.3	5907.3	77.8	274.6	71.0	193.5
CV%	1.2	1.2	1.3	1.4	1.4	1.2	1.3	1.2
n	62	62	62	62	62	62	62	62

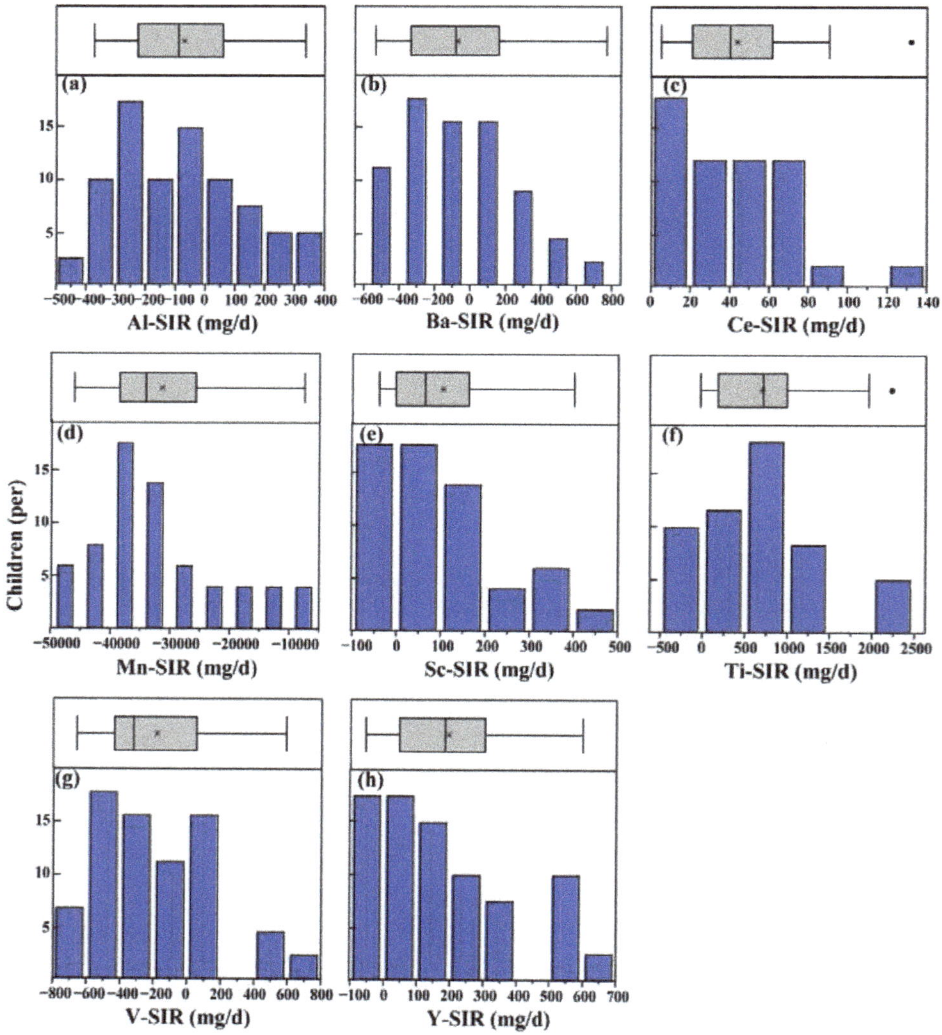

Figure 1. Frequency distribution histogram and outlier box of soil ingestion rate (SIR) based on tracer Al (**a**), Ba (**b**), Ce (**c**), Mn (**d**), Sc (**e**), Ti (**f**), V (**g**), and Y (**h**). The * represents mean values, and dot sign represents outliers.

3.2. Soil Ingestion Rate Based on the BTM

The complex metabolism of the human body can lead to different behaviors between the tracers. This is an important reason why different tracers give rise to different SIR values. The reliable estimation of soil absorption for each tracer can improve inter-tracer consistency in soil absorption values [22]. Unfortunately, there are substantial differences between the

SIR values estimated by different tracers. Doyle et al. [29] found that this variability may be partly due to measurement error, source error, or transit time misalignment. Regardless of its origin, there is a clear need to identify a reliable tracer for SIR estimation. The BTM method was developed for this purpose [29]. This method depends on the Food-to-Soil (F/S) ratio, which is the ratio from the mass of tracer elements taken from food to the mass of tracer elements in 1g of soil within one day. The most suitable tracer elements are identified by the F/S ratio; the lower an element's F/S ratio, the closer the estimated SIR value is to the true value [30]. The average F/S ratios determined were 0.000121, 0.001118, 0.000001, 0.012732, 0.000035, 0.000024, 0.000658, and 0.000028 for Al, Ba, Ce, Mn, Sc, Ti, V, and Y in this work. Accordingly, Al, Ce, Sc, Ti, and Y were identified as the best tracer elements. However, the SIR values based on these five tracers still show differences. Both the mean and median SIR values determined using Ti as the tracer were significantly higher than the SIR values obtained for other tracers.

The estimate based on the best five tracer elements (Al, Ce, Sc, Ti, and Y) was found to be the best approximation of the SIR, i.e., the one expected to be closest to the true value. The SIR determined using the BTM approach are shown in Figure 2 (including the frequency distribution and basic statistical parameters). It is clear that the distribution remained skewed (Figure 3). The SIR observed for the children ranged from −76.8 to 1725.0 mg/day, with 47.9, 148.3, and 383.3 mg/day for median, mean, and 95th percentile values, respectively. These data lie in the reference intervals (Ris; USEPA, 2011), which range from upper (URL) to lower (LRL) reference bounds. The LRL is −112.4–100.9 mg/day, which is considered the lower limit of the 90% confidence interval (CI) of the 2.5th percentile (P2.5), whereas the URL is 516.9–730.2 mg/day, which is considered the upper limit of the 90% CI of the 97.5th percentile (P97.5) [10]. However, since negative SIR values are physically meaningless, the RI of the SIR for children living in e-waste dismantling sites is 0–730.4 mg/day. In this study, 95th percentile values (383.3 mg/g) would be the recommended value.

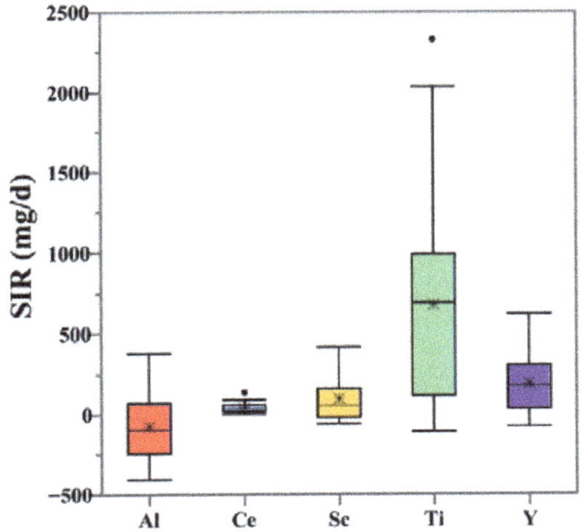

Figure 2. The outlier box plot, minimum, mean, median, and maximum, of SIRs based on the best tracer Al, Ce, Sc, Ti, and Y. The * represents mean values, and dot sign represents outliers.

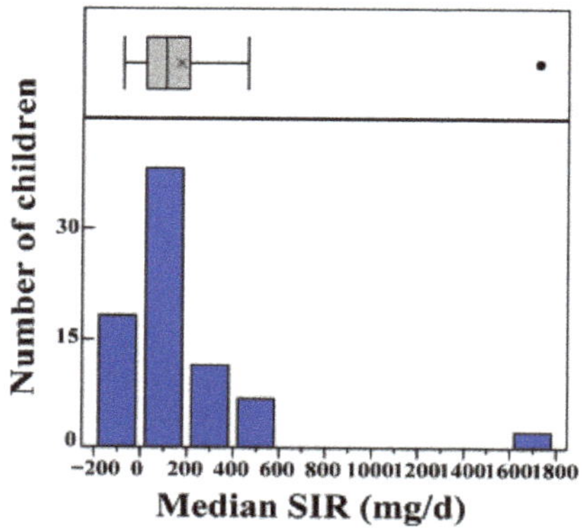

	SIR	95%CI
Max	1725.0	
99.5%	1504.7	
97.5%	623.5	516.9-730.2
95.0%	383.3	
90.0%	247.3	
75.0%	89.0	
Median	45.9	
25.0%	−5.8	−112.4-100.9
10.0%	−39.6	
5.0%	−61.6	
2.5%	−66.5	
0.5%	−74.7	
Min	−76.8	
Mean	148.3	
SD	305.6	
CV%	2.1	
n	62	

Figure 3. Frequency distribution histogram based on the BTM. All basic statistical parameters of SIR results are listed by the side of histogram. The * represents mean values, and dot sign represents outliers.

3.3. Comparison of SIR Results

The SIR measurements obtained in this work are compared to those reported in Table 2, which also specifies the regions, ages, and tracer elements considered in each study. In most cases, Al, Ti and Ba emerged as the usual tracer elements for SIR estimation. The mean SIR values obtained for Al range from 2.7 [22] to 154.0 [31] mg/day, while the medians range from −3.3 [22] to 33.3 [32] mg/day. The median SIR value obtained for Al in previous studies is higher than that reported in this study, but the mean is slightly lower. The mean SIR values for Ti reported in the literature range from −544.4 [22] to 3368.0 [25] mg/day, while the medians range from 11.9 [22] to 1861.0 [25] mg/day. Our mean and median SIR values are within these ranges. The SIR means for Ba range from 29.0 [31] to 368.0 [25] mg/day, while the medians range from −19.0 [31] to 394.0 [25] mg/day. The SIR mean that we calculated is in the middle of this range, and the median value is lower than those reported previously, as was also the case for Al.

The Ce- and Sc-based mean SIR values obtained in this work are more similar to those reported previously than the medians, while the median Mn-based SIR values obtained in this study are more widely dispersed than those for the other tracers. Additionally, the median SIR value for Mn was large and negative (−22,532.8 mg/day). The median Mn-based SIR determined by Calabrese et al. was −340.0 mg/day, which is the lowest value reported in the literature [31]. The V-based SIR means range from −182.0 [25] to 456.0 [31] mg/day, and the medians range from −185.0 [25] to 123.0 [31] mg/day. As was the case for Al and Ba, our V-SIR median is low. The Y-based SIR obtained in this study is higher than that in some studies, but the means and medians are similar to those reported previously.

Table 2. SIR (mean, median, SD, and SIR recommended, mg/day) comparisons between published studies and this study.

Reference	Age (Years)	n	Region	Tracer Element	Mean	Median	SD	SIR Recommended
Calabrese et al., 1989 [31]	1–4	64	America	Al	154.0	30.0	629.0	154.0
				Ti	170.0	30.0	691.0	
				Ba	29.0	−19.0	868.0	
				Mn	−496.0	−340.0	1974.0	
				V	456.0	123.0	1013.0	
				Y	65.0	11.0	717.0	
Calabrese et al., 1997 [22]	1–4	10	America	Al	2.7	−3.3	95.8	
				Ti	−544.4	11.9	2509.0	
				La	8.6	84.5	1377.2	
				BTM	6.8	−2.4	74.5	
Davis, et al., 2006 [32]	3–8	12	Canada	Al	36.7	33.3	35.4	
				Ti	206.9	46.7	277.5	
Irvine et al., 2014 [25]	Adult	9	Canada	Al	33.0	32.0	55.0	32.0
				Ti	3368.0	1861.0	4277.0	
				Ba	368.0	394.0	725.0	
				Ce	11.0	10.0	34.0	
				La	12.0	11.0	36.0	
				Mn	1363.0	1408.0	5359.0	
				V	−182.0	−185.0	144.0	
				Y	−13.0	1.0	67.0	
Chien et al., 2017 [24]	0.5–3	66	Taiwan, China	Ti	957.1		477.0	
				Si	9.6		19.2	
Lin et al., 2017 [14]	2.5–11.9	177	China	Al	47.7	27.8	59.8	60.8
				Ti	81.9	36.7	177.6	
				Ba	63.1	36.5	125.9	
				Ce	53.5	34.8	48.8	
				Mn	230.8	146.6	617.6	
				Sc	77.7	54.8	68.8	
				V	106.4	92.1	64.6	
				Y	79.8	59.1	68.3	
				BTM	73.5	51.7	63.7	
This study	2–16	61	China	Al	112.5	−124.3	138.0	383.3
				Ba	192.6	−210.2	225.9	
				Ce	18.8	27.1	24.3	
				Mn	4282.5	−22532.8	5907.3	
				Sc	57.5	23.9	77.8	
				Ti	227.9	175.5	274.6	
				V	166.4	−263.2	193.5	
				Y	53.0	39.2	71.0	
				BTM	148.3	47.9	306.5	

Based on the above, using a single tracer is not enough to obtain an accurate SIR for children. Therefore, in this study, the Best-Tracer Method (BTM) was employed. This method has been used before. Based on the BTM approach, Al, Ti, and La were selected as the best tracers in [22]. Lin et al. [14] selected V, Y, Sc, Ce, and Al as the tracers with the best F/S radio values. We similarly chose the elements with the lowest F/S ratios, namely Al, Ce, Sc, Ti, and Y. The median SIR derived by Calabrese et al. using the BTM approach is negative [22], and the mean is low; this is a consequence of their tracer element selections. The BTM-based SIR values obtained by Lin et al. [14] and in our study are all positive, although the median and mean values obtained in this work are higher than those reported by Lin et al.

It is clear that the SIR values for different regions and ages differ depending on the considered tracer. Indeed, even when the SIR is calculated using the same set of elements using the BTM approach, there are pronounced differences between countries and regions in terms of children's SIR. It appears that the SIR of children is higher in China in this area (median: 171.5 mg/day) than in America and Canada (median: 100 mg/day and 32.0 mg/day, respectively) [25,28]. This may be attributable to differences in lifestyle. Some researchers observed and disciplined the behavior of children to control for the activity factor. The reference hand-to-mouth and object-to-mouth contact frequencies in children specified in the USEPA exposure factor handbook [28] are 3 times/h inside and 7 times/h outside for hand-to-mouth contact and 1 time/h inside and 1 time/h outside for object-to-mouth contact. The durations of these contacts are not specified. However, the hand-/object-to-mouth contact durations for children in Taiwan are 0.34 and 0.46 min/h, respectively, and the corresponding indoor contact frequencies are 8.91 and 11.39 times/h, respectively [24]. Thus, in theory, the SIR in China would be expected to be higher than in America.

In China, different provinces also have different SIR values. In Taiwan, the average SIR for children aged from 24 to 36 months were 90.7 and 29.8 mg/day in the sand and clay groups, respectively [11], whereas in Hubei, Guangdong, and Gansu provinces these values were 51.7 mg/day [14]. Our study collected children from e-waste dismantling areas in South China, and we estimated that the SIR median was 148.3 mg/day. It is possible that the children studied in Taiwan were younger children who spend most of their time indoors without outside activity, resulting in relatively low SIR values. However, our study and that of Lin et al. [14] focused on children with higher levels of outside activity, who would be expected to have higher SIR values than children who spend most of their time indoors. In general, as reflected in these results, older children (6–17 years old) who have more outside activities have a higher SIR (median: 202.9 mg/day) than younger children (3–6 years old, median: 53.9 mg/day) ($p < 0.05$). No activity data were gathered during this study, but such data would facilitate interpretation of the determined SIR values and would, therefore, be useful to obtain in future.

3.4. Health Risk Assessment of Heavy Metals Based on SIR Results

SIR is an important parameter for environmental health risk assessment, not least because hand-/object-to-mouth ingestion is the heavy metal exposure pathway associated with the greatest health risk [33]. Therefore, this study assessed the oral ingestion health risk to children in the studied e-waste dismantling area based on the heavy metal contamination of the area's soils and the calculated SIR values. In resident and park green areas, the highest concentration of heavy metals is found for Zn and the lowest for Cd. The concentrations of heavy metals in park green areas were slightly higher than that in residential area, which may be attributable to the difference in soil environmental quality management and control policy between these areas. After integrating data from the resident and park green areas, the median concentrations (mg/kg) of Cr, Ni, Cu, Zn, As, Cd, and Pb were 48.8, 63.9, 128, 413, 6.30, 0.513, and 115, respectively. Cr showed higher values than the risk screening values for soil contamination in development land in China (GB36600-2018). That means the living spaces in this area have high levels of Cr pollution.

The corresponding carcinogenic risk, and non-carcinogenic risk of the different heavy metals are shown in Table 3; only Cr and As pose a carcinogenic risk to children. Despite the contamination of the area's soils, non-carcinogenic risk assessments showed that there was no appreciable oral non-carcinogenic risk to children due to heavy metal contamination. Conversely, HQ values between 1 and 10 indicate likely damage to human health [29], and HQ values above 10 are associated with serious chronic risks. The 95th HQ values based on SIR mean value decreased in the order of Pb > As > Cr > Cu > Ni > Zn > Cd, and all of them were below 1. Even though the results regarding SIR recommended values seem to indicate a low risk, Pb is still the main heavy metal source of non-carcinogenic risk to children in the studied area, and 2.5% of children are exposed to Pb health risks. Zhang et al. [34]

also found high concentrations of Pb and Cd in the blood of children in our sampling area, which were much higher than the concentrations in the blood of children in the control area. Due to the influence of Pb on the human nervous system and immune system, this is a matter of concern. The calculated carcinogenic risks for the different heavy metals varied widely. Carcinogenic risk values below 10^{-6} are considered safe; however, the carcinogenic risk due to Cr and As calculated using the SIR recommended value in this study were 3.73×10^{-5}, and 1.44×10^{-5}, respectively. In addition, 75% and 50% children were suffering from a high carcinogenic risk caused by Cr and As, respectively.

Table 3. The estimated children's health risk assessment results for heavy metal soil pollution in this area based on the data on children's soil intake obtained from this study.

	Non-Carcinogenic Risk							Carcinogenic Risk ($\times 10^{-6}$)	
	Cr	Ni	Cu	Zn	As	Cd	Pb	Cr	As
Max	1.42	0.275	0.278	0.120	1.83	0.045	2.87	168	65.0
99.50%	1.24	0.240	0.243	0.105	1.60	0.039	2.50	146	56.7
97.50%	0.512	0.099	0.101	0.043	0.661	0.016	1.04	60.6	23.5
95.00%	0.315	0.061	0.062	0.027	0.406	0.010	0.637	37.3	14.4
90.00%	0.279	0.054	0.055	0.024	0.361	0.009	0.565	33.1	12.8
75.00%	0.209	0.041	0.041	0.018	0.270	0.007	0.423	24.8	9.59
Median	0.087	0.017	0.017	0.007	0.113	0.003	0.177	10.4	4.01
25.00%	0.018	0.003	0.003	0.001	0.023	0.001	0.036	2.091	0.810
10.00%	−0.020	−0.004	−0.004	−0.001	−0.026	−0.001	−0.041	−2.41	−0.934
5.00%	−0.034	−0.007	−0.007	−0.003	−0.043	−0.001	−0.068	−3.98	−1.54
2.50%	−0.044	−0.009	−0.009	−0.004	−0.057	−0.001	−0.089	−5.20	−2.02
0.50%	−0.059	−0.012	−0.012	−0.005	−0.076	−0.002	−0.120	−7.01	−2.72
Min	−0.063	−0.012	−0.012	−0.005	−0.081	−0.002	−0.128	−7.47	−2.89
Mean	0.141	0.027	0.028	0.012	0.182	0.004	0.285	16.7	6.46
SD	0.252	0.049	0.049	0.021	0.325	0.008	0.509	29.8	11.5
n	62	62	62	62	62	62	62	62	62

If SIR is not localized but adopts the recommended value of USEPA (100 mg/day), the health risk to local children will be assessed as being low. Overall, the obtained results suggest that the two elements posing the greatest health risk to children in the studied e-waste dismantling site are Cr and As. Continuous monitoring of their concentration in the area's soil is required. Otherwise, children are at greater risk of exposure to soil pollution than adults. To reduce the risk of As and Cr, schools should keep desks and teaching aids clean and tidy and urge children to clean up after outdoor activities. Due to the relationship between soil heavy metal pollution and e-waste disposal sites [9], new schools should be built as far away from e-waste disposal sites as possible.

4. Conclusions

Al, Ce, Sc, Ti, and Y had the lowest F/S ratios of the elements included in our analysis and were, therefore, better tracer elements for SIR calculation than Ba, Mn, and V. The mean, median, and 95th percentile SIR values calculated based on measurements of these five elements were 148.3, 47.9, and 383.3 mg/day, respectively. Our estimate of children's SIR in South China was slightly higher than the values reported for America, Canada, and other areas of China. These differences may be due to the regional differences in children's lifestyles. Thus, the children's behavior patterns associated with soil intake warrant further investigation. We should also sample a greater number of individuals to improve the accuracy of the subsequent experiments.

The calculated SIR values were used in conjunction with soil pollution data to assess the risks to children's health due to heavy metal exposure via the hand-/object-to-mouth intake pathway in the studied region. The overall health risk was found to be high. Although the non-carcinogenic risk is within the lowest range specified in the relevant guidelines, the carcinogenic risk for Cr and As was over the acceptable range of below 10^{-6}. It should be noted that there are more than 75% children in this area living with a carcinogenic health risk. To our knowledge, this study is the first to apply the tracer mass-balance method to determine the SIR for children living in e-waste dismantling sites in Southern China, and to use the calculated SIR values for health risk assessment in children.

Supplementary Materials: The following supporting information can be downloaded at: https://www.mdpi.com/article/10.3390/ijerph19127332/s1. The information and data associated with the present study can be found in Tables S1–S6 and Figures S1–S4.

Author Contributions: Data curation, C.L. and Q.C.; Formal analysis, M.Z., H.C. and C.L.; Investigation, T.L.; Methodology, Q.C. and T.L.; Project administration, Y.Y.; writing—original draft preparation, Y.Y., M.Z., H.C. and Z.Q.; writing—review and editing, Y.Y., M.Z. and Z.Q. All authors have read and agreed to the published version of the manuscript.

Funding: This research was funded by The National Key R&D Program of China (2018YFC1801102), The Chemistry and Chemical Engineering Guangdong Laboratory (Grant No. 2032008), and The Science and Technology Planning Project of Guangdong Province (200106105876892, 190807115560881, 190325224778589).

Institutional Review Board Statement: The study was conducted in accordance with the Declaration of Helsinki, and approved by the Institutional Review Board (or Ethics Committee) of Guangdong University of Technology (19 March 2019).

Informed Consent Statement: Informed consent was obtained from all subjects involved in the study.

Data Availability Statement: The study did not report any data.

Acknowledgments: We would like to acknowledge The National Key R&D Program of China (2018YFC1801102), The Chemistry and Chemical Engineering Guangdong Laboratory (Grant No. 2032008), and The Science and Technology Planning Project of Guangdong Province (200106105876892, 190807115560881, 190325224778589).

Conflicts of Interest: The authors declare that there are no competing financial interests.

References

1. Rene, E.R.; Sethurajan, M.; Ponnusamy, V.K.; Kumar, G.; Pugazhendhi, A. Electronic waste generation, recycling and resource recovery: Technological perspectives and trends. *J. Hazard. Mater.* **2021**, *416*, 125664. [CrossRef] [PubMed]
2. Alabi, O.A.; Adeoluwa, Y.M.; Huo, X.; Xu, X.; Bakare, A.A. Environmental contamination and public health effects of electronic waste: An overview. *J. Environ. Health Sci. Eng.* **2021**, *19*, 1209–1227. [CrossRef] [PubMed]
3. Houessionon, M.G.; Karel, O.E.D.; Bouland, C.; Takyi, S.A.; Kedote, N.M.; Fayomi, B.; Fobil, J.N.; Basu, N. Environmental Heavy Metal Contamination from Electronic Waste (E-Waste) Recycling Activities Worldwide: A Systematic Review from 2005 to 2017. *Int. J. Environ. Res. Public Health* **2021**, *18*, 3517. [CrossRef] [PubMed]
4. Ngo, H.T.T.; Watchalayann, P.; Li, L.; Nguyen, D.B. Environmental health risk assessment of heavy metal exposure among children living in an informal e-waste processing village in Viet Nam. *Sci. Total Environ.* **2021**, *763*, 142982. [CrossRef]
5. Janicki, T.; Długoński, J.; Krupiński, M. Detoxification and simultaneous removal of phenolic xenobiotics and heavy metals with endocrine-disrupting activity by the non-ligninolytic fungus Umbelopsis isabellina. *J. Hazard. Mater.* **2018**, *360*, 661–669. [CrossRef]
6. Duan, W.W.; Xu, C.; Liu, Q.; Xu, J.; Weng, Z.K.; Zhang, X.; Basnet, T.B.; Dahal, W.; Gu, A.H. Levels of a mixture of heavy metals in blood and urine and all-cause, cardiovascular disease and cancer mortality: A population-based cohort study. *Environ. Pollut.* **2020**, *263*, 114630. [CrossRef]
7. Tshala-Katumbay, D.; Mwanza, J.C.; Rohlman, D.S.; Maestre, G.; Oriá, R.B. A global perspective on the influence of environmental exposures on the nervous system. *Nature* **2015**, *527*, S187–S192. [CrossRef]
8. Feng, L.F.; Du, J.; Yao, C.J.; Jiang, Z.Q.; Liao, T.; Zhang, Q.; Guo, X.N.; Yua, M.; Xia, H.L.; Shi, L.; et al. Ribosomal DNA copy number is associated with P53 status and levels of heavy metals in gastrectomy specimens from gastric cancer patients. *Environ. Int.* **2020**, *138*, 105593. [CrossRef]
9. Yang, H.; Huo, X.; Yekeen, T.A. Effects of lead and cadmium exposure from electronic waste on child physical growth. *Environ. Sci. Pollut. Res.* **2013**, *20*, 4441–4447. [CrossRef]
10. Huang, W.L.; Shi, X.L.; Wu, K.S. Human Body Burden of Heavy Metals and Health Consequences of Pb Exposure in Guiyu, an E-Waste Recycling Town in China. *Int. J. Environ. Res. Public Health* **2021**, *18*, 12428. [CrossRef]

11. Wang, Y.L.; Tsou, M.C.M.; Pan, K.H.; Özkaynak, H.; Dang, W.; His, H.C.; Chien, L.C. Estimation of Soil and Dust Ingestion Rates from the Stochastic Human Exposure and Dose Simulation Soil and Dust Model for Children in Taiwan. *Environ. Sci. Technol.* **2021**, *55*, 11805–11813. [CrossRef] [PubMed]
12. Moya, J.; Phillips, L. A review of soil and dust ingestion studies for children. *J. Expo. Sci. Environ. Epidemiol.* **2014**, *24*, 545–555. [CrossRef] [PubMed]
13. Ma, J.; Pan, L.B.; Wang, Q.; Lin, C.Y.; Duan, X.L.; Hou, H. Estimation of the daily soil/dust (SD) ingestion rate of children from Gansu Province, China via hand-to-mouth contact using tracer elements. *Environ. Geochem. Health* **2018**, *40*, 295–301. [CrossRef] [PubMed]
14. Lin, C.; Wang, B.; Cui, X.; Xu, D.; Cheng, H.; Wang, Q.; Ma, J.; Chai, T.; Duan, X.; Liu, X. Estimates of Soil Ingestion in a Population of Chinese Children. *Environ. Health Perspect.* **2017**, *125*, 077002. [CrossRef] [PubMed]
15. Hogan, K.; Marcus, A.; Smith, R.; White, P. Integrated exposure uptake biokinetic model for lead in children: Empirical comparisons with epidemiologic data. *Environ. Health Perspect.* **1998**, *106*, 1557–1567. [CrossRef]
16. Özkaynak, H.; Xue, J.; Zartarian, V.G.; Glen, G.; Smith, L. Modeled estimates of soil and dust ingestion rates for children. *Risk Anal. Off. Publ. Soc. Risk Anal.* **2011**, *31*, 592–608. [CrossRef]
17. Tsou, M.C.; Özkaynak, H.; Beamer, P.; Dang, W.; Hsi, H.C.; Jiang, C.B.; Chien, L.C. Mouthing activity data for children aged 7 to 35 months in Taiwan. *J. Expo. Sci. Environ. Epidemiol.* **2015**, *25*, 388–398. [CrossRef]
18. Tsou, M.C.; Özkaynak, H.; Beamer, P.; Dang, W.; Hsi, H.C.; Jiang, C.B.; Chien, L.C. Mouthing activity data for children age 3 to <6 years old and fraction of hand area mouthed for children age <6 years old in Taiwan. *J. Expo. Sci. Environ. Epidemiol.* **2018**, *28*, 182–192. [CrossRef]
19. Ian, V.L.; Susan, S.; Lin, S.M.; Wichers, S.L.; Casey, B. Estimating Children's Soil/Dust Ingestion Rates through Retrospective Analyses of Blood Lead Biomonitoring from the Bunker Hill Superfund Site in Idaho. *Environ. Health Perspect.* **2016**, *124*, 1462–1470. [CrossRef]
20. White, P.D.; Leeuwen, P.V.; Davis, B.D.; Maddaloni, M.; Hogan, K.A.; Marcus, A.H.; Elias, R.W. The conceptual structure of the integrated exposure uptake biokinetic model for lead in children. *Environ. Health Perspect.* **1998**, *106*, 1513–1530. [CrossRef]
21. Binder, S.; David, S.; David, M. Estimating Soil Ingestion: The Use of Tracer Elements in Estimating the Amount of Soil Ingested by Young Children. *Arch. Environ. Health Int. J.* **1986**, *41*, 341–345. [CrossRef] [PubMed]
22. Calabrese, E.J.; Stanek, E.J., III; Pekow, P.; Barnes, R.M. Soil Ingestion Estimates for Children Residing on a Superfund Site. *Ecotoxicol. Environ. Saf.* **1997**, *36*, 258–268. [CrossRef] [PubMed]
23. Clausing, P.; Brunekreef, B.; Wijnen, J.H. A method for estimating soil ingestion by children. *Int. Arch. Occup. Environ. Health* **1987**, *59*, 73–82. [CrossRef] [PubMed]
24. Chien, L.C.; Tsou, M.C.; His, H.C.; Beamer, P.; Bradham, K.; Hseu, Z.Y.; Jien, S.H.; Jiang, C.B.; Dang, W.; Özkaynak, H. Soil ingestion rates for children under 3 years old in Taiwan. *J. Expo. Sci. Environ. Epidemiol.* **2017**, *27*, 33–40. [CrossRef] [PubMed]
25. Irvine, G.; Doyle, J.R.; White, P.A.; Blais, J.M. Soil ingestion rate determination in a rural population of Alberta, Canada practicing a wilderness lifestyle. *Sci. Total Environ.* **2014**, *470*, 138–146. [CrossRef] [PubMed]
26. Jang, J.Y.; Kim, S.Y.; Kim, S.J.; Lee, K.E.; Cheong, H.K.; Kim, E.H.; Choi, K.H.; Kim, Y.H. General factors of the Korean exposure factors handbook. *J. Prev. Med. Public Health* **2014**, *47*, 7–17. [CrossRef]
27. Wang, B.B.; Lin, C.Y.; Zhang, X.; Duan, X.L.; Xu, D.Q.; Cheng, H.G.; Wang, Q.; Liu, X.T.; Ma, J.; Ma, J.W.; et al. A soil ingestion pilot study for teenage children in China. *Chemosphere* **2018**, *202*, 40–47. [CrossRef]
28. Moya, J. *Exposure Factors Handbook*, 2011 ed.; US EPA National Center for Environmental Assessment, Immediate Office: Washington, DC, USA, 2011.
29. Doyle, J.R.; Blais, J.M.; White, P.A. Mass balance soil ingestion estimating methods and their application to inhabitants of rural and wilderness areas: A critical review. *Sci. Total Environ.* **2010**, *408*, 2181–2188. [CrossRef]
30. Doyle, J.R.; Blais, J.M.; White, P.A. A soil ingestion pilot study of a population following a traditional lifestyle typical of rural or wilderness areas. *Sci. Total Environ.* **2021**, *424*, 110–120. [CrossRef]
31. Calabrese, E.J.; Barnes, R.; Stanek, E.J.; Pastides, H.; Gilbert, C.E.; Veneman, P.; Wang, X.R.; Lasztity, A.; Kostecki, P.T. How much soil do young children ingest: An epidemiologic study. *Regul. Toxicol. Pharmacol.* **1989**, *10*, 123–137. [CrossRef]
32. Davis, S.; Mirick, D.K. Soil ingestion in children and adults in the same family. *J. Expo. EnceEnviron. Epidemiol.* **2006**, *16*, 63–75. [CrossRef] [PubMed]
33. Ackah, M. Soil elemental concentrations, geoaccumulation index, non–carcinogenic and carcinogenic risks in functional areas of an informal e–waste recycling area in Accra, Ghana. *Chemosphere* **2019**, *235*, 908–917. [CrossRef] [PubMed]
34. Zheng, L.; Wu, K.; Li, Y.; Qi, Z.; Han, D.; Zhang, B.; Gu, C.; Chen, G.; Liu, J.; Chen, S.; et al. Blood lead and cadmium levels and relevant factors among children from an e–waste recycling town in China. *Environ. Res.* **2008**, *108*, 15–20. [CrossRef] [PubMed]

International Journal of
Environmental Research and Public Health

MDPI

Article

Performance and Mechanisms of Sulfidated Nanoscale Zero-Valent Iron Materials for Toxic TCE Removal from the Groundwater

Yue Lang [1], Yanan Yu [2], Hongtao Zou [1,*], Jiexu Ye [2] and Shihan Zhang [2]

1 College of Land and Environment, Shenyang Agricultural University, Shenyang 110866, China; langyue@sjzu.edu.cn
2 Key Laboratory of Microbial Technology for Industrial Pollution Control of Zhejiang Province, College of Environment, Zhejiang University of Technology, Hangzhou 310014, China; 1112127014@zjut.edu.cn (Y.Y.); yejiexu@zjut.edu.cn (J.Y.); shihanzhang@zjut.edu.cn (S.Z.)
* Correspondence: hongtaozou208@126.com

Abstract: Trichloroethylene (TCE) is one of the most widely distributed pollutants in groundwater and poses serious risks to the environment and human health. In this study, sulfidated nanoscale zero-valent iron (S-nZVI) materials with different Fe/S molar ratios were synthesized by one-step methods. These materials degraded TCE in groundwater and followed a pathway that did not involve the production of toxic byproducts such as dichloroethenes (DCEs) and vinyl chloride (VC). The effects of sulfur content on TCE dechlorination by S-nZVI were thoroughly investigated in terms of TCE-removal efficiency, H_2 evolution, and reaction rate. X-ray diffraction (XRD) and X-ray Photoelectron Spectroscopy (XPS) characterizations confirmed Fe(0) levels in S-nZVI were larger than for zero-valent iron (nZVI). An Fe/S molar ratio of 10 provided the highest TCE-removal efficiencies. Compared with nZVI, the 24-h TCE removal efficiencies of S-nZVI (Fe/S = 10) increased from 30.2% to 92.6%, and the Fe(0) consumed during a side-reaction of H_2 evolution dropped from 77.0% to 12.8%. This indicated the incorporation of sulfur effectively inhibited H_2 evolution and allowed more Fe(0) to react with TCE. Moreover, the pseudo-first-order kinetic rate constants of S-nZVI materials increased by up to 485% compared to nZVI. In addition, a TCE degradation was proposed based on the variation of detected degradation products. Noting that acetylene, ethylene, and ethane were detected rather than DCEs and VC confirmed that TCE degradation followed β-elimination with acetylene as the intermediate. These results demonstrated that sulfide modification significantly enhanced nZVI performance for TCE degradation, minimized toxic-byproduct formation, and mitigated health risks. This work provides some insight into the remediation of chlorinated-organic-compound-contaminated groundwater and protection from secondary pollution during remediation by adjusting the degradation pathway.

Keywords: S-nZVI; sulfidation; trichloroethylene; pathway; groundwater safety

Citation: Lang, Y.; Yu, Y.; Zou, H.; Ye, J.; Zhang, S. Performance and Mechanisms of Sulfidated Nanoscale Zero-Valent Iron Materials for Toxic TCE Removal from the Groundwater. *Int. J. Environ. Res. Public Health* **2022**, *19*, 6299. https://doi.org/10.3390/ijerph19106299

Academic Editor: Cheng Yan

Received: 10 April 2022
Accepted: 17 May 2022
Published: 22 May 2022

1. Introduction

Chlorinated hydrocarbons are common industrial solvents. Accidental spills and improper disposal result in these organic compounds entering the soil, drinking water, and groundwater [1–5]. Trichloroethene (TCE) is an intensively detected chlorinated hydrocarbon in groundwater [6,7]. It readily migrates, is persistent in abiotic and biotic degradation, and poses serious threats to groundwater [8–12]. TCE poses a significant carcinogenic risk to human health by all routes of exposure [13] and is classified as "carcinogenic to humans" by the International Agency for Research on Cancer [14]. TCE exposure is associated with kidney cancer, liver cancer, and lymphatic-system cancer. Groundwater serves as a primary source of drinking water in China, which underscores the importance of developing an environmentally friendly technology to remediate the TCE-contaminated groundwater in order to mitigate its health risks.

Methods such as adsorption, oxidation/reduction, and biologically reductive dechlorination have been used to eliminate TCE in groundwater [2,15,16]. Among these methods, nanoscale zero-valent iron (nZVI) is promising for the in situ chemical remediation of TCE-contaminated groundwater, readily reacting and dechlorinating TCE into non-toxic end products [17]. However, nZVI readily reacts with dissolved oxygen, water, and other groundwater substrates to form iron oxides and/or hydroxides. The continuously thickened iron oxide or hydroxide shell passivates the nZVI surface, which decreases the electron-transfer efficiency between Fe(0) and the pollutants in the inner core, and decreases nZVI activity [18–22]. Furthermore, nZVI shows relatively poor selectivity for TCE reduction over water in an aqueous solution [23].

In response to these problems, many groups have reported their research regarding ZVI modification to improve its reactivity, selectivity, and longevity. A few recent studies have reported that sulfidation could improve nZVI performance in the reduction of chlorinated contaminants. During sulfidation, FeS forms on the nZVI surface, thereby enhancing electron transfer between Fe(0) and chlorinated hydrocarbons and restraining hydrogen (H_2) evolution between Fe(0) and water [24–26]. Kim et al. reported that 'Fe/FeS' nanoparticles, synthesized via dithionite addition, enhanced TCE removal as compared to nZVI [27]. Rajajayavel and Ghoshal reported that sulfidation of nZVI facilitated TCE dechlorination and suggested the Fe/S molar ratio played a key role in the reactivity of sulfidated nZVI (S-nZVI) [28]. Han and Yan reported that enhanced TCE dechlorination rates with increasing sulfur levels were only significant at low sulfur loadings, which differed from the findings of Rajajayavel and Ghoshal [23,28]. Thus, although incorporating sulfur into nZVI has been demonstrated, S-nZVI needs additional refinement for better TCE dechlorination, and the critical variables such as sulfur content and speciation need a systematic investigation. Moreover, the abiotic TCE-degradation pathway typically follows one of three pathways such as hydrogenolysis, α-elimination, and β-elimination [17]. Among them, β-elimination is promising since it produces acetylene as an intermediate rather than highly toxic DCEs and VC [28]. Therefore, in-depth research on the pathway of TCE dechlorination is specifically needed for S-nZVI.

In this study, S-nZVI materials with different Fe/S molar ratios were prepared using sodium dithionite as the sulfidation agent. The effects of Fe/S molar ratios on the morphology and structure were characterized by scanning electron microscopy (SEM), Brunner–Emmet–Teller measurements (BET), X-ray diffraction (XRD), and X-ray photoelectron spectroscopy (XPS). The reactivities of different S-nZVI materials and bare nZVI to TCE and water were assessed with respect to TCE-removal efficiency, H_2 evolution, and the kinetics of TCE degradation. Furthermore, dechlorination products and pathways were identified in order to gain insight into the mechanism of S-nZVI for TCE degradation and the mitigation of health risks. This work provides some guidance to develop green and environmentally friendly technologies to remediate chlorinated-organic-compound-contaminated groundwaters and protect them from secondary pollution and potential health risks.

2. Materials and Methods

2.1. Reagents

$FeSO_4 \cdot 7H_2O$ (>99%, Aanalytical Reagent), ethanol (99.8%, Guaranteed Reagent) and NaOH (95%, Analytical Reagent) were purchased from Rhawn (Shanghai, China). $NaBH_4$ (98%, Analytical Reagent) was purchased from Sinopharm Chemical Reagents Ltd., (Shanghai, China). TCE (C_2HCl_3, >99.5%, Guaranteed Reagent) and $Na_2S \cdot 9\ H_2O$ (>98.0%, Analytical Reagent) were purchased from Aladdin (Shanghai, China). N_2 (purity 99.99%), air (purity 99.99%), Ar (purity 99.99%) and H_2 (purity 99.99%) were purchased from Hangzhou Jingong Special Gas Co., Ltd. (Hangzhou, China). H_2 standard gas (10%) and

Alkane olefin standard gas (purity 99.99%) were purchased from Dalian Date Gas Co., Ltd. (Dalian, China).

2.2. Preparation of nZVI

A total of 4.9643 g of $FeSO_4 \cdot 7H_2O$ was dissolved in 50 mL deoxygenated water and transferred into a 250 mL three-necked flask. Absolute ethanol (20 mL) was added and stirred to dilute the dispersed solute and make the synthesized iron particles smaller. $NaBH_4$ (2.0266 g) (excess Fe^{2+}:BH_4^- = 1:3) was dissolved in 20 mL of 0.1% NaOH solution (inhibiting $NaBH_4$ hydrolysis). The mixture was added dropwise using a constant-pressure funnel to a three-neck flask at 2 drops/s and stirred at 1000 rpm. After the dropwise addition was completed, stirring was continued for 0.5 h. Then, the nZVI particle was separated from the suspension using a magnet, and washed with deoxygenated deionized water and pure ethanol three times. After that, the material was sequentially dried in an argon-filled glove box for 24 h and aged in the transition chamber of the argon-filled glove box for 8 h. Finally, it was sealed and stored in the anaerobic glove box. All the reactant solutions used in the experiment were stripped with N_2 for at least 0.5 h to remove the dissolved oxygen.

2.3. One-Step Preparation of S-nZVI

A total of 3.723 g of $FeSO_4 \cdot 7H_2O$ was added into 110 mL of deoxygenated water and stirred for 0.5 h under N_2 atmosphere in a 250 mL three-necked flask. Then, 7.5 mL of aq. NaOH (1 M) was added dropwise into the $FeSO_4$ solution at a rate of 1 drop/s while stirring. A total of 1.0165 g of $NaBH_4$ was dissolved into 40 mL of deoxygenated water, and a certain amount of 0.1 M $Na_2S \cdot 9H_2O$ solution was added in order to prepare a synthesis mixture. After that, the mixture of $NaBH_4$ and Na_2S solution were introduced into the three-necked flask at a rate of 0.2 mL/s by a peristaltic pump, followed by 0.5 h of mechanically stirring under an atmosphere of N_2. The dosages of $Na_2S \cdot 9H_2O$ solution in the synthesis mixture were 52.23, 26.12, 17.41 and 13.06 mL to achieve the Fe/S molar ratios of 2.5, 5, 7.5 and 10, respectively. The resulting S-nZVI suspension was washed at least three times with the deoxygenated deionized water and pure ethanol, and dried in an argon-filled glove box for 24 h. Finally, the obtained S-nZVI particles were placed in the transition chamber of the anaerobic glove box for 8 h of aging, and then sealed and stored in the argon-filled glove box. All the reactant solutions used in the experiment were purged with N_2 for at least 0.5 h to remove the dissolved oxygen.

2.4. TCE Degradation and H_2 Evolution

Inside an anaerobic glove box, 0.026 g of S-nZVI or nZVI was added to a 52 mL crimp-top vial containing 26 mL of a deoxygenated 4-(2-hydroxyethyl)-1-piperazine ethane sulfonic acid (HEPEs) buffer (50 mM, pH = 7), and sealed with an aluminum cover and a polytetrafluoroethylene (PTFE) septum. Outside the glove box, 15 μL of a 0.136 M TCE stock solution was injected to ensure an initial TCE concentration of 10 ppm. Finally, the crimp-top vial was placed on a constant-temperature shaker and maintained at 30 °C. At certain intervals, 100 μL of headspace gas in the bottle was extracted with a micro-sampling needle; the TCE concentration, H_2 concentration, and the TCE-degradation products were measured by GC. The solution pH in the bottle was measured after the reaction.

2.5. Characterization

A physical adsorption instrument (Mike 2460 type, Norcross, GA, USA) measured the adsorption isotherm of the material on N_2 at 77 K using N_2 as the adsorbate to calculate the specific surface area and pore size of the material. The samples were degassed at 120 °C for 6 h before testing, and the specific BET surface area of the tested material was determined based on the N_2 adsorption and desorption of BJH models at 77 K. The surface morphology of the material was analyzed by SEM (Zeiss Sigma 300, Oberkochen, Germany). The surface composition of the material was analyzed using XPS (K-AlpHa, Thermo Scientific, Waltham,

MA, USA). The initial binding energy of the XPS spectrum was corrected based on the C1s peak at 284.8 eV, and the energy spectrum was analyzed using XPS-peak fit software (V.4.1, USA). The crystal type of the material was qualitatively analyzed by XRD (X'Pert'3 Powder, Panalytical, Almelo, The Netherlands). The test conditions of the instrument were as follows: the X-ray source was Cu target Kα rays (λ = 0.154056 nm), with a scanning range of 10–80° at a scanning rate of 5° min^{-1}.

2.6. Kinetic Modelings

The degradation of TCE followed the pseudo-first-order reaction kinetics (Equation (1))

$$y = Ae^{-k_{sa}t} \quad (1)$$

where k_{sa} ($L/(m^2 \cdot h)$) refers to the pseudo-first-order reaction-rate constant, y (mg/L) represents the TCE concentration of headspace sampling in the crimp-top vial, t (h) represents the reaction time, and A (mg/L) represents the initial TCE concentration.

2.7. Analytical Methods

Concentrations of TCE and its degradation products were measured by a GC (Agilent 7890B, Santa Clara, CA, USA) equipped with a flame ionization detector (FID). The capillary column was a GS-Q, 30 m long with a 0.53 mm diameter. The inlet temperature was 200 °C, the detector temperature was 230 °C, the injection volume was 100 μL, and the injection port split ratio was 10:1. The initial column oven temperature was 50 °C, maintained for 7 min, then increased to 230 °C at a heating rate of 20 °C/min for 10 min. During this process, TCE and its degradation products in the samples were separated.

The concentration of H_2 was measured by a gas chromatograph (FULI-9790, Taizhou, China) equipped with a thermal conductive detector (TCD). A TXD-01 packed column (3 mm × 2 m, Lanzhou Institute of Chemical Physics, Chinese Academy of Sciences, Lanzhou, China) was used, with Ar as the carrier gas. The inlet temperature was 140 °C, the column oven was 60 °C, the detector temperature was 140 °C, and the injection volume was 100 μL. Since H_2 solubility in water is negligible, H_2 in the reaction-vial headspace was the total H_2 produced during the reaction.

3. Results and Discussion

3.1. Materials Characterization

Figure 1 shows the surface morphology of nZVI and S-nZVI materials with different Fe/S molar ratios. Figure 1a shows spherical nZVI particles with relatively uniform sizes (50–200 nm) that aggregate into chains due to electrostatic and magnetic forces. The S-nZVI contained spherical and amorphous structures as shown in Figure 1b–e. By comparing S-nZVI materials with different Fe/S ratios, the spherical structure continuously decreased, and the main structure gradually changed from a relatively regular sphere to an irregular amorphous structure as sulfur was added. These changes illustrated that sulfidation increased material dispersion and improved its specific surface area, which could be related to the formation of an FeS_x layer on the nZVI surface.

Figure 2 shows XRD spectra of nZVI and S-nZVI with different Fe/S ratios. The characteristic diffraction peaks at 2θ = 44.6° and 65.0° corresponded to the (110) and (200) faces of Fe(0) (PDF#00-006-0696), respectively. On the other hand, there was no obvious iron-oxide peak, which indicated that the degree of oxidation was relatively light during the preparation and preservation of the material. Additionally, S-nZVI had a strong diffraction peak at the same position, which indicated high crystallinity Fe(0) in S-nZVI, and that sulfur introduction promoted the formation of Fe(0) crystals, and the peak intensity increased with the increase of Fe/S ratio.

Figure 1. SEM images of nZVI and S-nZVI materials with different Fe/S molar ratios (**a**) nZVI; (**b**) S-nZVI (Fe/S = 10); (**c**) S-nZVI (Fe/S = 7.5); (**d**) S-nZVI (Fe/S = 5); (**e**) S-nZVI (Fe/S = 2.5).

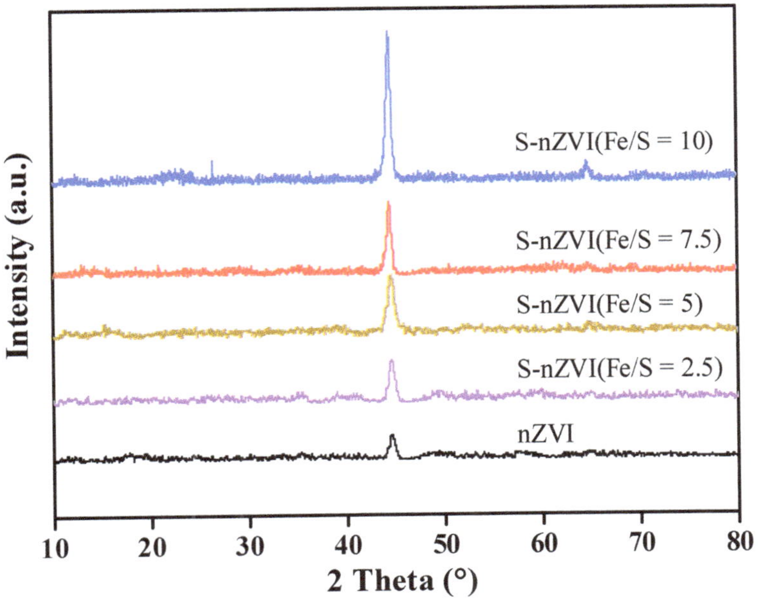

Figure 2. XRD spectra of nZVI and S-nZVI materials with different Fe/S molar ratios.

The pore structure and size of different materials were analyzed to explore surface-structure differences among nZVI and S-nZVI materials with different Fe/S ratios. As shown in Table 1, the specific surface area of nZVI was 8.76 m^2/g, and the specific surface areas of S-nZVI with different Fe/S ratios (2.5, 5, 7.5, and 10) were 95.66, 53.52, 57.60, and 27.11 m^2/g, respectively. The specific surface area and pore volume increased with added sulfur, perhaps due to the presence of FeS_x, which inhibited material agglomeration and increased the surface roughness and surface area. For Fe/S = 5 and 2.5, the average pore size and volume both increased significantly. This might be due to increased FeS_x; the amorphous structure became the primary structure, which resulted in larger pore size and volume. This conclusion agreed with SEM image data. These results indicated the Fe/S ratio significantly impacted the pore size and characteristics of S-nZVI as well as its reaction activity.

Table 1. Pore-structure parameters of nZVI and S-nZVI materials with different Fe/S ratios.

Sample	S_{BET} (m^2/g)	$D_{average}$ (nm)	V_{total} (cm^3/g)
nZVI	8.76	3.41	0.01
S-nZVI (Fe/S = 10)	27.11	3.82	0.05
S-nZVI (Fe/S = 7.5)	57.60	3.41	0.12
S-nZVI (Fe/S = 5)	53.52	22.17	0.31
S-nZVI (Fe/S = 2.5)	95.66	18.91	0.48

XPS analysis determined the elemental composition and valence information on the material surface. Fe 2p peak fitting (Figure 3a) yielded binding energies for Fe 2p 3/2, Fe(III), Fe(II), and Fe(0) at 711.0, 709.0, and 706.7 eV, respectively, while Fe 2p 1/2, Fe(III), Fe(II), and Fe(0) had binding energies of 724.8, 722.8, and 719.8 eV, respectively. Shoulder peaks for Fe(III) and Fe(II) came at 718.9 and 733.4 eV, and 714.6 and 729.5 eV, respectively. Additionally, no obvious elemental-iron peaks were observed because the XPS analysis determined the nanomaterial components on the surface. Fitting data revealed the S-nZVI surface was covered by iron sulfide, so the ZVI content was not detected. As shown in Figure 3c, the one-step sulfidation of nZVI showed a drop in the Fe(III) content with additional sulfur, while levels of Fe(II) remained steady. Peak fitting of S 2p in the material is shown in Figure 3b, in which the binding energies of monosulfide (S^{2-}), disulfide ($S_2{}^{2-}$), and sulfate ($SO_4{}^{2-}$) were 161.5, 162.2, and 168.5 eV, respectively. As shown in Figure 3d, S in S-nZVI mainly existed as S^{2-}, $S_2{}^{2-}$, or $SO_4{}^{2-}$. With added sulfur, S^{2-} levels decreased, and the content of $S_2{}^{2-}$ increased, possibly due to excessive sulfur (from sodium sulfide) that reduced to $S_2{}^{2-}$, and the electron-transfer utilization of FeS_2 was much smaller than FeS [29], so the generation of excessive $S_2{}^{2-}$ reduced its electron-transfer efficiency and reduced dechlorination of S-nZVI on TCE.

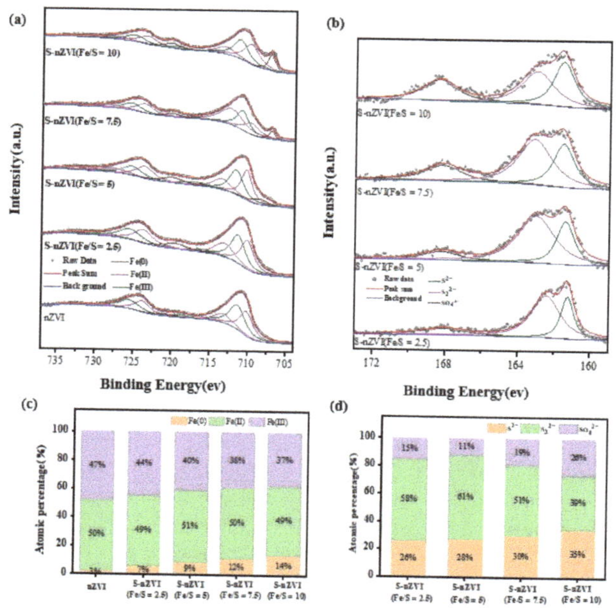

Figure 3. XPS spectra of nZVI and S-nZVI materials with different Fe/S ratios: (**a**) XPS spectrum of Fe2 p; (**b**) XPS spectrum of S 2p; (**c**) Fe levels in different valences; (**d**) S levels in different valences.

3.2. Performance of TCE Degradation

Figure 4 shows the performances of nZVI and S-nZVI materials with different Fe/S ratios (Fe/S = 2.5, 5, 7.5, and 10) in TCE degradation. As shown in Figure 4a, the 24 h TCE-removal efficiency of nZVI was 30.2%, while that of S-nZVI (Fe/S = 10) exceeded 92.6%, which indicated that nZVI did not degrade TCE, while the dechlorination of S-nZVI improved significantly. Meanwhile, the removal efficiency of TCE increased as the Fe/S molar ratio increased. To confirm whether Fe (not S) degraded TCE in S-nZVI, Na_2S was mixed with TCE to monitor its degradation effect. As shown in Figure 4a, Na_2S did not degrade TCE. A quantitative test of H_2 in the crimp-top vial was performed (Figure 4b) to explore how much Fe(0) was used for H_2 evolution. That test indicated that the side reaction of H_2 evolution consumed 77.0% of the initial Fe(0) in nZVI. However, only 12.8% of the initial Fe(0) was consumed for H_2 evolution in the S-nZVI at an Fe/S ratio of 10. This indicated that sulfidation greatly inhibited side reactions so that a large amount of Fe(0) could react fully with TCE. Some researchers reported that H_2 evolution mainly occurred in the iron oxide layer rather than the FeS_x layer, while the main reaction site for TCE occurred in the FeS_x layer [30,31]. Since the FeS_x layer had strong electrical conductivity [32], electron transfer from the surface of the FeS_x layer was used for TCE degradation. This was the reason S-nZVI simultaneously suppressed H_2 evolution and promoted TCE degradation.

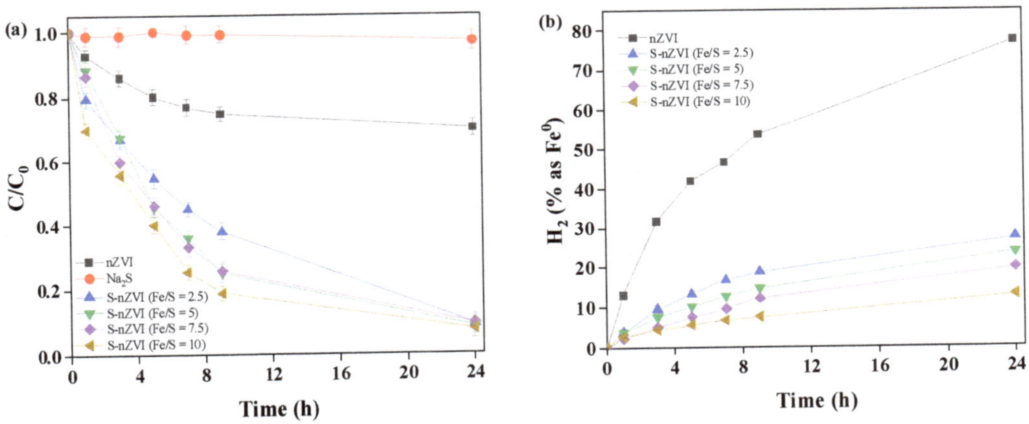

Figure 4. Performance of nZVI and S-nZVI materials with different Fe/S molar ratios in TCE degradation, (**a**) TCE removal; (**b**) H_2 evolution (pH: 7, t: 30 °C; S-nZVI: 0.026 g, TCE: 10 ppm).

3.3. Kinetics of TCE Degradation

An evaluation of degradation rate constants for nZVI and S-nZVI materials was conducted and is shown in Figure 5. Rate constants were obtained by fitting the 24 h removal-efficiency curve for TCE using pseudo-first-order kinetics. As shown in Figure 5, the incorporation of sulfur into nZVI greatly increased the degradation rate constant, which was likely due to the FeS shell of S-nZVI improving electron-transfer efficiency. The degradation activity of S-nZVI maximized for Fe/S = 10, with a rate constant of 0.117 L/($m^2 \cdot h$). However, when the Fe/S molar ratio varied between 2.5–10, the rate constant for S-nZVI materials changed slightly. This may be related to excess Fe(0) in the S-nZVI system, where the initial TCE concentration was as low as 10 ppm.

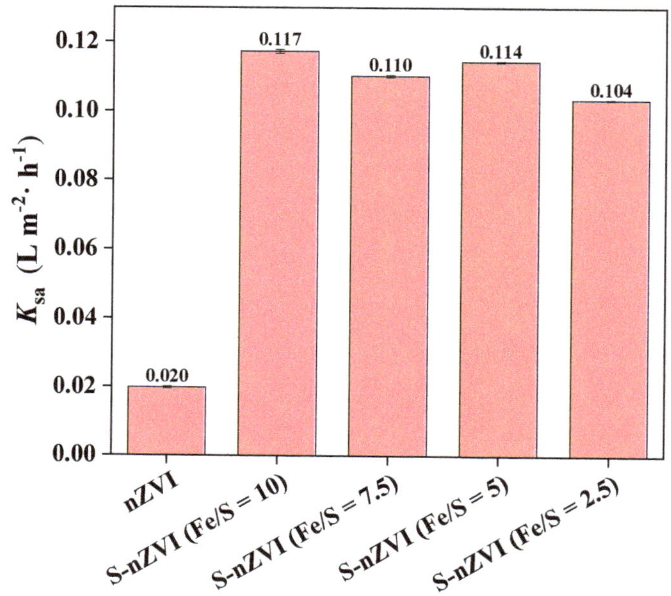

Figure 5. Reaction rates of TCE degradation by nZVI and S-nZVI materials with different Fe/S molar ratios. (pH: 7, t: 30 °C; nZVI and S-nZVI: 0.026 g, TCE: 10 ppm).

3.4. Mechanism of TCE Degradation

As shown in Figure 6, toxic products such as DCEs and VC were not detected in either neat nZVI or S-nZVI. Moreover, both materials primarily produced ethylene, which confirmed that TCE degradation followed a β-elimination mechanistic pathway with acetylene as the intermediate. In this work, primary products of TCE degradation such as acetylene, ethylene, and ethane were measured and are shown in Figure 6. Compared with the neat nZVI, a small amount of acetylene was detected with the S-nZVI, which indicated that sulfidated modification of nZVI enhanced the conversion of TEC to form acetylene. Moreover, the ethylene produced using S-nZVI was 11.3 times greater than nZVI, which confirmed that sulfidated modification enhanced the reactivity of each step during TCE degradation.

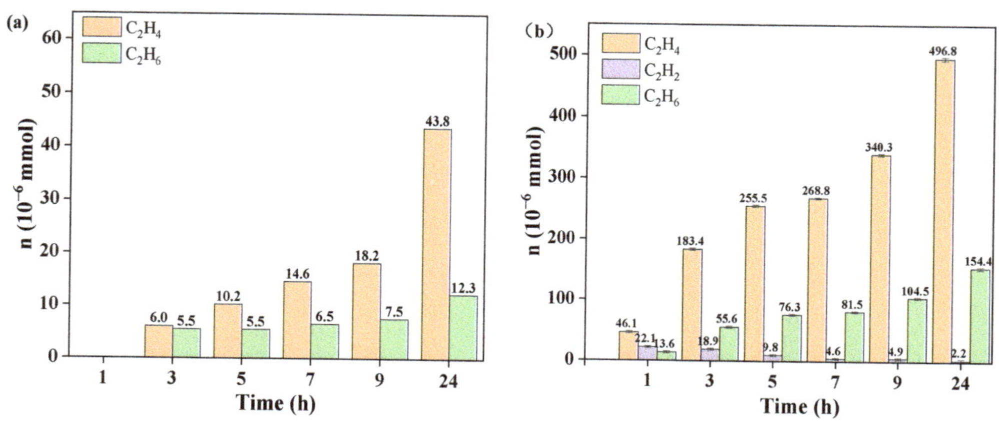

Figure 6. Gas phase production during TCE degradation with: (**a**) neat nZVI; and (**b**) S-nZVI with a Fe/S ratio of 10 (pH: 7, T: 30 °C; nZVI and S-nZVI: 0.026 g).

Overall, dechlorination of TCE occurred via β-elimination through chloroacetylene, followed by hydrogenolysis of chloroacetylene to acetylene, which agreed with previous reports [23,33]. A larger dose of S-nZVI produced ethylene and ethane via hydrogenation of acetylene, and may generate C3-C6 hydrocarbons via the polymerization of acetylene (Figure 7).

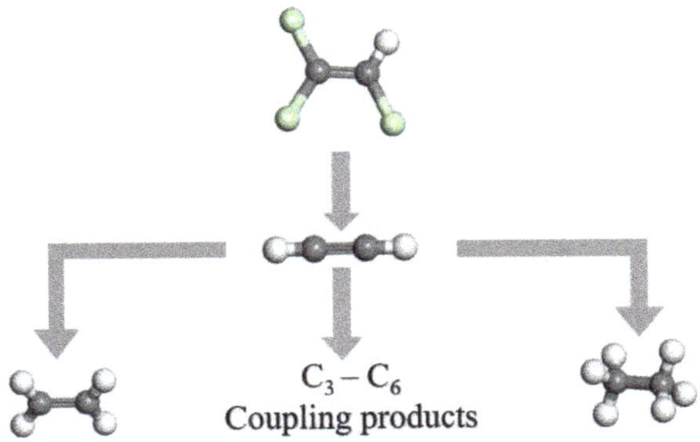

Figure 7. A possible mechanism for the TCE degradation with the S-nZVI.

4. Conclusions

This study found that sulfidation significantly improved TCE dechlorination via nZVI. XRD and XPS results illustrated that sulfidation caused the material to store more Fe(0), and Fe(0) increased with the decreasing sulfur ratio (Fe/S $\leq$ 10). The 24 h TCE-removal efficiency of S-nZVI with an Fe/S ratio of 10 increased from 30.2% to 92.6% compared to nZVI. This was attributed to a reduction in H_2 evolution from nZVI after sulfidation. At an Fe/S molar ratio of 10, Fe(0) consumed by H_2 evolution was only 12.8%, which suggested a large amount of reactive Fe(0) was available to react with TCE. Compared with nZVI, the pseudo-first-order kinetic rate constants of S-nZVI materials increased significantly. Additionally, the reaction of TCE degradation by S-nZVI was explored. Acetylene was the intermediate product, followed by the production of ethylene and ethane. C_3–C_6 hydrocarbons may be generated via the polymerization of acetylene. Overall, this work provides insights into developing green and environmentally friendly technologies to remediate chlorinated organic compound-contaminated groundwater and protect them from secondary pollution as well as potential health risks in groundwater utilization.

Author Contributions: Y.L.: Writing—Original Draft, Investigation; Y.Y.: Investigation, Data Analysis. H.Z.: Conceptualization, Supervision, Validation; J.Y.: Formal Analysis; S.Z.: Writing—Review & Editing. All authors have read and agreed to the published version of the manuscript.

Funding: This work was supported by the Central Government Guided Local Science and Technology Development Fund (2021ZY1022), and National Natural Science Foundation of China (Nos: 21876157).

Institutional Review Board Statement: Not applicable.

Informed Consent Statement: Not applicable.

Data Availability Statement: The data presented in this study are available on request from the corresponding author. The data are not publicly available due to privacy.

Conflicts of Interest: The authors declare no conflict of interest.

References

1. Seyama, T.; Adachi, K.; Yamazaki, S. Kinetics of photocatalytic degradation of trichloroethylene in aqueous colloidal solutions of TiO_2 and WO_3 nanoparticles. *J. Photochem. Photobiol. A* **2012**, *249*, 15–20. [CrossRef]
2. Wei, Z.S.; Seo, Y. Trichloroethylene (TCE) adsorption using sustainable organic mulch. *J. Hazard. Mater.* **2010**, *181*, 147–153. [CrossRef] [PubMed]
3. Dong, H.R.; He, Q.; Zeng, G.M.; Tang, L.; Zhang, L.H.; Xie, Y.K.; Zeng, Y.L.; Zhao, F. Degradation of trichloroethene by nanoscale zero-valent iron (nZVI) and nZVI activated persulfate in the absence and presence of EDTA. *Chem. Eng. J.* **2017**, *316*, 410–418. [CrossRef]
4. Huang, B.B.; Lei, C.; Wei, C.H.; Zeng, G.M. Chlorinated volatile organic compounds (Cl-VOCs) in environment-sources, potential human health impacts, and current remediation technologies. *Environ. Int.* **2014**, *71*, 118–138. [CrossRef] [PubMed]
5. Valentín, L.; Nousiainen, A.; Mikkonen, A. Introduction to Organic Contaminants in Soil: Concepts and Risks. In *Emerging Organic Contaminants in Sludges: Analysis, Fate and Biological Treatment*; Springer: Berlin/Heidelberg, Germany, 2013; pp. 1–29. [CrossRef]
6. Yan, J.C.; Han, L.; Gao, W.G.; Xue, S.; Chen, M.F. Biochar supported nanoscale zerovalent iron composite used as persulfate activator for removing trichloroethylene. *Bioresour. Technol.* **2015**, *175*, 269–274. [CrossRef]
7. Dong, H.R.; Zhang, C.; Deng, J.M.; Jiang, Z.; Zhang, L.H.; Cheng, Y.J.; Hou, K.J.; Tang, L.; Zeng, G.M. Factors influencing degradation of trichloroethylene by sulfide-modified nanoscale zero-valent iron in aqueous solution. *Water Res.* **2018**, *135*, 1–10. [CrossRef]
8. Dong, H.R.; Zhang, C.; Hou, K.J.; Cheng, Y.J.; Deng, J.M.; Jiang, Z.; Tang, L.; Zeng, G.M. Removal of trichloroethylene by biochar supported nanoscale zero-valent iron in aqueous solution. *Sep. Purif. Technol.* **2017**, *188*, 188–196. [CrossRef]
9. Starr, R.C.; Orr, B.R.; Lee, M.H.; Delwiche, M. *Final Project Report-Coupled Biogeochemical Process Evaluation for Con-Ceptualizing Trichloriethylene Co-Metabolism: Co-Metabolic Enzyme Activity Probes and Modeling Co-Metabolism and Attenuation*; North Wind Inc.: Idaho Falls, ID, USA, 2010.
10. Slater, G.F.; Dempster, H.S.; Sherwood Lollar, B.; Ahad, J. Headspace analysis: A new application for isotopic characterization of dissolved organic contaminants. *Environ. Sci. Technol.* **1999**, *33*, 190–194. [CrossRef]
11. Meng, Q.J.; Li, P.F.; Qu, J.H.; Liu, Y.; Wang, Y.F.; Chen, Z.B.; Zhang, Y. Study on the community structure and function of anaerobic granular sludge under trichloroethylene stress. *Ecotoxicology* **2021**, *30*, 1408–1418. [CrossRef]
12. Chiu, W.A.; Jinot, J.; Scott, C.S.; Makris, S.L.; Cooper, G.S.; Dzubow, R.C.; Bale, A.S.; Evans, M.V.; Guyton, K.Z.; Keshava, N.; et al. Human health effects of trichloroethylene: Key findings and scientific issues. *Environ. Health Persp.* **2013**, *121*, 303–311. [CrossRef]
13. Luo, Y.-S.; Furuya, S.; Soldatov, V.Y.; Kosyk, O.; Yoo, H.S.; Fukushima, H.; Lewis, L.; Iwata, Y.; Rusyn, I. Metabolism and Toxicity of Trichloroethylene and Tetrachloroethylene in Cytochrome P450 2E1 Knockout and Humanized Transgenic Mice. *Toxicol. Sci.* **2018**, *164*, 489–500. [CrossRef]
14. Guha, N.; Loomis, D.; Grosse, Y.; Lauby-Secretan, B.; El Ghissassi, F.E.; Bouvard, V.; Benbrahim-Tallaa, L.; Baan, R.; Mattock, H.; Straif, K. Carcinogenicity of trichloroethylene, tetrachloroethylene, some other chlorinated solvents, and their metabolites. *Lancet Oncol.* **2012**, *13*, 1192–1193. [CrossRef]
15. Ahmad, M.; Lee, S.S.; Dou, X.M.; Mohan, D.; Sung, J.-K.; Yang, J.E.; Ok, Y.S. Effects of pyrolysis temperature on soybean stover- and peanut shell-derived biochar properties and TCE adsorption in water. *Bioresour. Technol.* **2012**, *118*, 536–544. [CrossRef] [PubMed]
16. Shao, H.B.; Butler, E.C. The Relative Importance of Abiotic and Biotic Transformation of Carbon Tetrachloride in Anaerobic Soils and Sediments. *Soil Sediment. Contam. Int. J.* **2009**, *18*, 455–469. [CrossRef]
17. Pasinszki, T.; Krebsz, M. Synthesis and Application of Zero-Valent Iron Nanoparticles in Water Treatment, Environmental Remediation, Catalysis, and Their Biological Effects. *Nanomaterials* **2020**, *10*, 917. [CrossRef]
18. Mu, Y.; Jia, F.L.; Ai, Z.H.; Zhang, L.Z. Iron oxide shell mediated environmental remediation properties of nano zero-valent iron. *Environ. Sci. Nano* **2017**, *4*, 27–45. [CrossRef]
19. Zhang, W.-X. Nanoscale Iron Particles for Environmental Remediation: An Overview. *J. Nanoparticle Res.* **2003**, *5*, 323–332. [CrossRef]
20. Nurmi, J.T.; Tratnyek, P.G.; Sarathy, V.; Baer, D.R. Characterization and properties of metallic iron nanoparticles: Spectroscopy, electrochemistry, and kinetics. *Environ. Sci. Technol.* **2005**, *39*, 1221–1230. [CrossRef]
21. Fan, D.M.; Anitori, R.P.; Tebo, B.M.; Tratnyek, P.G.; Lezama Pacheco, J.S.; Kukkadapu, R.K.; Engelhard, M.H.; Bowden, M.E.; Kovarik, L.; Arey, B.W. Reductive sequestration of pertechnetate (99TcO4–) by nano zerovalent iron (nZVI) trans-formed by abiotic sulfide. *Environ. Sci. Technol.* **2013**, *47*, 5302–5310. [CrossRef]
22. Park, S.W.; Kim, S.K.; Kim, J.B.; Choi, S.W.; Inyang, H.I.; Tokunaga, S. Particle surface hydrophobicity and the dechlo-rination of chloro-compounds by iron sulfides. *Water Air Soil Pollut.* **2006**, *6*, 97–110. [CrossRef]
23. Han, Y.L.; Yan, W.L. Reductive dechlorination of trichloroethene by zero-valent iron nanoparticles: Reactivity enhancement through sulfidation treatment. *Environ. Sci. Technol.* **2016**, *50*, 12992–13001. [CrossRef] [PubMed]
24. Fan, D.; Lan, Y.; Tratnyek, P.G.; Johnson, R.L.; Filip, J.; O'Carroll, D.M.; Garcia, A.N.; Agrawal, A. Sulfidation of Iron-Based Materials: A Review of Processes and Implications for Water Treatment and Remediation. *Environ. Sci. Technol.* **2017**, *51*, 13070–13085. [CrossRef] [PubMed]

25. Li, J.X.; Zhang, X.Y.; Sun, Y.K.; Liang, L.P.; Pan, B.C.; Zhang, W.M.; Guan, X.H. Advances in Sulfidation of Zerovalent Iron for Water Decontamination. *Environ. Sci. Technol.* **2017**, *51*, 13533–13544. [CrossRef] [PubMed]
26. Garcia, A.N.; Zhang, Y.Y.; Ghoshal, S.; He, F.; O'Carroll, D.M. Recent advances in sulfidated zerovalent iron for contaminant transformation. *Environ. Sci. Technol.* **2021**, *55*, 8464–8483. [CrossRef]
27. Kim, E.J.; Kim, J.H.; Azad, A.M.; Chang, Y.S. Facile synthesis and characterization of Fe/FeS nanoparticles for environ-mental applications. *ACS Appl. Mater. Interfaces* **2011**, *3*, 1457–1462. [CrossRef]
28. Rajajayavel, S.R.C.; Ghoshal, S. Enhanced reductive dechlorination of trichloroethylene by sulfidated nanoscale zerovalent iron. *Water Res.* **2015**, *78*, 144–153. [CrossRef]
29. Fang, H.Y.; Huang, T.Z.; Mao, J.F.; Dinesh, M.M.; Sun, Y.; Liang, D.; Qi, L.; Yu, J.M.; Jiang, Z.K. Investigation on the cat-alytic performance of reduced-graphene-oxide-interpolated FeS2 and FeS for oxygen reduction reaction. *ChemistrySelect* **2018**, *3*, 10418–10427. [CrossRef]
30. Mangayayam, M.; Dideriksen, K.; Ceccato, M.; Tobler, D.J. The Structure of Sulfidized Zero-Valent Iron by One-Pot Synthesis: Impact on Contaminant Selectivity and Long-Term Performance. *Environ. Sci. Technol.* **2019**, *53*, 4389–4396. [CrossRef]
31. Cao, Z.; Liu, X.; Xu, J.; Zhang, J.; Yang, Y.; Zhou, J.L.; Xu, X.H.; Lowry, G.V. Removal of Antibiotic Florfenicol by Sulfide-Modified Nanoscale Zero-Valent Iron. *Environ. Sci. Technol.* **2017**, *51*, 11269–11277. [CrossRef]
32. Guo, Y.N.; Park, T.; Yi, J.W.; Henzie, J.; Kim, J.; Wang, Z.L.; Jiang, B.; Bando, Y.; Sugahara, Y.; Tang, J.; et al. Nanoarchitectonics for transition-metal-sulfide-based electrocatalysts for water splitting. *Adv. Mater.* **2019**, *31*, 1807134. [CrossRef]
33. Gu, Y.W.; Wang, B.B.; He, F.; Bradley, M.J.; Tratnyek, P.G. Mechanochemically Sulfidated Microscale Zero Valent Iron: Pathways, Kinetics, Mechanism, and Efficiency of Trichloroethylene Dechlorination. *Environ. Sci. Technol.* **2017**, *51*, 12653–12662. [CrossRef] [PubMed]

International Journal of
Environmental Research and Public Health

MDPI

Article

Aerosolization Behaviour of Fungi and Its Potential Health Effects during the Composting of Animal Manure

Ruonan Wang [1,2,†], Aoyuan Yu [1,2,†], Tianlei Qiu [2], Yajie Guo [2], Haoze Gao [2], Xingbin Sun [1,*], Min Gao [2,*] and Xuming Wang [2]

1 College of Forestry, Northeast Forestry University, Harbin 150040, China; loumeier@163.com (R.W.); yuaoyuan36@126.com (A.Y.)
2 Beijing Key Laboratory of Agricultural Genetic Resources and Biotechnology, Beijing Agro-Biotechnology Research Center, Beijing Academy of Agriculture and Forestry Sciences, Beijing 100097, China; qiutianlei@baafs.net.cn (T.Q.); guoyajie2007@163.com (Y.G.); ghz220414@163.com (H.G.); wangxuming@baafs.net.cn (X.W.)
* Correspondence: sunxingbin1025@163.com (X.S.); gm_baafs@126.com (M.G.); Tel.: +86-(451)-82192120 (X.S.); +86-(10)-51503803 (M.G.)
† These authors contributed equally to this work.

Abstract: Compost is an important source of airborne fungi that can adversely affect occupational health. However, the aerosol behavior of fungi and their underlying factors in composting facilities are poorly understood. We collected samples from compost piles and the surrounding air during the composting of animal manure and analyzed the aerosolization behavior of fungi and its potential health effects based on the fungal composition and abundance in two media using high-throughput sequencing and ddPCR. There were differences in fungal diversity and richness between the air and composting piles. Ascomycota and Basidiomycota were the two primary fungal phyla in both media. The dominant fungal genera in composting piles were Aspergillus, Thermomyces, and Alternaria, while the dominant airborne fungal genes were Alternaria, Cladosporium, and Sporobolomyces. Although the communities of total fungal genera and pathogenic/allergenic genera were different in the two media, fungal abundance in composting piles was significantly correlated with abundance in air. According to the analysis on fungal composition, a total of 69.10% of the fungal genera and 91.30% of pathogenic/allergenic genera might escape from composting pile into the air. A total of 77 (26.64%) of the fungal genera and six (20%) of pathogenic/allergenic genera were likely to aerosolize. The influence of physicochemical parameters and heavy metals on the aerosol behavior of fungal genera, including pathogenic/allergenic genera, varied among the fungal genera. These results increase our understanding of fungal escape during composting and highlight the importance of aerosolization behavior for predicting the airborne fungal composition and corresponding human health risks in compost facilities.

Keywords: composting facility; airborne fungi; pathogenic/allergenic genera; aerosolization behaviour; factor analysis

Citation: Wang, R.; Yu, A.; Qiu, T.; Guo, Y.; Gao, H.; Sun, X.; Gao, M.; Wang, X. Aerosolization Behaviour of Fungi and Its Potential Health Effects during the Composting of Animal Manure. *Int. J. Environ. Res. Public Health* **2022**, *19*, 5644. https://doi.org/10.3390/ijerph19095644

Academic Editor: Cheng Yan

Received: 10 March 2022
Accepted: 14 April 2022
Published: 6 May 2022

1. Introduction

The increase of centralized animal feeding operations (CAFOS) in many locations has increased the amount of manure produced [1]. Composting is an important method for recycling and stabilizing animal manure [2]. This biochemical process involves the interaction of diverse microbial communities to convert organic wastes into nutrient-rich, safe, and stable fertilizers and soil amendments [3,4]. Fungi play an important role in composting due to their ability to use many carbon substrates as a food source and attack organic residues that are too dry, acidic, or low in nitrogen for bacterial decomposition [5]. However, some operations involving vigorous movement during composting, such as moving and handling the compost material, are associated with the release of large amounts of bioaerosols [5–8]. Exposure to aerosolized microorganisms can have adverse effects on human health [9]. Fungal bioaerosols consist of spores, mycelium fragments, and debris which are easily inhaled by workers and cause numerous symptoms including allergies, irritation, and opportunistic infections. Long-term lung exposure to fungal bioaerosols can be associated with chronic diseases while the effects of short-term exposure range from irritation of the eyes and nose to coughing and a sore throat [10,11]. Given the inhalation of fungi can promote human health issues [10–15], there have been many studies on the abundance and composition of fungi in the air of composting sites [6,8,14,16].

However, there is little information on airborne fungi generated during the composting of animal manure [17,18] and this hampers the assessment of the environmental health risks under these conditions. The occupational health effects of airborne fungi may be underestimated since the worst fungal infections are usually caused by only a few species [19]. Traditional culture-based methods, fluorescence, and scanning electron microscopy cannot be used to effectively describe the entire fungal profile, and undetected fungal pathogens make it difficult to establish a definitive link between fungal exposure and respiratory problems [20]. High-throughput sequencing technologies have been used for the analysis of airborne fungi since they account for the relative abundance and full diversity of microorganisms present [17]. Accordingly, with the estimation of absolute abundance (EAA), we can obtain information on taxa abundance in the microbial communities [21–23]. The composition and quantification of fungi released from animal manure composting have not been previously studied, especially with regard to pathogens. This knowledge gap has prevented the quantitative assessment of human health risks from occupational exposure to airborne fungi.

Compost piles are important sources of airborne fungi [6]. Some fungi in piles are more prone to being aerosolized, and this preferential aerosolization could contribute to its widespread distribution in different air environments [24,25]. The aerosolization behavior of fungi was studied during the composting of vegetable waste [16] and during sewage sludge biostabilization [26] by comparing the fungal composition of air and its contributing sources. The aerosolization behavior of microorganisms in compost may be affected by their specific morphological, biochemical characteristics [24,27]. Other contributing conditions in compost piles include temperature, ventilation [28], as well as physico-chemical parameters [29], including heavy metals, exercising some influence on the microbial community [30]. However, the relationship between those potential factors and aerosolization of fungi, or fungal pathogens, is not known.

The objectives of this study were to investigate the aerosolization behavior of fungi and their potential factors during the composting of animal manure. The specific research goals included: (1) characterizing the concentration, diversity, and composition of fungi and pathogenic/allergenic genera and their potential health effects in air and composting piles; (2) describing the aerosol behavior of pathogenic/allergenic genera based on differences among key fungal genera, and connections of community between two media; (3) exploring the influence of physicochemical parameters and heavy metals on the aerosol index of fungal pathogenic/allergenic genera during composting. The present results will help us to understand the environmental risks and quantitatively assess human health issues posed by occupational exposure to airborne fungi during animal manure composting.

2. Materials and Methods

2.1. Sample Collection

We collected air and compost pile (C) samples from a commercial composting plant in Shouguang city, Shandong Province (118.50 N, 36.68 E) during thermophilic phases. Livestock manure, vegetables, and straws were the main raw materials used for compost production. In this study, totally 23 air samples and 12 solid samples were collected. Twelve and 11 air samples were collected from inside (I) (above the composting pile) and outside (O) the composting workshop, respectively, from 13 to 25 June 2020. Six solid samples were collected from the compost pile every two days during the same period.

We used total suspended particulate (TSP) air samplers (2030A, Laoying, Qingdao, China) to collect air onto a 90 mm (diameter) sterilized quartz-fiber filter (Ahlstrom Munktell, NO. 420065, Falun, Sweden) at a flow of 100 L/min for 24 h. Filters were baked in a muffle furnace at 500 °C for 5 h and stored in a sterilized plastic box until loaded into the sampler. Regarding each of the solid samples, 5 random samples were collected from the surface of compost pile using a sterile shovel before air collection. Those samples were mixed evenly and 100 g was taken for subsequent experiments. After sampling, both filters and composting piles were placed inside an ice chest, returned to the laboratory, and stored at −80 °C until analysis.

2.2. DdPCR and ITS Sequencing

For the TSP samples, 1/8 filter of each filter was used and cut into small pieces, which were loaded into Lysing Matrix E. A 0.3 g solid sample was loaded into Lysing Matrix E directly. DNA was extracted using the FastDNA® SPIN Kit for soil (MP Biomedicals, Santa Ana, CA, USA) using manufacturer instructions. A Qubit® dsDNA High Sensitivity Assay Kit (Life Technologies, Carlsbad, CA, USA) was used to measure the concentration of extracted DNA. All DNA samples were stored at −20 °C until analysis.

We used ddPCR to analyze the copy number of ITS in air and solids samples to determine the absolute abundance of fungi. ddPCR was run on a QX200 Droplet Digital™ PCR System (BioRad, Hercules, CA, USA). Each ddPCR reaction contained 20 μL of QX200 ddPCR EvaGreen Supermix (Bio-Rad), 100 nM of each primer, and 1 μL of sample DNA. The operating procedures and methods were the same as previously published [7], and the corresponding primers were ITS1F (5′-CTTGGTCATTTAGAGGAAGTAA-3′) and ITS2R (5′-GCTGCGTTCTTCATCGATGC-3′) [31]. PCR amplification of ITS was performed using the following conditions (heating rate 2.5 °C/s): 95 °C for 10 min, 40 cycles of 95 °C for 30 s, 55 °C annealing temperature for 60 s, and 72 °C 30 s at 4 °C, 5 min at 90 °C. After the PCR was completed, the 96-well plate was transferred to a Droplet Reader (Bio-Rad) for data collection. QuantaSoft™ software 1.7.4.0917 (Bio-Rad, Hercules, CA, USA) was used to automatically measure, record, and analyze the fluorescence per droplet and per well.

The internal transcribed spacer 1 (ITS1) region of the fungal rRNA gene was amplified using primers ITS1F and ITS2R through polymerase chain reactions (PCRs) [31]. The ITS sequencing was performed using Illumina MiSeq platform at Novogene Bio-Pharm Technology Co. Ltd. (Shanghai, China). Operational taxonomic units (OTUs) with a 97% similarity cutoff were clustered by UPARSE (Uparse v7.0.1001, http://drive5.com/uparse/, accessed on 24 May 2020) [32]. UCHIME was used to identify and remove chimeric sequences [33]. QIIME (version 1.9.0) [34] was used to assign taxonomy to OTUs based on the UNITE fungal ITS reference training data set for taxonomic assignment and to generate an OTU table.

2.3. Determination of Heavy Metals and Physical and Chemical Parameters

Concentrations of Cd were measured by a Graphite Furnace Atomic Absorption Spectrophotometer (GFAAS, HITACHI Z5000). Other heavy metals were determined using an Inductively Coupled Plasma-Atomic Emission Spectrometry (ICP-AES, PerkinElmer Optima 5300DV). Total nitrogen (N) was measured using an automatic N meter and Kjeldahl determination [35]. Total potassium (TK) and total phosphorus (TP) were measured according to NYT87-1988 and NYT88-1988 standard experimental methods, respectively. The organic matter (OM) in composting piles was determined as previously reported [36]. The pH was measured using a pH meter (pHS2F, Shanghai Precision and Scientific Instrument Co., Ltd., Shanghai, China). For air-dried moisture (AD), the solid samples were baked in a 105 °C incubator for about 2 h, moved to a desiccator to cool to room temperature. They were then weighed, and the AD value was calculated. A thermometer was used to determine the temperature of the compost pile every 48 h (Kedengbao Energy Technology TM-902C).

2.4. Data Analysis

The estimated absolute abundance (EAA) of certain fungal genera was calculated as the product of its relative abundance multiplied by the corresponding total copy number of ITS according to previous research on bacteria [21]. The relative abundance (*RA*) of a fungal taxon was determined by ITS sequencing and the copy number of airborne fungi was measured by ddPCR. The aerosolization behavior of microorganisms was quantified by the Bioaerosolization Index (*BI*) [24].

$$BI = \frac{RA_{aerosol}}{RA_{compost}},$$

where *RA* aerosol was the relative abundance (the percentage of a certain fungal genus among the total fungi) of certain fungi in the air and *RA* compost was the relative abundance of the corresponding fungus in the compost piles.

Box plots were drawn using the Origin Pro 8.5. One-way ANOVA was used to analyze the differences across the sampling settings by SPSS 26.0, and all statistical tests were considered significant at $p < 0.05$. A website (Draw Venn Diagram (ugent.be, accessed on 9 January 2022) was used to draw Venn diagrams. STAMP analysis was conducted using the Tukey–Kramer test to compare the statistical differences in fungal genera between different samples [37]. The correlation of fungal communities among the three media was analyzed using conditioned constrained principal coordinate analysis (CPCoA) and non-metric multidimensional scaling (NMDS), which were performed by the package "ape" and "ggplot2" based on the Bray–Curtis distance with R 4.1.2 (http://www.r-project.org/; accessed on 22 July 2021). The redundancy analysis diagram (RAD) was drawn using the webpage (http://www.cloud.biomicroclass.com/, accessed on 9 January 2022). Both Procrustes analysis ("vegan" and "labdsv" packages) and Heatmaps ("pheatmap" package) were drawn in the R environment.

3. Results and Discussion

3.1. Fungal Abundance and Diversity in Air and Compost Piles

The fungal concentration (copies of ITS) and diversity (OTU number and Shannon index) in compost piles and air (inside and outside of the composting workshop) are presented in Figure 1A. The estimation of absolute abundance (EAA) was used for the quantitative analysis of 30 pathogenic/allergenic genera, and their biological indices are shown in Figure 1B.

The concentration of fungi in composting pile was 4.71×10^4–2.99×10^3 copies/g, and the corresponding concentration of airborne fungi from inside and outside composting workshop were 6.8×10^3 and 9.721×10^3 copies/m^3, with concentration ranges of 2.72×10^2–2.57×10^4 and 24.88–3.41×10^4 copies/m^3, respectively (Figure 1A). The values of both richness and diversity of fungi in composting piles were higher than those in inside and outside composting workshop. Overall, the OTUs number (fungal richness) ranged from 258 to 458. Although the average value of OTUs observed in composting piles (385 ± 12.95) was higher than that in inside workshop (357 ± 58.04) and outside composting working shop (334 ± 42.84), there was no significant difference between the two air samples. The Shannon index of composting piles (4.31 ± 0.32) was significantly higher than that of inside (2.73 ± 0.26) and outside composting workshop (3.06 ± 0.21) (one-way ANOVA, $p < 0.05$), which is consistent with research on the composting of green waste (5 ± 0.1) [31].

The EAA value of airborne pathogenic/allergenic genera had a similar trend with the total fungi. A higher value was detected outside the composting workshop ($2.05 \times 10^3 \pm 0.92 \times 10^3$ copies/m^3) compared to the value inside ($1.25 \times 10^3 \pm 0.31 \times 10^3$ copies/m^3) (Figure 1B), but the difference was not statistically significant. The EAA of airborne pathogenic/allergenic genera was similar to biochemical and culture method test data from other composting facilities [18,38]. Our results confirmed the exposure risk of pathogenic/allergenic genera in the surrounding air environment during composting of animal manure. The number of pathogenic/allergenic genera from inside composting workshop (18) was slightly higher than that in composting piles (17) and outside composting workshop (16). No significant difference was detected on the Shannon index of pathogenic/allergenic genera among the three sampling settings in this study.

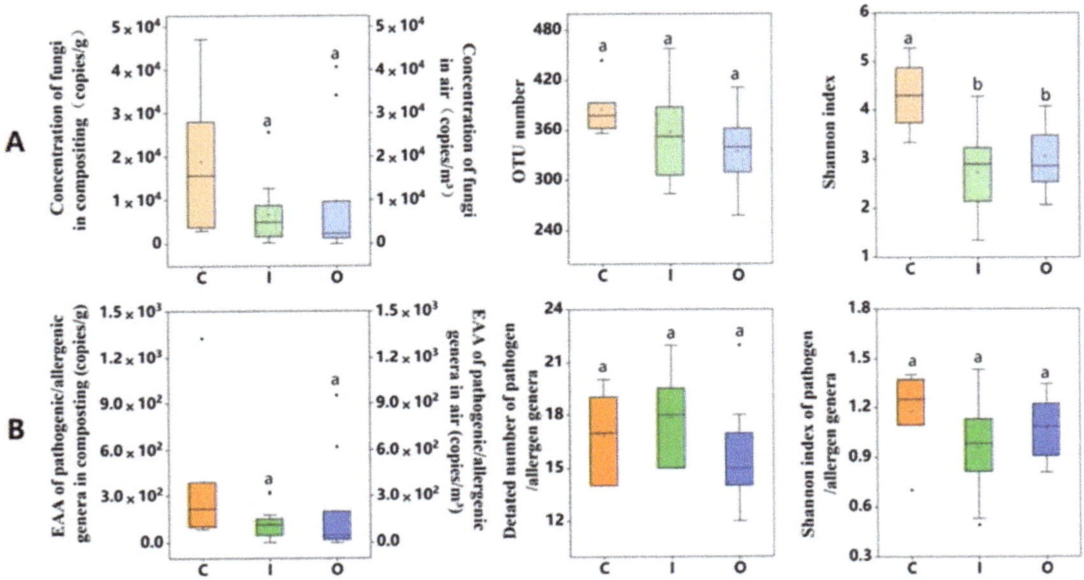

Figure 1. Abundance and diversity of fungal genera and pathogenic/allergenic genera in composting piles (C), inside air (I), and outside air (O) of composting workshop. (**A**) Concentration, OTU number, and Shannon indices of total fungal genera; (**B**) the estimation of absolute abundance (EAA), detected number, and Shannon indices of pathogenic/allergenic genera. (There is a significant difference between "a" and "b" above the box diagram; $p < 0.05$).

We analyzed the dominant fungal phyla in the air (inside and outside air of composting workshop) and composting piles (Figure S1) and found that Ascomycota and Basidiomycota were consistently the two primary phyla in all three sample settings during thermophilic phases (Figure S2). The same dominant fungal phyla were detected in the composting of pig carcasses [6] and green waste [8], as well as sewage sludge biostabilization [26]. Given that Ascomycetes are major contributors to the degradation of holocellulose [39], they are commonly found in other composting processes as well as in this study due to the presence of straw in the raw materials. The relative abundance of Ascomycota in compost piles (86.96 ± 4.56%) was higher than their abundance in the air (61.77 ± 26.85%). However, a higher relative abundance of Basidiomycota was detected in the air (23.43 ± 22.29%) compared with abundance in compost piles (3.83 ± 2.75%). We analyzed the dominant fungal genera in 29 samples from compost piles, inside and outside the composting workshop (Figure S3). The average values of dominant fungal genera by relative abundance (heat map) and concentration (radar map) are shown in Figure 2A. The dominant fungi in compost piles were *Aspergillus* (45.28 ± 16.64%), *Thermomyces* (14.98 ± 11.27%), *Alternaria* (13.98 ± 11.27%), *Diutina* (10.30 ± 21.70%), and *Cladosporium* (1.26 ± 0.56%), while the dominant airborne fungal genes were from *Alternaria* (47.41 ± 9.62%), *Cladosporium* (11.57 ± 2.67%), *Sporobolomyces* (7.47 ± 1.37%), *Schizophyllum* (6.78 ± 1.44%), and *Aspergillus* (1.25 ± 0.42%). Distinctive distributions of airborne fungal compositions compared to those in compost have been previously reported [6,26]. The differences may be due to the different aerosolization behavior of fungi and different environmental pressures in the two media [40]. This difference may also be caused by differences in the biological mechanisms of spore ontogeny and spore release [41,42].

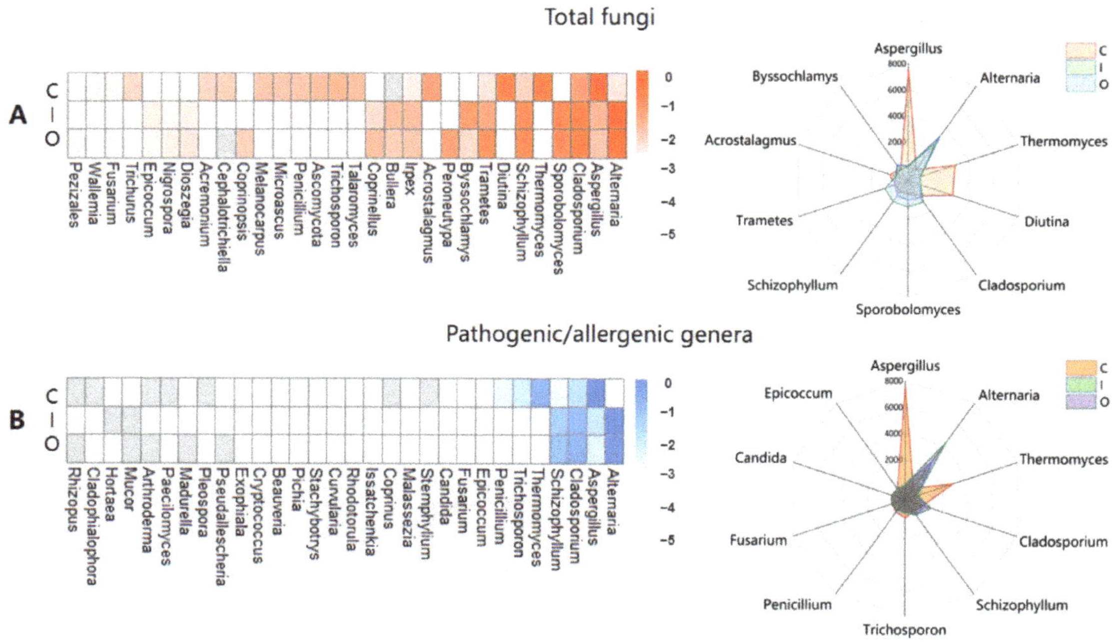

Figure 2. The dominant fungi and pathogenic/allergenic genera in compost piles (C), inside (I), and outside air (O) of composting workshop. (**A**), the relative abundance (Log value) of the top 30 dominant fungi by heat map and the estimation of absolute abundance (EAA) of 10 dominant fungi by radar map; (**B**), the relative abundance (Log value) of 30 pathogenic/allergenic genera by heat map and the EAA of 10 dominant pathogenic/allergenic genera by radar map.

Advances in high-throughput sequencing and bioinformatics enable the study of fungal pathogenic/allergenic genera. *Alternaria*, *Aspergillus*, *Cladosporium*, *Schizophyllum*, and *Thermomyces* were the top five fungi with the highest relative abundance among the target 30 fungal pathogenic/allergenic genera (Figure 2B, heat map). Combined with the concentration of ITs, the quantitative information (expressed as EAA) of the top five pathogenic/allergenic genera were also analyzed (Figure 2B, radar map). According to the EAA, the highest concentration of airborne pathogenic/allergenic genera was detected from *Alternaria* ($3.43 \times 10^3 \pm 5.80 \times 10^3$ copies/m^3). The EAA of *Cladosporium* in the air was ($8.82 \times 10^2 \pm 1.08 \times 10^3$ copies/m^3), which was similar to previous studies [18,38,43,44]. *Aspergillus fumigatus* was the only specific taxon for which UK legislation requires compost site testing [45]. *A. fumigatus* has also been detected in the air of other composting facilities [8,27,31]. Because of infections caused by *A. fumigatus* [46], the considerable relative abundance ($1.26 \pm 0.86\%$) and EAA value ($0.86 \times 10^2 \pm 0.98 \times 10^2$ copies/m^3) of *Aspergillus* in the compost facility air is a health concern. The quantitative and qualitative data on pathogenic/allergenic genera in this study reveal possible human health risks from the air environment during the composting of animal manure.

The physicochemical parameters in the composting piles over 12 days are shown in Table 1. During this thermophilic stage, the temperature ranged from 68.5 and 72.5 °C, with an average value of 70.5 ± 1.35. The content of total nitrogen (TN) increased from $1.58 \pm 0.00\%$ to $2.02 \pm 0.01\%$. Usually, the variation of TN was caused by the degradation of organic compounds and the metabolic processes of microorganisms. The similar trend was also detected in previous research [26]. Meanwhile, the increasing biological metabolism during the composting process [47] might also have resulted in the increase of organic matter as we found that OM% increased from 19.65 ± 0.12% to 30.05 ± 0.04%. Both AD and pH were relatively stable in the thermophilic stage, with mean values of 1.9 ± 0.10 and 8.72 ± 0.13, respectively. Seven heavy metals in the compost piles were also analyzed (Table 2). The highest heavy metal concentration was Zn (255.61 ± 1.56 mg/kg) and the lowest was Hg (0.04 ± 0.00 mg/kg). The concentration of Cu is within the range of the published literature [30].

Table 1. Physicochemical parameters in composting piles.

Time	*TK (%)*	*TN (%)*	*TP (%)*	*OM (%)*	*pH*	*AD (%)*	*T (°C)*
2020.06.14	6.49 ± 0.01	1.58 ± 0.00	0.73 ± 0.00	19.65 ± 0.12	8.93 ± 0.03	1.90 ± 0.00	72.50 ± 0.00
2020.06.16	6.31 ± 0.00	1.80 ± 0.01	0.70 ± 0.00	33.70 ± 0.16	8.67 ± 0.03	2.00 ± 0.00	72.00 ± 0.00
2020.06.18	6.16 ± 0.02	1.81 ± 0.01	0.76 ± 0.00	28.85 ± 0.04	8.69 ± 0.00	1.90 ± 0.00	70.00 ± 0.00
2020.06.20	5.98 ± 0.00	1.90 ± 0.00	0.80 ± 0.00	27.30 ± 0.08	8.50 ± 0.02	1.70 ± 0.00	68.50 ± 0.00
2020.06.22	6.48 ± 0.03	2.01 ± 0.01	0.81 ± 0.00	28.10 ± 0.08	8.78 ± 0.04	1.90 ± 0.00	70.00 ± 0.00
2020.06.24	6.01 ± 0.02	2.02 ± 0.01	0.67 ± 0.00	30.05 ± 0.04	8.74 ± 0.03	2.00 ± 0.00	70.00 ± 0.00
Mean value	6.24 ± 0.21	1.85 ± 0.15	0.75 ± 0.05	27.94 ± 4.24	8.72 ± 0.13	1.90 ± 0.00	70.50 ± 0.00

Organic matter (OM); air-dried moisture (AD); Total nitrogen (TN); Total potassium (TK); Total phosphorus (TP); The organic matter (OM); Air-dried moisture (AD); Temperature (T). "±" stands for "SD".

Table 2. Concentration of heavy metals in composting piles (mg/kg).

Time	*Pb*	*Cd*	*Cr*	*Cu*	*Zn*	*Hg*	*As*
2020.06.14	18.05 ± 0.58	1.37 ± 0.02	20.42 ± 0.03	41.80 ± 0.08	231.26 ± 4.68	0.06 ± 0.00	2.03 ± 0.07
2020.06.16	14.09 ± 0.87	1.12 ± 0.04	16.66 ± 0.25	44.32 ± 0.24	223.14 ± 0.91	0.06 ± 0.00	1.80 ± 0.01
2020.06.18	14.81 ± 0.81	1.13 ± 0.02	20.60 ± 0.15	49.62 ± 0.04	277.25 ± 0.35	0.03 ± 0.00	1.72 ± 0.06
2020.06.20	15.93 ± 0.12	1.33 ± 0.01	31.56 ± 0.09	49.55 ± 0.18	278.00 ± 2.48	0.03 ± 0.00	1.33 ± 0.01
2020.06.22	14.55 ± 0.58	1.26 ± 0.01	22.83 ± 0.13	48.27 ± 0.01	253.77 ± 0.92	0.03 ± 0.00	1.30 ± 0.02
2020.06.24	13.81 ± 0.99	1.26 ± 0.05	19.36 ± 0.19	46.39 ± 0.19	270.23 ± 0.01	0.01 ± 0.00	1.20 ± 0.01
Mean value	15.20 ± 0.66	1.25 ± 0.02	21.90 ± 0.14	46.66 ± 0.12	255.61 ± 1.56	0.04 ± 0.00	1.56 ± 0.03

3.2. Connection and Difference of the Fungal Community and Abundance between Composting Piles and Air

We examined microbial community correlations between air and composting piles by total fungal genera and pathogenic/allergenic genera (Figure 3A,B).

Fungal populations in different sampling settings were clustered using non-metric multidimensional scaling (NMDS) analysis at the genus level based on the Bray–Curtis dissimilarity coefficient. Both the community of total fungal genera and pathogenic/allergenic genera in the compost piles were distinguished from those in the two air samples (Figure 3A). CPCoA also demonstrated the significance of this difference ($p = 1 \times 10^{-4}$ in Figure S4). The airborne fungal communities of air sampling settings were clustered closer compared with those in the compost piles. A similar distribution of clusters was also observed in pathogenic/allergenic fungal communities from the three sampling settings. This dissimilarity in fungal communities between air and compost was confirmed using Procrustes and Mantel analyses (Figure 3B).

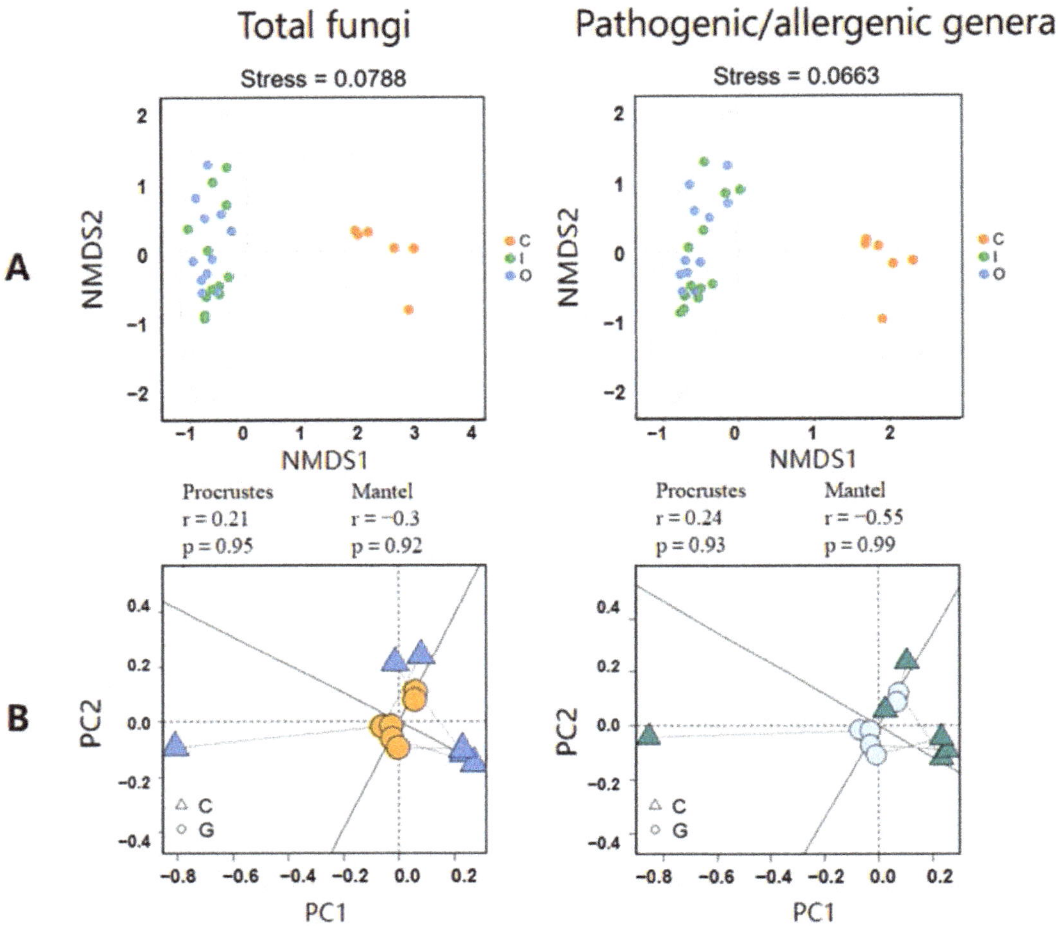

Figure 3. Relation of fungal community between composting piles and air by total fungal genera and pathogenic/allergenic genera. (**A**), fungal lineages used non-metric multidimensional scaling analysis (the orange point represents compost piles, the green represents the inside, and the blue represents outside air of composting workshop); (**B**), fungal community correlation used Procrustes and Mantel analysis (the triangle represents compost piles, and the circle represents the air sample).

Our study data suggest that the aerosol process of fungal communities is not a synchronized behavior of the overall fungi, and the aerosolization behavior of specific fungal genera may vary. In addition, the environmental stresses to which fungi are exposed in air and compost piles might also contribute to their different abundances and compositions [40]. Since the composition in compost piles was not sufficiently accurate to predict the escaped airborne fungal community and subsequent potential health risk, it is crucial to develop a targeted assessment of the aerosolization behaviors of specific fungi.

To explain the above distinguishing community, we analyzed specific fungal genera between air and compost piles, including the number of common and endemic fungal genera (Venn diagram in Figure 4A), and specific differences among fungal genera (Stamp analysis in Figure 4B).

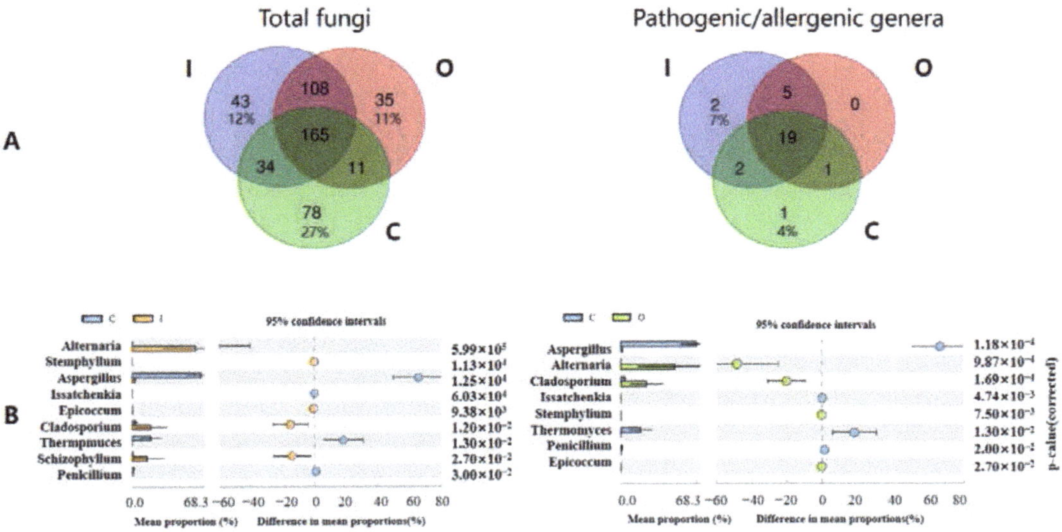

Figure 4. Common and endemic fungi in air and compost piles. (**A**) Number of common and endemic total fungi and pathogenic/allergenic genera in composting piles (C), inside (I), and outside air (O) of composting workshop using a Venn diagram, (**B**) Identification of different pathogenic/allergenic genera between C and I, or C and O using Stamp analysis. The bar chart shows the average proportion of differential pathogenic/allergenic genera predicted using Picrust 2. The difference in proportions between any two groups is presented with 95% confidence intervals. Only $p < 0.05$ (Welch's t test, FDR adjusted) are depicted.

A total of 78 endemic genera were detected in composting piles, followed by inside (43), and outside of composting workshop (35), with corresponding proportions (ratio of endemic genera to the total fungal genera) of 27%, 12%, and 11%, respectively. A total of 165 airborne fungal genera were shared across the three sampling settings. Compost piles shared more common fungal genera with inside (199) than outside workshops (176), and the most common fungal genera were detected in the two air sampling areas (273). Based on the fungal composition, 69.10% of fungal genera in compost piles could become airborne. Among the pathogenic/allergenic genera, only one endemic genus was detected in composting piles (*Mucor*). Consistent with fungal genera, the number of shared pathogenic/allergenic genera between composting piles and I was greater than between composting piles and O. Most (91.30%) of the pathogenic/allergenic genera in composting piles were shared with inside air, suggesting that most could aerosolize and emit into the air.

The key fungal genera driving the significantly different proportions between composting piles and I (Figure S5), and composting piles and O (Figure S6) were analyzed using Stamp analysis. More different genera were detected between composting piles and inside air (38) than between composting piles and outside air from composting workshop (33). Stamp analysis on the key pathogenic/allergenic genera (Figure 4B) showed that the key fungal genera from composting piles to inside air was slightly higher than that to outside air. Most of the critical genera were identical, except for *Schizophyllum* with significantly high relative abundance in inside air of workshop.

To quantitatively assess fungal aerosolization behavior, we analyzed the correlation of fungal concentrations and relative abundances in compost piles and air using linear regression analysis. The total fungal concentration in air was significantly correlated with that in composting workshop ($R^2 = 0.102$, $p < 0.01$) (Figure 5). We also detected significant correlations between pathogenic/allergenic genera in air and compost media based on EAA values ($R^2 = 0.35$, $p < 0.01$). The relative abundance of both total fungi and pathogenic/allergenic genera in the air also increased with the increase of the corresponding value in compost piles ($p < 0.01$). Most research on fungal aerosolization behavior has focused on the relative abundance of specific fungal genera between air and compost [6,26] Based on concentrations and relative abundances, we established a link between the biomass of airborne fungi (including the pathogenic/allergenic genera) and its potential source (compost). This can help predict the quantitative trends of airborne fungi according to their levels in compost piles.

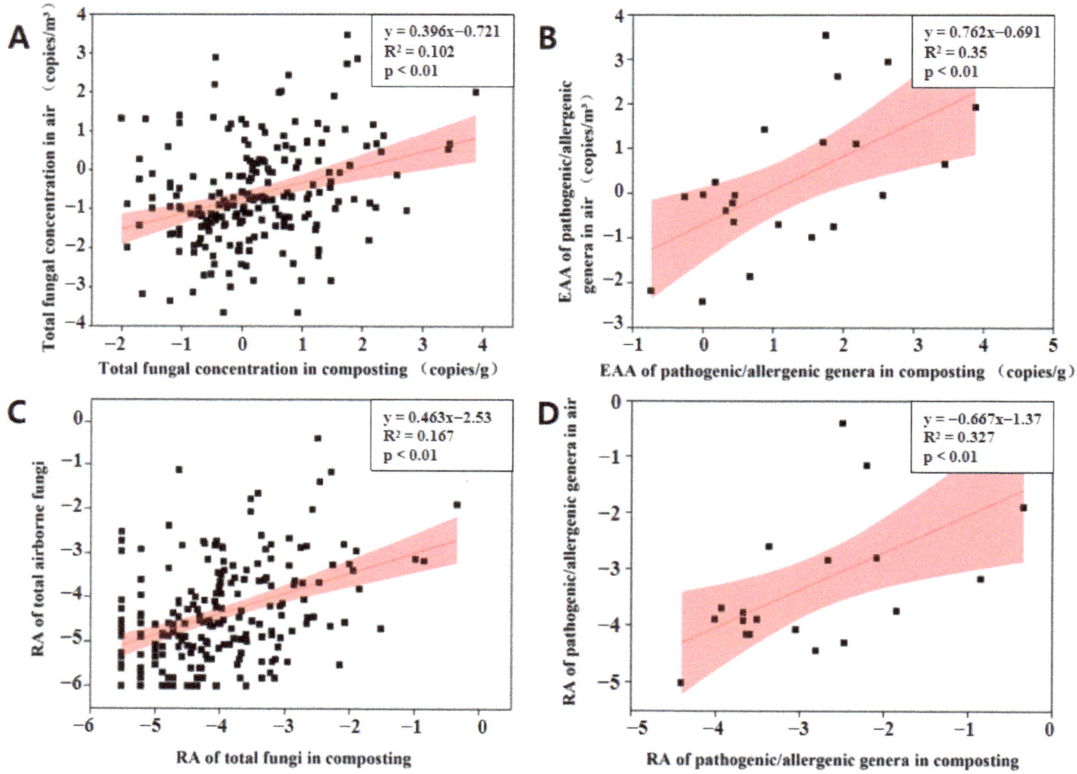

Figure 5. Correlation of fungal concentrations and relative abundances between air and compost piles using linear regression analysis. (**A**), total fungal concentrations; (**B**), the estimation of absolute abundance (EAA) of pathogenic/allergenic genera; (**C**), Relative abundance of total fungi; (**D**), Relative abundance of pathogenic/allergenic genera. Data were log-transformed.

3.3. Potential Factors Affecting the Aerosolization Behavior of Airborne Fungi and Pathogenic/Allergenic Genera

The aerosolization behavior of each specific fungal phylum or genus was described using the Bioaerosolization Index (BI). A log BI value greater than 0 indicates that the fungus is more likely to spread from the compost into the air and vice versa. The aerosolization behavior of fungal phyla was analyzed and the results are shown in Figure S7. Although the Ascomycota and Basidiomycota were the most dominant fungal phyla in both air and compost piles, they have different aerosolization behaviors. Basidiomycota (log BI = 0.798 ± 0.42) had preferential aerosolization while Ascomycota (log BI = -0.16 ± 0.22) represented passively aerosolized microorganisms. A similar trend in the aerosolization behavior of the same fungal phyla was observed during the composting of vegetable waste [16] and sewage sludge biostabilization [26].

The results on aerosolization behavior of fungal genera (Figure S8) indicated that 26.64% of all fungi were easily spread from composting piles into the surrounding air with log BI value being higher than zero. The highest value of log BI (*Sporobolomyces* of 3.94) was comparable to previous studies [16,26]. The number of preferential fungus genera (77) was higher than that in the composting of vegetable waste (7) [16] and sewage sludge biostabilization (11) [26], suggesting that a greater number of fungi genera in manure compost are likely to aerosolize. The five most easily aerosolized fungi were *Sporobolomyces* (BI = 3.52 ± 1.24), *Dioszegia* (BI = 2.97 ± 0.89), *Pyrenophora* (BI = 2.77 ± 0.51), *Tilletiopsis* (BI = 2.53 ± 0.24), and *Coprinopsis* (BI = 2.53 ± 0.24). Similar to the results of fungal phyla, the five dominant fungal genera (Figure 2) were not completely consistent with the top five genera with preferential aerosolization (Figure 6) (except *Sporobolomyces*). This discrepancy here, as well as in previous studies [16,26], indicates that detailed studies are needed to determine the underlying mechanisms of fungal aerosolization behavior.

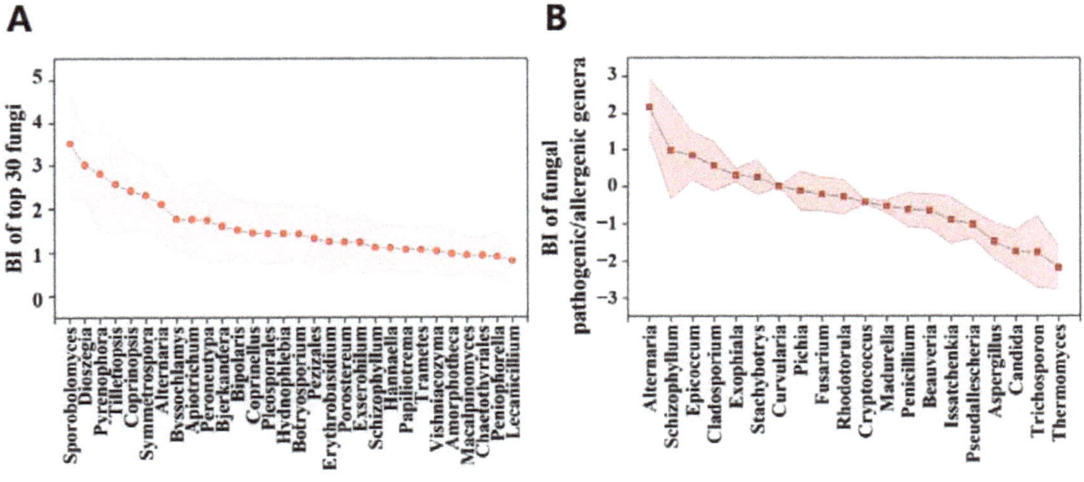

Figure 6. Bioaerosolization Index (BI) of the 30 top fungal genera (**A**) and pathogenic/allergenic genera (**B**) (BI value was t log-transformed).

The aerosolization behavior of pathogenic/allergenic genera is shown in Figure 6B. Among the 30 target pathogenic/allergenic genera, the log BI values of six genera (20%) were greater than 0. There was species-specific aerosolization behavior in the pathogenic/allergenic genera (Figure 6B). Although the enrichment of *Aspergillus* in the air was reported in both animal and domestic composting facilities [6,16], its log BI (log BI = −1.56 ± −1.53) value here was lower than 0. This result indicates that, in addition to the characteristics of the fungus itself, other factors might play a non-ignored role on fungal aerosolization behavior. Extensive exposure to mycotoxins produced by *Fusarium* may cause respiratory symptoms in humans [48]. *Fusarium* (log BI around 3) was the most preferred fungal genus for aerosolization during biostabilization [26]. In this study, we found it to be a passively aerosolized microorganism with log BI = −0.17 ± 0.093. Our results suggest that the aerosolization behavior of some fungal genera might differ between different composting facilities.

To explore the potential factors related to the aerosolization behavior of fungi, we analyzed the relation of physicochemical parameters (Table 1) and specific heavy metals (Table 2) with the log BI of preferentially aerosolized fungi using redundancy analysis (RDA). According to the RDA result on the fungal phylum (Figure 7), AD, TN, TP, and OM had a positive effect on the aerosolization behavior of most fungal phyla, including *Ascomycota*, *Basidiomycota*, *Chytridiomycota*, and *Neocallimastigomycota*. The heavy metals Cd, Cr, Pb, Zn, and Cu promote the aerosolization behavior of *Ascomycota*, *Basidiomycota*,

Chytridiomycota, and *Neocallimastigomycota*. The effect of both physicochemical parameters and heavy metals on the above four fungal phyla were different from the effects on *Mucoromycota* and *Mortierellomycota*. Although the heavy metals, such as Cu and As, have been reported to limit the growth and diversity of the microbial communities during manure composting [30], we found that Cu promotes the aerosolization behavior of certain fungal phylum. Current results highlight that future study should be conducted regarding the detailed correlation between heavy metal and specific fungal phylum in composting piles, as well as their influence on the aerosolization behavior.

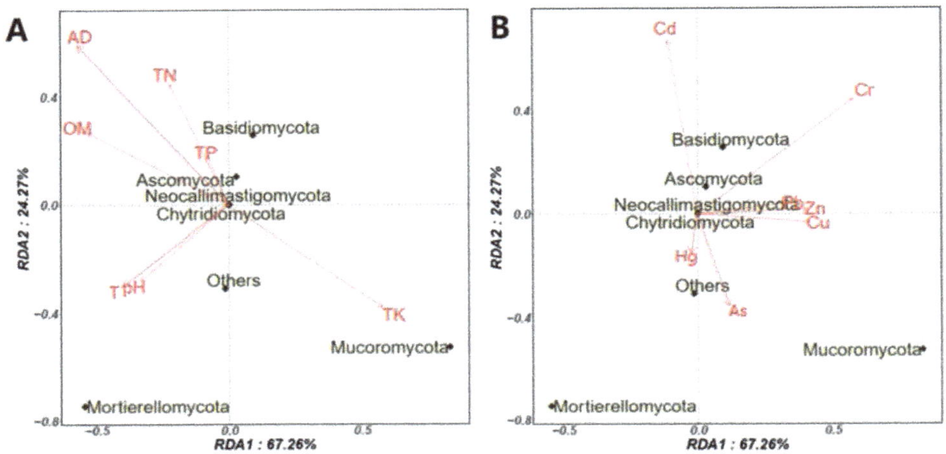

Figure 7. Factors involved in the aerosolization behavior of fungal phylum using redundancy analysis (RDA). (**A**), physicochemical parameters on log BI of fungal phylum; (**B**), Specific heavy metals on log BI of the fungal phylum. OM, Organic matter; AD, air-dried moisture; TN, Total nitrogen; TK, Total potassium; TP, Total phosphorus; OM, The organic matter; AD, Air-dried moisture; T, Temperature.

Figure 8 shows the potential influence of physicochemical parameters and heavy metals on the aerosolization behavior of the five fungal genera with the highest log BI value. These included *Sporobolomyces*, *Dioszegia*, *Pyrenophora*, *Tilletiopsis*, and *Coprinopsis* (Figure 6). Most previous studies focused on the effects of physicochemical factors on microorganisms in compost piles [49,50]. To our best knowledge, no studies determined their influences on aerosolization behavior. We found that the TK or TN was positively correlated with at least three fungi (Figure 8A), and a similar trend was found in pathogenic/allergenic genera (Figure 8C). TN is an important nitrogen source that supports microbial activity and increases the growth of fungal communities in piles [50], which might have a potential effect on their aerosolization behavior. Although the adjustment of moisture content (AD) might eliminate the bioaerosol emission of hydrophilic microorganisms (e.g., fungi) [26], both the positive and negative effect of AD on fungal aerosolization behavior were detected here and depended on the specific fungal genus. The concentration of airborne bacteria emissions in the composting of swine manure was positively correlated with the temperature in the compost pile [51]. We also found that temperature facilitated aerosolization of *Tilletiopsis*, *Coprinopsis*, and *Epicoccum*. The heavy metals Zn and Cu seem to promote the aerosolization process of *Dioszegia*, *Pyrenophora*, *Tilletiopsis*, and *Coprinops* (Figure 8B) as well as the pathogenic/allergenic genera, including *Schizophyllum*, *Cladosporium*, and *Exophiala* (Figure 8D).

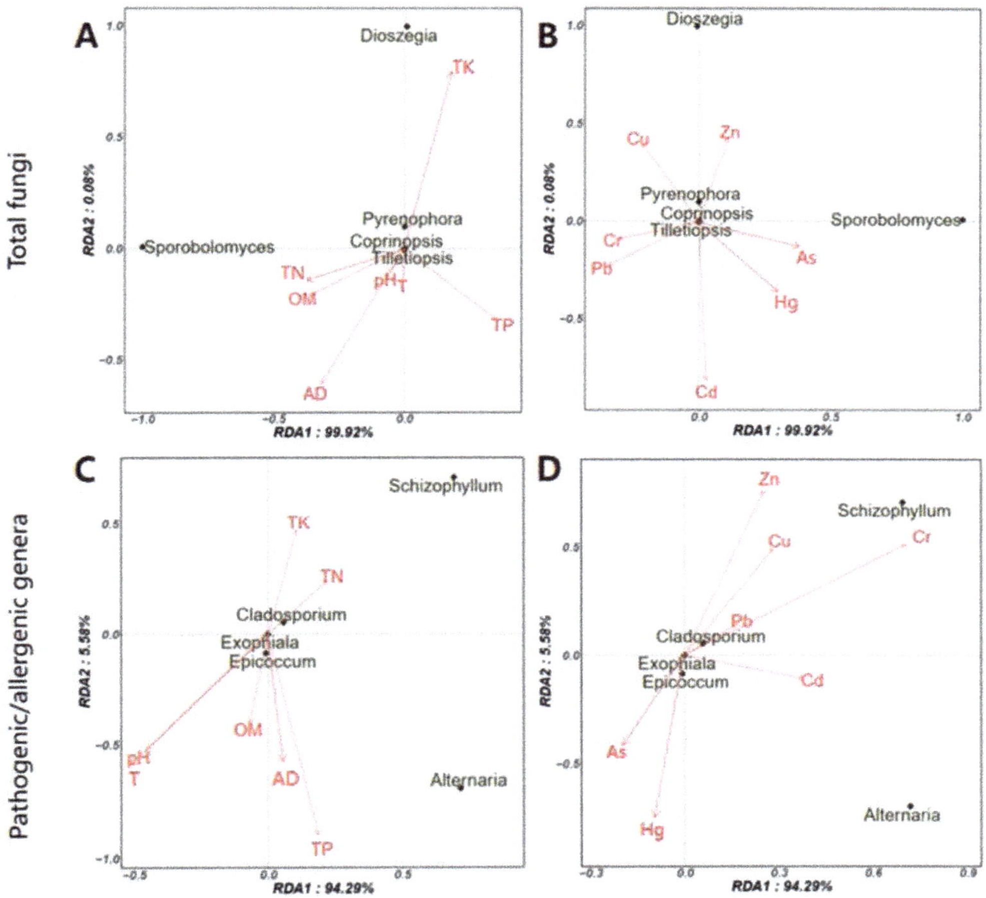

Figure 8. Factors associated with the aerosolization behavior of fungal genera using redundancy analysis. (**A**), physicochemical parameters on top 5 fungi with the highest log BI; (**B**), Specific heavy metal on top 5 fungi with the highest log BI; (**C**), physicochemical parameters on top 5 pathogenic/allergenic genera with the highest log BI; (**D**), Specific heavy metal on top 5 pathogenic/allergenic genera with highest log BI.

We found no consistent influence of a specific factor on fungal aerosolization including physicochemical parameters or heavy metals. This may be because factors other than those tested here may be involved in fungal aerosolization. Understanding the mechanisms involved during aerosol emission will require studies on a broader range of potential factors, including compost age, activity, or material composition [25]. The capability of fungi to become aerosolized may also depend on their morphological characteristics (form, structure [52], and size [16]), and on their biochemical characteristics, such as hydrophobicity, which vary significantly among species [24,25,27]. These factors also need to be evaluated in future research.

4. Conclusions

Fungi in compost can be aerosolized, resulting in adverse effects on occupational health. In this study, we analyzed the aerosol behavior of fungi and their underlying factors in composting facilities. Our results indicate that, due to differences in the community structure in air and compost, the fungal composition in compost piles could not accurately predict the composition of airborne fungi. However, most fungal genera and pathogen/allergen containing genera in compost might potentially aerosolize into the air, and about 25% of these are likely to aerosolize. The influence of physicochemical parameters and heavy metals on the aerosol behavior of fungi, including pathogenic/allergenic genera, varied depending on the fungal genus evaluated. Current results on the compositions and abundance of pathogenic/allergenic genera highlight the possible risks to human health posed by the air environment during animal manure composting. The link between the biomass of airborne fungi (including the pathogenic/allergenic genera) and its potential source (compost) can help predict the quantitative trends of airborne fungi and corresponding health risk according to their levels in compost piles.

Supplementary Materials: The following supporting information can be downloaded at: https://www.mdpi.com/article/10.3390/ijerph19095644/s1, Figure S1: Top 10 dominant fungal phyla in compost piles (C), inside (I) and outside (O) composting workshop; Figure S2: Heatmap of relative abundance of the top 10 dominant fungal phyla in compost piles (C), inside (I) and outside (O) composting workshop; Figure S3: Heatmap of relative abundance of the top 30 dominant genera of fungi in composting piles (C), inside (I) and outside (O) composting workshop; Figure S4: Cluster analysis results of fungal community in samples from compost piles (C), inside (I) and outside (O) composting workshop; Figure S5: Identification of different fungal genera between compost piles (C) and inside composting workshop (I) using Stamp analysis. The bar chart depicts the average proportion of differential pathogen/allergen containing genera predicted using Picrust2. The difference in proportions between any two groups is presented with 95% confidence intervals. Only p values < 0.05 (Welch's t-test, FDR adjusted) are depicted; Figure S6: Identification of different fungal genera between compost piles (C) and outside composting workshop (O) using Stamp analysis. The bar chart depicts the average proportion of differential pathogen/allergen containing genera predicted using Picrust2. The difference in proportions between any two groups is presented with 95% confidence intervals. Only p values < 0.05 (Welch's t-test, FDR adjusted) are depicted; Figure S7: Bioaerosolization Index (BI) of fungal phylum. (BI value was transformed by log); Figure S8: Bioaerosolization Index (BI) of 77 fungal genera with a BI value greater than 0. (BI value was transformed by log).

Author Contributions: Methodology and data curation, writing—original draft preparation R.W. and A.Y.; Software, formal analysis, T.Q., H.G. and Y.G.; Conceptualization, writing—review and editing, project administration and funding acquisition, X.S., M.G. and X.W. All authors have read and agreed to the published version of the manuscript.

Funding: This research was funded by the National Natural Science Foundation of China (grant numbers 41961134033, 51878053) and the Research Foundation of BAAFS (grant number KJCX20200402).

Data Availability Statement: Not applicable.

Conflicts of Interest: The authors declare that they have no known competing financial interests or personal relationships that could have appeared to influence the work reported in this paper.

References

1. Zhang, B.; Tian, H.; Lu, C.; Dangal, S.R.S.; Yang, J.; Pan, S. Global manure nitrogen production and application in cropland during 1860–2014: A 5 arcmin gridded global dataset for Earth system modeling. *Earth Syst. Sci. Data* **2017**, *9*, 667–678. [CrossRef]
2. Li, J.; Bao, H.; Xing, W.; Yang, J.; Liu, R.; Wang, X.; Lv, L.; Tong, X.; Wu, F. Succession of fungal dynamics and their influence on physicochemical parameters during pig manure composting employing with pine leaf biochar. *Bioresour. Technol.* **2020**, *297*, 122377. [CrossRef] [PubMed]
3. Robertson, S.; Douglas, P.; Jarvis, D.; Marczylo, E. Bioaerosol exposure from composting facilities and health outcomes in workers and in the community: A systematic review update. *Int. J. Hyg. Environ. Health* **2019**, *222*, 364–386. [CrossRef] [PubMed]
4. Wery, N. Bioaerosols from composting facilities-a review. *Front. Cell. Infect. Microbiol.* **2014**, *4*, 42. [PubMed]
5. Langarica-Fuentes, A.; Zafar, U.; Heyworth, A.; Brown, T.; Fox, G.; Robson, G.D. Fungal succession in an in-vessel composting system characterized using 454 pyrosequencing. *FEMS Microbiol. Ecol.* **2014**, *88*, 296–308. [CrossRef]
6. Mbareche, H.; Veillette, M.; Bonifait, L.; Dubuis, M.-E.; Benard, Y.; Marchand, G.; Bilodeau, G.J.; Duchaine, C. A next generation sequencing approach with a suitable bioinformatics workflow to study fungal diversity in bioaerosols released from two different types of composting plants. *Sci. Total Environ.* **2017**, *601–602*, 1306–1314. [CrossRef]
7. Gao, M.; Qiu, T.; Sun, Y.; Wang, X. The abundance and diversity of antibiotic resistance genes in the atmospheric environment of composting plants. *Environ. Int.* **2018**, *116*, 229–238. [CrossRef]

8. Ferguson, R.M.W.; Neath, C.E.E.; Nasir, Z.A.; Garcia-Alcega, S.; Tyrrel, S.; Coulon, F.; Dumbrell, A.J.; Colbeck, I.; Whitby, C. Size fractionation of bioaerosol emissions from green-waste composting. *Environ. Int.* **2021**, *147*, 106327. [CrossRef]
9. Domingo, J.L.; Nadal, M. Domestic waste composting facilities: A review of human health risks. *Environ. Int.* **2009**, *35*, 382–389. [CrossRef]
10. Fung, F.; Hughson, W.G. Health effects of indoor fungal bioaerosol exposure. *Appl. Occup. Environ. Hyg.* **2003**, *18*, 535–544. [CrossRef]
11. Eduard, W. A health-based criteria document on fungal spore exposure in the working population. Is it relevant for the general population? *Indoor Air* **2008**, *18*, 257–258. [CrossRef] [PubMed]
12. Pearson, C.; Littlewood, E.; Douglas, P.; Robertson, S.; Gant, T.W.; Hansell, A.L. Exposures and health outcomes in relation to bioaerosol emissions from composting facilities: A systematic review of occupational and community studies. *J. Toxicol. Environ. Health Part B* **2015**, *18*, 43–69. [CrossRef] [PubMed]
13. Taha, M.P.M.; Pollard, S.J.T.; Sarkar, U.; Longhurst, P. Estimating fugitive bioaerosol releases from static compost windrows: Feasibility of a portable wind tunnel approach. *Waste Manag.* **2005**, *25*, 445–450. [CrossRef]
14. Sanchez-Monedero, M.A.; Stentiford, E.I.; Mondini, C. Biofiltration at Composting Facilities: Effectiveness for Bioaerosol Control. *Environ. Sci. Technol.* **2003**, *37*, 4299–4303. [CrossRef] [PubMed]
15. Huffnagle, G.B.; Noverr, M.C. The emerging world of the fungal microbiome. *Trends Microbiol.* **2013**, *21*, 334–341. [CrossRef] [PubMed]
16. He, P.; Wei, S.; Shao, L.; Lü, F. Aerosolization behavior of prokaryotes and fungi during composting of vegetable waste. *Waste Manag.* **2019**, *89*, 103–113. [CrossRef]
17. Le Goff, O.; Bru-Adan, V.; Bacheley, H.; Godon, J.J.; Wery, N. The microbial signature of aerosols produced during the thermophilic phase of composting. *J. Appl. Microbiol.* **2010**, *108*, 325–340. [CrossRef]
18. Gutarowska, B.; Skóra, J.; Stępień, Ł.; Szponar, B.; Otlewska, A.; Pielech-Przybylska, K. Assessment of microbial contamination within working environments of different types of composting plants. *J. Air Waste Manag.* **2015**, *65*, 466–478. [CrossRef]
19. Peccia, J.; Hernandez, M. Incorporating polymerase chain reaction-based identification, population characterization, and quantification of microorganisms into aerosol science: A review. *Atmos. Environ.* **2006**, *40*, 3941–3961. [CrossRef]
20. Tischer, C.G.; Heinrich, J. Exposure assessment of residential mould, fungi and microbial components in relation to children's health: Achievements and challenges. *Int. J. Hyg. Environ. Health* **2013**, *216*, 109–114. [CrossRef]
21. Zhang, Z.; Qu, Y.; Li, S.; Feng, K.; Wang, S.; Cai, W.; Liang, Y.; Li, H.; Xu, M.; Yin, H. Soil bacterial quantification approaches coupling with relative abundances reflecting the changes of taxa. *Sci. Rep.* **2017**, *7*, 4837. [CrossRef] [PubMed]
22. Finnegan, S.; Droser, M.L. Relative and absolute abundance of trilobites and rhynchonelliform brachiopods across the Lower/Middle Ordovician boundary, eastern Basin and Range. *Paleobiology* **2005**, *31*, 480–502. [CrossRef]
23. Stämmler, F.; Gläsner, J.; Hiergeist, A.; Holler, E.; Weber, D.; Oefner, P.J.; Gessner, A.; Spang, R. Adjusting microbiome profiles for differences in microbial load by spike-in bacteria. *Microbiome* **2016**, *4*, 28. [CrossRef] [PubMed]
24. Moletta-Denat, M.; Bru-Adan, V.; Delgenes, J.-P.; Hamelin, J.; Wéry, N.; Godon, J.-J. Selective microbial aerosolization in biogas demonstrated by quantitative PCR. *Bioresour. Technol.* **2010**, *101*, 7252–7257. [CrossRef]
25. Veillette, M.; Bonifait, L.; Mbareche, H.; Marchand, G.; Duchaine, C. Preferential aerosolization of Actinobacteria during handling of composting organic matter. *J. Aerosol Sci.* **2018**, *116*, 83–91. [CrossRef]
26. Lu, F.; Hu, T.; Wei, S.; Shao, L.; He, P. Bioaerosolization behavior along sewage sludge biostabilization. *Front. Environ. Sci. Eng.* **2020**, *15*, 45. [CrossRef]
27. Perrott, P.; Turgeon, N.; Gauthier-Levesque, L.; Duchaine, C. Preferential aerosolization of bacteria in bioaerosols generated in vitro. *J. Appl. Microbiol.* **2017**, *123*, 688–697. [CrossRef]
28. Ekinci, K.; Keener, H.; Akbolat, D. Effects of feedstock, airflow rate, and recirculation ratio on performance of composting systems with air recirculation. *Bioresour. Technol.* **2006**, *97*, 922–932. [CrossRef]
29. Zhang, J.; Zeng, G.; Chen, Y.; Yu, M.; Yu, Z.; Hui, L.; Yu, Y.; Huang, H. Effects of physico-chemical parameters on the bacterial and fungal communities during agricultural waste composting. *Bioresour. Technol.* **2011**, *102*, 2950–2956. [CrossRef]
30. Shehata, E.; Cheng, D.-m.; Ma, Q.-q.; Li, Y.-l.; Liu, Y.-w.; Feng, Y.; Ji, Z.-y.; Li, Z.-j. Microbial community dynamics during composting of animal manures contaminated with arsenic, copper, and oxytetracycline. *J. Integr. Agric.* **2021**, *20*, 1649–1659. [CrossRef]
31. Mbareche, H.; Veillette, M.; Bilodeau, G.; Duchaine, C. Comparison of the performance of ITS1 and ITS2 as barcodes in amplicon-based sequencing of bioaerosols. *PeerJ* **2020**, *8*, e8523. [CrossRef] [PubMed]
32. Edgar, R.C. UPARSE: Highly accurate OTU sequences from microbial amplicon reads. *Nat. Methods* **2013**, *10*, 996–998. [CrossRef] [PubMed]
33. Edgar, R.C.; Haas, B.J.; Clemente, J.C.; Quince, C.; Knight, R. UCHIME improves sensitivity and speed of chimera detection. *Bioinformatics* **2011**, *27*, 2194–2200. [CrossRef] [PubMed]
34. Caporaso, J.G.; Kuczynski, J.; Stombaugh, J.; Bittinger, K.; Bushman, F.D.; Costello, E.K.; Fierer, N.; Pena, A.G.; Goodrich, J.K.; Gordon, J.I. QIIME allows analysis of high-throughput community sequencing data. *Nat. Methods* **2010**, *7*, 335–336. [CrossRef] [PubMed]
35. Bremner, J.M. Determination of nitrogen in soil by the Kjeldahl method. *J. Agric. Sci.* **1960**, *55*, 11–33. [CrossRef]

36. Guo, H.; Nasir, M.; Lv, J.; Dai, Y.; Gao, J. Understanding the variation of microbial community in heavy metals contaminated soil using high throughput sequencing. *Ecotoxicol. Environ. Saf.* **2017**, *144*, 300–306. [CrossRef]
37. Parks, D.; Tyson, G.; Hugenholtz, P.; Beiko, R. STAMP: Statistical analysis of taxonomic and functional profiles. *Bioinformatics* **2014**, *30*, 3123–3124. [CrossRef]
38. Bonifait, L.; Marchand, G.; Veillette, M.; M'Bareche, H.; Dubuis, M.-E.; Pépin, C.; Cloutier, Y.; Bernard, Y.; Duchaine, C. Workers' exposure to bioaerosols from three different types of composting facilities. *J. Occup. Environ. Hyg.* **2017**, *14*, 815–822. [CrossRef]
39. Glass, N.L.; Schmoll, M.; Cate, J.H.D.; Coradetti, S. Plant Cell Wall Deconstruction by Ascomycete Fungi. *Annu. Rev. Microbiol.* **2013**, *67*, 477–498. [CrossRef]
40. Pal, C.; Bengtsson-Palme, J.; Kristiansson, E.; Larsson, D.G.J. The structure and diversity of human, animal and environmental resistomes. *Microbiome* **2016**, *4*, 54. [CrossRef]
41. Sebők, F.; Dobolyi, C.; Bobvos, J.; Szoboszlay, S.; Kriszt, B.; Magyar, D. Thermophilic fungi in air samples in surroundings of compost piles of municipal, agricultural and horticultural origin. *Aerobiologia* **2015**, *32*, 255–263. [CrossRef]
42. Feeney, P.; Rodríguez, S.F.; Molina, R.; McGillicuddy, E.; Hellebust, S.; Quirke, M.; Daly, S.; O'Connor, D.; Sodeau, J. A comparison of on-line and off-line bioaerosol measurements at a biowaste site. *Waste Manag.* **2018**, *76*, 323–338. [CrossRef] [PubMed]
43. Alvarado, C.S.; Gandara, A.; Flores, C.; Perez, H.R.; Green, C.F.; Hurd, W.W.; Gibbs, S.G. Seasonal changes in airborne fungi and bacteria at a dairy cattle concentrated animal feeding operation in the southwest United States. *J. Environ. Health* **2009**, *71*, 40–44. [PubMed]
44. O'Connor, D.; Daly, S.M.; Sodeau, J.R. On-line monitoring of airborne bioaerosols released from a composting/green waste site. *Waste Manag.* **2015**, *42*, 23–30. [CrossRef]
45. Agency, E. *M9: Environmental Monitoring of Bioaerosols at Regulated Facilities*; Environment Agency: Bristol, UK, 2018.
46. Janbon, G.; Quintin, J.; Lanternier, F.; d'Enfert, C. Studying fungal pathogens of humans and fungal infections: Fungal diversity and diversity of approaches. *Microbes Infect.* **2019**, *21*, 237–245. [CrossRef]
47. Lü, F.; Shao, L.-M.; Zhang, H.; Fu, W.-D.; Feng, S.-J.; Zhan, L.-T.; Chen, Y.-M.; He, P.-J. Application of advanced techniques for the assessment of bio-stability of biowaste-derived residues: A minireview. *Bioresour. Technol.* **2018**, *248*, 122–133. [CrossRef]
48. Niculita-Hirzel, H.; Hantier, G.; Storti, F.; Plateel, G.; Roger, T. Frequent Occupational Exposure to Fusarium Mycotoxins of Workers in the Swiss Grain Industry. *Toxins* **2016**, *8*, 370. [CrossRef]
49. Hashemi, H.; Abbasi, F.; Samaei, M.R.; Khodadadi, H. Determination of Fungi Species Variety in Thermal Phases of Compost Production and Related Operational Parameters. *J. Environ. Eng.* **2018**, *144*, 04018065. [CrossRef]
50. Qiu, X.; Zhou, G.; Zhang, J.; Wang, W. Microbial community responses to biochar addition when a green waste and manure mix are composted: A molecular ecological network analysis. *Bioresour. Technol.* **2019**, *273*, 666–671. [CrossRef]
51. Kang, W.; Kim, I.-H.; Lee, T.-J.; Kim, K.-Y.; Kim, D. Effect of temperature on bacterial emissions in composting of swine manure. *Waste Manag.* **2014**, *34*, 1006–1011. [CrossRef]
52. Galès, A.; Bru-Adan, V.; Godon, J.-J.; Delabre, K.; Catala, P.; Ponthieux, A.; Chevallier, M.; Birot, E.; Steyer, J.-P.; Wéry, N. Predominance of single bacterial cells in composting bioaerosols. *Atmos. Environ.* **2015**, *107*, 225–232. [CrossRef]

International Journal of
Environmental Research and Public Health

MDPI

Article

Factors Driving Microbial Community Dynamics and Potential Health Effects of Bacterial Pathogen on Landscape Lakes with Reclaimed Water Replenishment in Beijing, PR China

Junzhi Zhang [1,*], Xiao He [1], Huixin Zhang [1], Yu Liao [1], Qi Wang [2,3], Luwei Li [1] and Jianwei Yu [2,3,*]

1 Beijing Climate Change Response Research and Education Center, Beijing University of Civil Engineering and Architecture, Beijing 100044, China; hexiao0365@126.com (X.H.); zhanghuixin_bucea@163.com (H.Z.); yuliao_1994@126.com (Y.L.); li_hany@163.com (L.L.)
2 Key Laboratory of Drinking Water Science and Technology, Research Center for Eco-Environmental Sciences, Chinese Academy of Sciences, Beijing 100085, China; qiwang_st@rcees.ac.cn
3 University of Chinese Academy of Sciences, Beijing 100019, China
* Correspondence: zhangjunzhi@bucea.edu.cn (J.Z.); jwyu@rcees.ac.cn (J.Y.); Tel.: +86-010-6120-9207 (J.Z.); +86-010-6284-9149 (J.Y.)

Abstract: Assessing the bacteria pathogens in the lakes with reclaimed water as major influents are important for public health. This study investigated microbial communities of five landscape lakes replenished by reclaimed water, then analyzed driven factors and identified health effects of bacterial pathogens. 16S rRNA gene sequence analysis demonstrated that *Proteobacteria*, *Actinobacteria*, *Cyanobacteria*, *Firmicutes*, and *Verrucomicrobia* were the most dominant phyla in five landscape lakes. The microbial community diversities were higher in Jun. and Jul. than that in other months. Temperature, total nitrogen and phosphorus were the main drivers of the dominant microbial from the Redundancy analysis (RDA) results. Various potential bacterial pathogens were identified, including *Pseudomonas*, *GKS98_freshwater_group*, *Sporosarcina*, *Pseudochrobactrum*, *Streptomyces* and *Bacillus*, etc, some of which are easily infectious to human. The microbial network analysis showed that some potential pathogens were nodes that had significant health effects. The work provides a basis for understanding the microbial community dynamics and safety issues for health effects in landscape lakes replenished by reclaimed water.

Keywords: reclaimed water; health effects; microbial community dynamics; diversity and richness; driven factors; pathogens; networks

Citation: Zhang, J.; He, X.; Zhang, H.; Liao, Y.; Wang, Q.; Li, L.; Yu, J. Factors Driving Microbial Community Dynamics and Potential Health Effects of Bacterial Pathogen on Landscape Lakes with Reclaimed Water Replenishment in Beijing, PR China. *Int. J. Environ. Res. Public Health* **2022**, *19*, 5127. https://doi.org/10.3390/ijerph19095127

Academic Editor: Cheng Yan

Received: 13 March 2022
Accepted: 20 April 2022
Published: 22 April 2022

Publisher's Note: MDPI stays neutral with regard to jurisdictional claims in published maps and institutional affiliations.

1. Introduction

Reclaimed water is a special water source has received concentrated concerns in many countries and regions [1–3]. Replenishing landscape lakes with reclaimed water is already one of the effective ways to improve water utilization [4]. Over a quarter of the reclaimed water was used to replenish landscape lakes [3,5,6]. However, there still have some obvious differences in the water quality of reclaimed water as compared to natural lakes [4]. Especially, high levels of N, P, disinfection by-products and bacterial pathogens could still remain in reclaimed waters after extensive wastewater treatment, which would cause health issues of water safety easily [6–10]. Thus, the pathogens cause potential threats on the public health, especially to the visitors exposed to these waters such as swimmers or fishers [11–13].

Recently, the impact of reclaimed water as a water source has attracted much attention [14]. For the landscape lakes receiving reclaimed water, the water quality and biological community are largely changed. Although the driving factors for microorganism variation in different water environments has been studied [14,15], little is known regarding microorganism dynamics in surface waters with reclaimed water influents. Microbial communities reflect the structure and function of water ecosystems, which are strongly correlated with environmental factors. For example, temperature, pH, light, total nitrogen (TN), and total phosphorus (TP) affect the microbial community greatly [16,17]. Moreover, algal blooms and reproduction may also affect microbial community through nutrients consumptions [18]. Meanwhile, for the reclaimed water, great attention has been paid to the hazard induced by bacterial pathogens [19–21]. A variety of bacterial pathogens have been identified and hardly removed by wastewater treatment [22,23], such as *Aeromonas*, *Pseudomonas*, *Mycobacterium*, *Legionella* and pathogenic *E. coli* [13,14]. Thus, together with chemical contaminants, the health effects of pathogens should be considered in the surface waters receiving reclaimed water [4,24,25]. The status of bacterial pathogens during treatment processes have been evaluated in previous studies [9,26,27]. For landscape lakes, it is unexplored that how reclaimed water influences microbial community structure, dynamics of bacterial pathogens.

In this study, the microbial community structure was explored with 16S rRNA gene sequencing for five landscape lakes replenished by reclaimed water. Environmental factor association analysis was employed to identify the environmental driven factors related to the dynamics of bacterial pathogens, which were visualized by redundancy analysis (RDA), Heatmap and microbial network. This study would contribute to the comprehensive evaluation on the hazards of pathogens in more receiving waters due reclaimed water replenishment, and the development of regulation strategy microbial community in reclaimed water.

2. Materials and Methods

2.1. Water Samples

The information about the five investigated landscapes (GBD, ALPK, NHZ, QN and LY in Beijing) are provided in Figure 1. GBD is located in the Tonghui River with an area of 142,900 m^2 and is replenished by the reclaimed wastewater treatment plant of GBD (GBDRWTP). NHZ as a wetland park was also chosen (300,000 m^2), which is mainly supplied by the XHM reclaimed wastewater treatment plant (XHMRWTP). ALPK is located on the Dragon River system of Olympic Forest Park with an area of 635,000 m^2, which is mainly replenished by the effluent of QH reclaimed wastewater treatment plant of (QHRWTP). QN and LY lakes are replenished by the reclaimed wastewater treatment stations of QNRWTS and LYRWTS, which are 611,000 m^2 and 700,000 m^2, respectively.

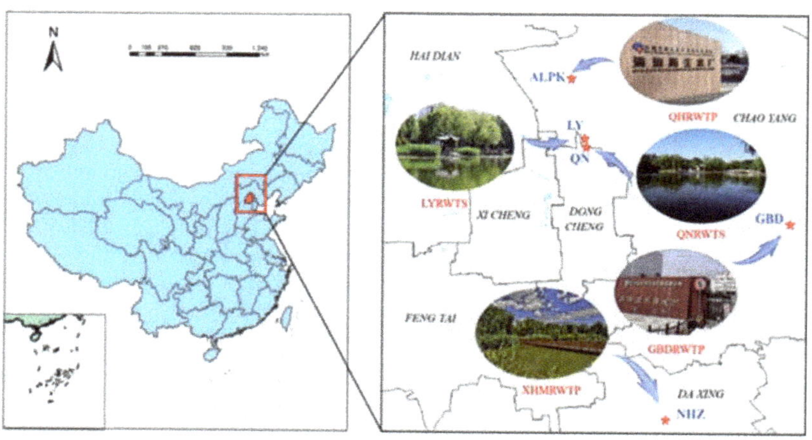

Figure 1. Five sampled landscape lakes in Beijing.

The water depth of the five lakes ranges from 0.5 to 2.5 m. The sampling was carried out once a month from May to September (on the same day for all five lakes), and the parallel samples from the surface and the bottom were collected at the same time. All 50 samples were stored in brown bottles (4 °C in the dark) before subsequent analysis.

2.2. Water Quality Analysis

All samples were analyzed as soon as possible within three days. Temperature (T) and pH were recorded simultaneously along with the sampling process. Other water quality parameters were determined according to the EPA standard methods including the total phosphorus (TP), total nitrogen (TN), orthophosphate (PO_4^{3-}), ammonium (NH_4^+), and nitrate (NO_3^-) [28]. Optical density 680 (OD_{680}) was employed to characterize algal cell density by a UV-vis spectrophotometer (Lambda 650s, PerkinElmer, Waltham, MA, USA) at 680 nm. Water quality characteristics are shown in Table 1, the temperature of the five lakes ranged from 23 °C to 30 °C and pH from 7.21 to 8.85, while average values with errors are included for other indicators.

Table 1. Water quality characteristics.

Parameters	Value	Parameters	Value
T (°C)	23.0~30.0	TN (mg/L)	3.70 ± 0.87
pH	7.21~8.85	TP (mg/L)	0.024 ± 0.009
NH_4^+-N (mg/L)	0.0851 ± 0.0012	PO_4^{3-}-P (mg/L)	0.0177 ± 0.0003
NO_3^- (mg/L)	5.31 ± 0.55	OD_{680} (cm^{-1})	0.457 ± 0.00

2.3. DNA Extraction and High-Throughput Sequencing

The method followed the procedure described in a previous work [29]. The E.Z.N.A.® Water DNA Kit (Omega Bio Tek, Norcross, GA, USA) was used to extract DNA from the samples. The V4-V5 region of the bacterial 16S ribosomal RNA genes was amplified by PCR reactions. The final extension phase was conducted at 72 °C and kept for 10 min until the PCR thermal cycler stopped at 10 °C [30]. PCR reactions were performed in triplicate using a combination of the three PCR products detected by 2% agarose gel electrophoresis. The amplified products were purified and quantified by the AxyPrep DNA Gel Extraction Kit (Axygen Biosciences, Union City, CA, USA) and QuantiFluor™-ST (Promega, Madison, WI, USA) according to the manufacturer's instructions, respectively.

Purified amplicons were pooled in equimolar amounts and paired-end sequencing (2 × 250) was performed on the Illumina MiSeq platform according to the standard protocol of Majorbio Bio-Pharm Technology Co., Ltd., Shanghai, China. The original reads of Accession Number: PRJNA759710 were readily available in the database NCBI Sequence Read Archive (SRA).

2.4. Data Processing and Statistical Analysis

The index sequence was used to distinguish the data for each sample and the QIIME2 (version 4.1) was employed to analyze the data. Sequencing was completed by the fastp software and FLASH software. The sequence de-hybridization method and parameters are shown in SI (Section S1).

The maximum mismatch rate allowed for overlapping regions of spliced sequences was 0.2 to screen out inconsistent sequences. Sequences with 97% similarity to any other sequence were classified as an operant classification unit (OTU) using UPARSE (version 7.1) software. Classification analysis of each 16S rRNA gene sequence was performed in the SILVA (SSU118) 16S rRNA database using a 70% confidence threshold.

Since the five lakes are all shallow and their stratification differences were insignificant, the data from 0.5 m and 2.5 m depth were averaged for analysis. OTU data was used for the statistical analysis of sequencing results. some rare taxa were removed to normalize the data. The microbial communities of landscapes lakes as well as their driving factors were analyzed in multiple ways. The richness (Sobs) and diversity (Shannon) indexes were calculated by the MOTHUR package (version 1.30.1), for describing the alpha diversity of the microbial communities. More environmentally relevant factors were selected using

Variance Inflation Factors (VIFs), calculated as $VIF_i = 1/(1 - R_i^2)$. Where R_i^2 represents the proportion of variance of the ith independent variable in the model that is correlated with the other independent variables. R_i was used to determine the covariance between the ith independent variable and the other independent variables. RDA and Spearman correlation Heatmap analysis was performed using pheatmap package based on R software to visualize the relationship between dominant community composition and environmental factors. BugBase was used to analyze the abundance of potential pathogens. The microbial network correlations were analyzed by microbial network analysis. A Correlation Network Figure was determined based on graph theory knowledge to analyze the correlation of the biological network using Networkx.

3. Results and Discussion

3.1. Microbial Community of Examined Lakes

A total of 52 classes belonging to 27 distinct phyla were recognized in the samples. Figure S2 gives the relative abundance of the dominant genera, indicating that the variation in relative abundance among the dominant genera varied considerably. The phylum of *Proteobacteria* was the most abundant phylum in the examined lakes, accounting for 41.78% of all the sample sequences. *Cyanobacteria* was the second most abundant phylum, accounted for 17.21% of the total phyla. The rest of the sequences belonged to *Firmicutes* (12.49%), *Actinobacteria* (12.31%), *Verrucomicrobia* (6.71%), *Bacteroidetes* (3.88%), *Planctomycetes* (3.11%) and others (2.51%, <1% for each). *Proteobacteria* has been identified as the dominant phylum either in natural water or reclaimed water, while the abundance of *Firmicutes* in reclaimed waters was higher than in natural waters [29,31–33]. Therefore, *Firmicutes* in the landscape lakes replenished by reclaimed water was one of the four dominant phyla (>10%), over the abundance of *Actinobacteria* [34,35].

3.2. Microbial Community Diversity and Potential Bacterial Pathogens

The microbiome survey yielded 2, 695, 332 reads from 50 samples. After normalization, a total of 1114 OTUs were obtained and verified by rarefaction curves. The end of the rarefaction curves gradually flattened out, suggesting that the detection rate of microbial communities was reaching saturation in all samples. The standardized Sobs index and Shannon index from normalized OTUs results were used to assess the richness (Figure 2a,c) and diversity (Figure 2b,d).

The microbial richness and diversity were not considerably varied across examined sample sites but diverged between months. From Figure 2a,b, the microbial Sob indexes were between 447 to 477 averagely and the Shannon indexes were between 3.22 to 3.76 averagely in different sites. As shown in Figure 2c,d, the Sobs indexes were 347 and Shannon indexes were 3.23 on average in Jun. and Jul. (28 °C~31 °C), while Sobs and Shannon indexes were 535.5 and 3.63 in May, August and September (20 °C~25 °C). The Sobs indexes in May, August and September were 1.54 times of Jun. and Jul. The diversities were higher and the richness was lower of microbial communities in June and July than them in other months. The index of similarities was calculated using ANOSIM/Adonis method (Figure S2), and the reliability was good (p values < 0.05). From the corresponding parameters listed in Table S2, it could be seen that sampling time (in month) ($R^2 = 0.27$) had a greater effect on microbial community composition than location ($R^2 = 0.16$).

The taxonomic of the microbial community in each month is shown in Figure 3. It could be seen that the proportion of *Firmicutes* was much higher than that in May, August and September In July, the proportion of *Firmicutes* was 28%, which equal to that of *Proteobacteria* and is the most abundant phylum. As shown in Figure 3, *Proteobacteria* and *Actinobacteria* were significantly reduced from Jun. to Jul. In previous studies, *Firmicutes* genera has been found to be capable of tolerating and thriving at high temperature [36]. During algal blooms in June and July, *Firmicutes* were better adapted to the dissolved oxygen depletion in water than *Proteobacteria* and *Actinobacteria*. Meanwhile, from Table 1, TN of the five investigated landscape lakes in this study was several times higher than that of the surface water quality standard, which might be due to the replenishment of reclaimed water [37,38]. High N concentrations has been found to be suitable for the growth and propagation of some *Firmicutes* genera, which could be a reason for the increase in abundance of *Firmicutes*. Additionally, as *Firmicutes*, contains several pathogens such as *Staphylococcus* and *Bacillus*, special attention should be paid for its abundance in waters. These pathogens induce health hazards through parasitizing the skin, nasal cavity, throat, gut, canker sores, abscesses. In the examined landscape lakes, the increased abundance of *Firmicutes* in Jun. and Jul. could elevate the hazards of diseases such as wound infections and diarrhea in humans [26,27,39].

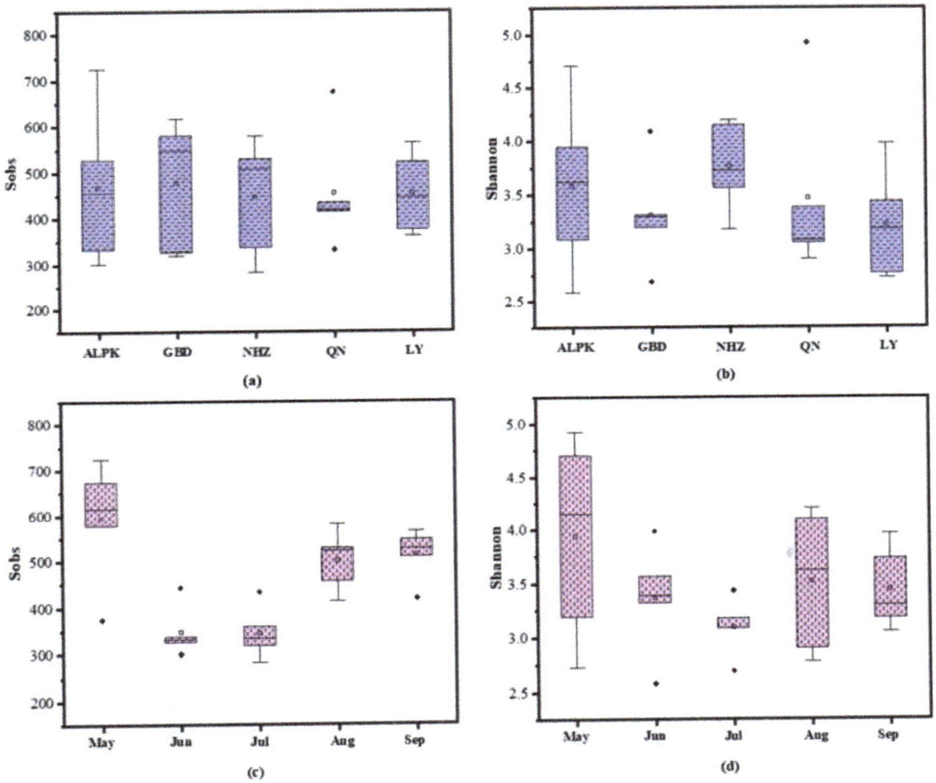

Figure 2. OTU richness (Sobs) and diversity (Shannon) of all samples in different locations (**a**,**b**) and different months (**c**,**d**).

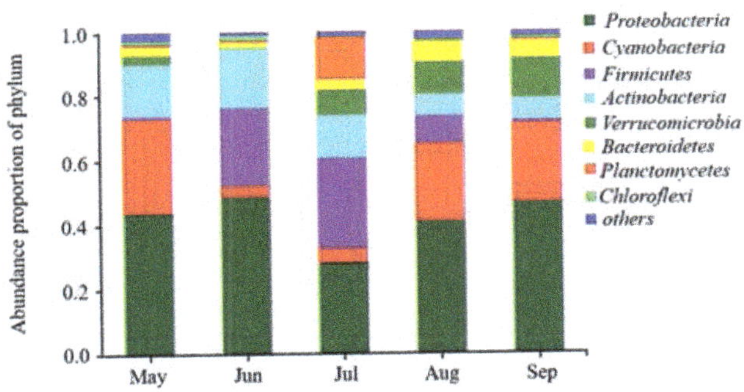

Figure 3. Taxonomic of microbial community in different months.

Compared to *Firmicutes*, *Proteobacteria* and *Actinobacteria* were commonly frequent phyla. *Proteobacteria* was the most dominant phylum each month (over 40% averagely). *Actinobacteria* was a common phylum and the average abundance in each month was over 10%. *Proteobacteria* and *Actinobacteria* have been widely detected in natural water, and could constitute more than half of the total bacteria in some surface waters, while the results of this study showed that the proportion of *Proteobacteria* and *Actinobacteria* in examined lakes was lower than in natural water. As the microbial composition is closely related to environmental factors, thus reclaimed water replenishing could affecting microbial community by changing environmental factors. Therefore, only further studies on the relationship between environmental factors and microbial community could better predict the dynamics. Moreover, the study found that bacteria such as *Legionella* exist in the *Proteobacteria* while *Rhodococcus* present in *Actinobacteria*, which have been recognized as emerging pathogens in reclaimed water sources [40–42]. Studies for health effects also found that these pathogens could lead to lung disease [43,44].

3.3. Responses of Community Changes and Potential Bacterial Pathogens to Key Environmental Factors

A vector plot of environmental variables correlated to the changes in microbial community (calculate using RDA analysis) is shown in Figure 4a, and the Spearman correlation Heatmap of environmental factors and major genera were shown in Figure 4b. To improve the accuracy of RDA analysis, several critical environmental factors were chosen based on VIF analysis (VIF < 10) [45], and Heatmap analysis was conducted on the genera to explore the microbial community and bacteria pathogen changes (Figure 4b).

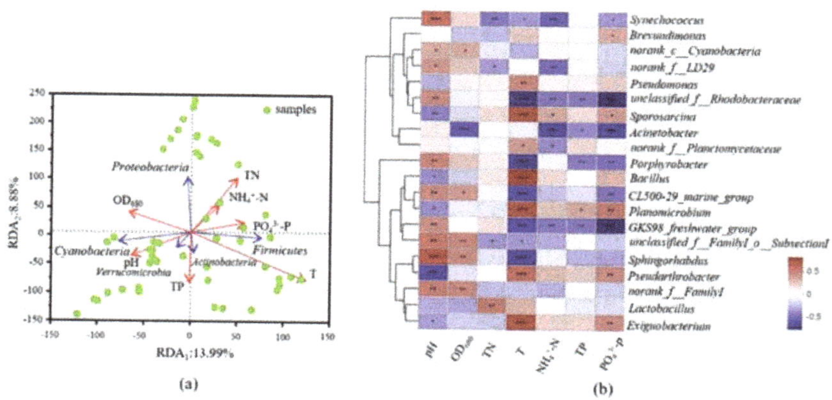

Figure 4. RDA analysis of environmental factors and the microbial community (**a**); Heatmap analysis of key environmental factors and microbial community with top 20 genera (**b**), * $0.01 < p \leq 0.05$, ** $0.001 < p \leq 0.01$, *** $p \leq 0.001$.

As shown in Figure 4a, RDA analysis showed that T, TN, TP, NH_4^+-N and PO_4^{3-}-P had considerable influences on the dominant phyla. T ($p = 0.001$) and TN ($p = 0.097$) were the two dominant driven factors affecting the dominant phyla. T was positively correlated with *Firmicutes*, and negatively correlated with *Proteobacteria*, which was aligned with the changes of the dominant phyla with month variations in Figure 3. It could be seen that the *Proteobacteria* were negatively correlated with T (Figure 4a), and its abundance in Jul. was significantly decreased (Figure 3), which could be attributed to the that *Firmicutes* became the dominant phylum and occupying the survival of *Proteobacteria*. Previous studies showed that some genera of *Firmicutes* have a strong ability to tolerate high temperatures [46], which could be the reason for the dominance of *Firmicutes* in Jun. and Jul. Nitrogen was positively correlated with *Firmicutes*. *Firmicutes*, as nitrogen-active strains, can adapt to higher nitrogen levels due to the influent of reclaimed water [46]. Compared to temperature, the nitrogen effect on *Firmicutes* was minor. This phenomenon could be associated to that nitrogen was the limiting factor for *Firmicutes* [34,35]. In Figure 4a, the temperature was positively correlated with *Actinobacteria,* and nitrogen was negatively correlated, indicating that *Actinobacteria* cannot well adapted to landscape lakes with high nitrogen levels. Similarly, the abundance of *Actinobacteria* in this study was lower than that in natural water [31,34,35].

Heatmap analysis was conducted on genera to further explore the microbial community structure. The top 20 genera were selected and analyzed with Heatmap in (Figure 4b). The important genera of *Firmicutes* were *Bacillus, Sporosarcina, Lactobacterium, Exiguobacterium,* and *Planomicrobium,* all of them were positively correlated with T. Several genera in *Proteobacteria* were negatively or poorly correlated with T and significantly negatively correlated with PO_4^{3-}-P. For *Proteobacteria,* PO_4^{3-}-P is usually a limiting factor [47,48]. In addition, several genera containing human pathogens were discovered, such as *Bacillus, Acinetobacter,* and *Pseudomonas*. Therefore, the analysis of the top 50 genera in relation to environmental factors and the identification of bacterial pathogens among them was continued (Figure S1). Many bacteria genera (*Acinetobacter, Bacillus, Pseudomonas, Mycobacterium, Escherichia-Shigella, Rhodococcus* and *Legionella*) have been found potential pathogenic. *Pseudomonas* are widely distributed in soil, water, food and air, including several bacterial pathogens such as *P. fluorescens* [49,50], which could cause wounds infections [50,51]. Since *Pseudomonas* was the dominant genus in landscape lakes replenished by reclaimed water, its potential hazard should be further noticed. Furthermore, *Mycobacterium, Rhodococcus* and *Legionella* are all lung pathogens that caused lung and respiratory infections [13,26,27,52]. Additionally, *Rhodococcus* and *Legionella* were positively correlated with T and PO_4^{3-}-P. As reclaimed waters usually have higher P level than natural waters, thus potential safety issues should be paid close attention. Previous studies showed that TN in landscape lakes influenced by the reclaimed water was usually high, which could make *Mycobacterium, Pseudomonas, Legionella sp.* and *Streptococcus* more active, especially during high temperature month when algal blooms were more possible to occur. Higher activity of bacterial pathogens increases the hazard of disease in humans. In addition, with similar levels of TN and other nutrients, *Mycobacterium, Pesudomonas, Legionella* and *Brevundimonas* have found to be dominantd in lakes with reclaimed water impacts, and these pathogens could attach to suspended particles to increase their pathogenic hazards [53,54].

3.4. Health Effects of Potential Pathogens and Microbial Co-Occurrence

The abundance of potential pathogens in the top 50 genera was analyzed using the BugBase in functional prediction. Figure 5 shows that the species and the abundance of bacteria pathogens were higher in Jun. and Jul. than that in the other three months. The abundance of potential pathogens in Jun. and Jul. was about 30% of total genera, while that in May, August and September were about 15%. The top five genera that contributed to potential pathogenicity were: *Pseudomonas, GKS98_freshwater_group, Sporosarcina, Pseudochrobactrum* and *Bacillus*. *GKS98_freshwater_group* (*Proteobacteria*) was the high contribution pathogenic genus in Aug. and Sep. *Pseudomonas* (*Proteobacteria*), *Pseudochrobactrum* (*Proteobacteria*) and *Sporosarcina* (*Firmicutes*) were the main pathogenic genera in Jun. and Jul. This result indicate that the microbial community composition was significant varied, which could be attributed to environmental factors changes in Jun. and Jul. In addition, several species of pathogens (*Kocuria rosea, Brevundimonas diminuta* and *Chryseobacteriun indologenes*) have been found in May to July [55]. *Kocuria rosea* colonizes oropharynx, skin and mucous membrane leading to endocarditis [56,57]. *Chryseobacterium indologenes* is an uncommon human pathogen, which causes severe infections such as septicaemia and ventilator-associated pneumonia in immunocompromised patients [58,59]. *Kocuria rosea* (p = 0.00015) and *Brevundimonas diminuta* (p = 0.00076) were positively correlated with T and *Chryseobacteriun indologenes* was also positively correlated with TN (p = 0.0003).

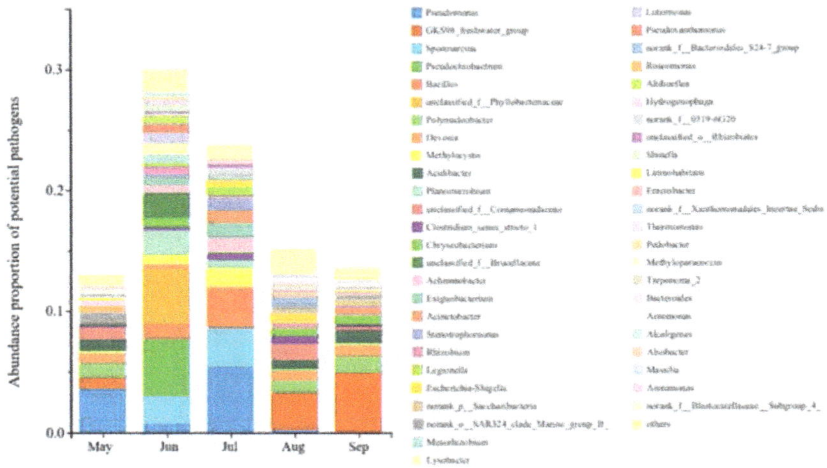

Figure 5. Abundance proportion of potential pathogens on genus level.

In addition to environmental factors, microbial co-occurrence patterns were an important driver shaping community. The main pathogenic contributing genera in May to July were *Pseudomonas*, *Sporosarcina* and *Bacillus* while those in Aug. to Sep. were *GKS98_freshwater_group* (Figure 5). Accordingly, microbial co-occurrence networks were constructed in May to July (Figure 6a) and from Aug. to Sep. (Figure 6b), respectively. There were 19 nodes with $p < 0.05$ in the group of May to July, 18 nodes with $p < 0.05$ in the group of Aug. to Sep. $p < 0.05$, suggesting that the nodes have relevance in the microbial network and they were necessary for the network. Most of the nodes in both groups belonged to *Proteobacteria, Firmicutes, Actinobacteria* and *Cyanobacteria*, consistent with the microbial community results. According to the Heatmap results, most of the important nodes were strongly correlated with T, TN and TP, which indicated that environmental factors could affect microbial networks through nodes. Focusing on the microbial network influenced by these nodes, it should be highlight that many of them were pathogenic [60,61]. As shown in Figure 6, potential genera pathogens with high abundance from May to July of microbial networks included *Sporosarcina, Bacillus, Pseudomonas, Streptomyces, Legionella* and *Rhodococcus*. In addition, from Aug. to Sep. included *GKS98_freshwater_group*, *Acinetobacter* and *Mycobacterium*.

More than highlighting important nodes microbial network also showed the co-occurrence relationship between the nodes. In Figure 6, both microbial networks had more promoting links relationships than inhibiting links. Genera belonging to the same phylum were mostly in a promoting relationship with each other, which was one reason why the rapid increase of a particular phylum in a month. However, some genera pathogens were also mostly promoting relationships with genera belonging to the same phylum (whether pathogenic or not), such as *Bacillus*, *Pseudomonas*, *Rhodococcus* and *Mycobacterium*. When environmental factors induce changes in the abundance of microbial communities, it is also an opportunity for bacteria pathogens to multiply. The higher abundance of nodal genera, the higher hazard for health effects due to genera pathogens would be posed. In terms of the number of links, there were much more connections in the second network (Figure 6b from Aug. to Sep.) than in the first network (Figure 6a from May to July), and the increased density of connections in the community made it a more complex network. Complex networks mean more stable relationships [31,62]. In a simple network, the structure changes are hard to predict, and the pathogenic hazard is higher [31,62]. Environmental factors were inextricably linked to microbial networks, and result of the current work highlights the importance of studying microbial networks based on environmental factors.

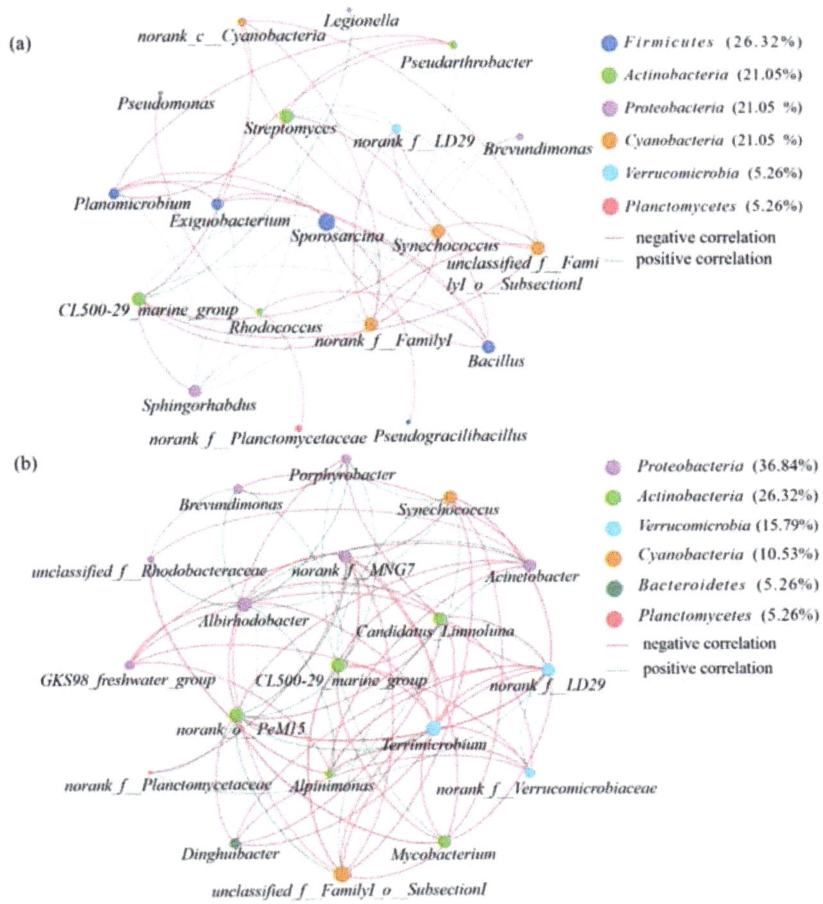

Figure 6. Co-occurrence networks for microbial communities in May, June and July (**a**); in August and September (**b**).

4. Conclusions

Microbial community composition of five landscape lakes replenished by reclaimed water were investigated. The conclusions are described as follows: the microbial community composition and diversity in the landscape lakes fluctuated in different months. *Proteobacteria*, *Cyanobacteria*, *Firmicutes* and *Actinobacteria* were the most dominant phyla, the diversity and richness of which were associated with time variations. The key environmental factors affecting the microbial community were temperature, nitrogen and phosphorus, and the temperature was the most significant. Many genera were found to be pathogenic, including *Acinetobacter*, *Mycobacterium*, *Pseudomonas*, *Rhodococcus* and *Legionella*, of which, *Kocuria rosea*, *Brevundimonas diminuta* and *Chryseobacteriun indologenes* were identified. Meanwhile, the network analysis indicated that genera pathogens already have a certain position in the network, indicating an increased potential hazard of pathogenicity in the microbial networks. Therefore, for the landscape lakes with reclaimed water influent, the potential health hazards induced by reclaimed water should be extensively assessed. The seasonal changes in microbial communities during May to September were identified here, however, the variation of microbial communities in the rest of the seasons still missing. In addition, the rRNA sequencing has low precision in identifying species-level microbial, making the available data are still scarce. Therefore, future studies may design a sampling throughout the whole year and some methodological updates and more highly accurate analysis tools should be considered.

Supplementary Materials: The following supporting information can be downloaded at: https://www.mdpi.com/article/10.3390/ijerph19095127/s1, Figure S1: Pie pilot of microbial community on phylum level. Figure S2: Analysis of similarities of microbial community in different groups on OTU level. The corresponding Adonis analysis results exhibited in Table S2. Figure S3: Spearman correlation Heatmap of primary environmental factors and bacterial community with top 50 bacterial community abundance. Table S1: TN in different lakes and different layers in five months (unit: mg/L). Table S2: Adonis analysis of microbial community in different factor on OTU level.

Author Contributions: Conceptualization, methodology and funding acquisition: J.Z.; investigation, data curation and writing-original draft preparation: X.H.; resources: H.Z.; writing—review and editing: Y.L.; investigation and writing—review and editing: Q.W.; visualization: L.L.; supervision, project administration and funding acquisition: J.Y. All authors have read and agreed to the published version of the manuscript.

Funding: This study was supported by Funds for the National Key R&D Program of China (2018YFE0204100, 2021YFC3200904), the National Natural Science Foundation of China (No. 52070185, 51408022), the special fund from the Key Laboratory of Drinking Water Science and Technology, Research Center for Eco-Environmental Sciences, Chinese Academy of Sciences (20K03KLDWST).

Data Availability Statement: All data and materials support their published claims and comply with field standards. The original reads of Accession Number: PRJNA759710 were readily available in the database NCBI Sequence Read Archive (SRA).

Acknowledgments: We are grateful for the suggestions from the reviewers and editor to improve the paper.

Conflicts of Interest: The authors declare no conflict of interest.

References

1. Chen, Z.; Tong, Y.; Ngo, H.H.; Yun, L.; Li, G.Q.; Wu, Q.Y.; Li, K.X.; Bai, Y.; Liu, S.M.; Hu, H.Y. Assimilable organic carbon (AOC) variation in reclaimed water: Insight on biological stability evaluation and control for sustainable water reuse. *Bioresour. Technol.* **2018**, *254*, 290–299. [CrossRef] [PubMed]
2. Ren, X.L.; Chen, H.B. Effect of residual chlorine on the interaction between bacterial growth and assimilable organic carbon and biodegradable organic carbon in reclaimed water. *Sci. Total Environ.* **2021**, *752*, 141223. [CrossRef] [PubMed]
3. Li, D.Q.; Huang, D.; Guo, C.F.; Guo, X.Y. Multivariate statistical analysis of temporal–spatial variations in water quality of a constructed wetland purification system in a typical park in Beijing, China. *Environ. Monit. Assess.* **2014**, *187*, 4219. [CrossRef] [PubMed]
4. Liu, W.; Xu, Z.Q.; Long, Y.J.; Feng, M.Q. Replenishment of urban landscape ponds with reclaimed water: Spatiotemporal variations of water quality and mechanism of algal inhibition with alum sludge. *Sci. Total Environ.* **2021**, *790*, 148052. [CrossRef] [PubMed]
5. Marks, J.S. Taking the public seriously: The case of potable and non-potable reuse. *Desalination* **2006**, *187*, 137–147. [CrossRef]
6. Zhu, L.B.; Torres, M.; Betancourt, W.Q.; Sharma, M.; Shirley, A.; Micallef, S.A.; Gerba, C.; Sapkota, A.R.; Sapkota, A.; Parveen, S.; et al. Incidence of fecal indicator and pathogenic bacteria in reclaimed and return flow waters in Arizona, USA. *Environ. Res.* **2019**, *170*, 122–127. [CrossRef]
7. Yue, L.Y.; Kong, W.D.; Ji, M.K.; Liu, J.B.; Morgan-Kiss, R.M. Community response of microbial primary producers to salinity is primarily driven by nutrients in lakes. *Sci. Total Environ.* **2019**, *696*, 134001. [CrossRef]
8. Malayil, L.; Ramachandran, P.; Chattopadhyay, S.; Cagle, R.; Hittle, L.; Ottesen, A.; Mongodin, E.F.; Sapkota, A.R. Metabolically-active bacteria in reclaimed water and ponds revealed using bromodeoxyuridine DNA labeling coupled with 16S rRNA and shotgun sequencing. *Water Res.* **2020**, *184*, 116185. [CrossRef]
9. Caicedo, C.; Rosenwinkel, K.H.; Exner, M.; Verstraete, W.; Suchenwirth, R.; Hartemann, P.; Nogueira, R. Legionella occurrence in municipal and industrial wastewater treatment plants and risks of reclaimed wastewater reuse: Review. *Water Res.* **2019**, *149*, 21–34. [CrossRef]
10. Kulkarni, P.; Olson, N.D.; Paulson, J.N.; Pop, M.; Maddox, C.; Claye, E.; Rosenberg Goldstein, R.E.; Sharma, M.; Gibbs, S.G.; Mongodin, E.F.; et al. Conventional wastewater treatment and reuse site practices modify bacterial community structure but do not eliminate some opportunistic pathogens in reclaimed water. *Sci. Total Environ.* **2018**, *639*, 1126–1137. [CrossRef]
11. Solaiman, S.; Micallef, S.A. Aeromonas spp. diversity in U.S. mid-Atlantic surface and reclaimed water, seasonal dynamics, virulence gene patterns and attachment to lettuce. *Sci. Total Environ.* **2021**, *779*, 146472. [CrossRef] [PubMed]

12. Pepper, I.L.; Gerba, C.P. Risk of infection from Legionella associated with spray irrigation of reclaimed water. *Water Res.* **2018**, *139*, 101–107. [CrossRef] [PubMed]
13. Hamilton, K.A.; Hamilton, M.T.; Johnson, W.; Jjemba, P.; Bukhari, Z.; LeChevallier, M.; Haas, C.N. Health risks from exposure to Legionella in reclaimed water aerosols: Toilet flushing, spray irrigation, and cooling towers. *Water Res.* **2018**, *134*, 261–279. [CrossRef] [PubMed]
14. Banda, J.F.; Zhang, Q.; Ma, L.Q.; Pei, L.X.; Du, Z.R.; Hao, C.B.; Dong, H.L. Both pH and salinity shape the microbial communities of the lakes in Badain Jaran Desert, NW China. *Sci. Total Environ.* **2021**, *791*, 148108. [CrossRef]
15. Aping, N.; Song, L.Y.; Xiong, Y.H.; Lu, C.J.; Junaid, M.; Pei, D.S. Impact of water quality on the microbial diversity in the surface water along the three Gorge reservoir (Tgr), China. *Ecotoxicol. Environ. Saf.* **2019**, *181*, 412–418.
16. Zhang, Y.T.; Shen, H.; He, X.H.; Thomas, B.W.; Lupwayi, N.Z.; Hao, X.Y.; Thomas, M.C.; Shi, X.J. Fertilization shapes bacterial community structure by alteration of soil pH. *Front. Microbiol.* **2017**, *8*, 1325. [CrossRef]
17. Guo, Z.B.; Liu, H.; Wan, S.X.; Hua, K.K.; Wang, D.Z.; Guo, X.S.; He, C.L. Fertilisation practice changes rhizosphere microbial community structure in the agroecosystem. *Ann. Appl. Biol.* **2019**, *174*, 123–132. [CrossRef]
18. Yang, Y.Y.; Chen, J.F.; Chen, X.L.; Jiang, Q.S.; Liu, Y.; Xie, S.G. Cyanobacterial Bloom Induces Structural and Functional Succession of Microbial Communities in Eutrophic Lake Sediments. *Environ. Pollut.* **2021**, *284*, 117157. [CrossRef]
19. Kevin, P.; Wang, J.; Fernando, S.C.; Thompson, J.R. Secondary metabolite gene expression and interplay of bacterial functions in a tropical freshwater cyanobacterial bloom. *ISME J.* **2014**, *8*, 1866–1878.
20. Wang, K.; Razzano, M.; Mou, X.Z. Cyanobacterial blooms alter the relative importance of neutral and selective processes in assembling freshwater bacterioplankton community. *Sci. Total Environ.* **2020**, *706*, 135724. [CrossRef]
21. Zhang, H.H.; Jia, J.Y.; Chen, S.N.; Huang, T.L.; Wang, Y.; Zhao, Z.F.; Feng, J.; Hao, H.Y.; Li, S.L.; Ma, X.X. Dynamics of Bacterial and Fungal Communities during the Outbreak and Decline of an Algal Bloom in a Drinking Water Reservoir. *Int. J. Environ. Res. Public Health* **2018**, *15*, 361. [CrossRef] [PubMed]
22. Gerba, C.P.; Betancourt, W.Q.; Kitajima, M.; Rock, C.M. Reducing uncertainty in estimating virus reduction by advanced water treatment processes. *Water Res.* **2018**, *133*, 282–288. [CrossRef] [PubMed]
23. Costán-Longares, A.; Montemayor, M.; Payán, A.; Méndez, J.; Jofre, J.; Mujeriego, R.; Lucena, F. Microbial indicators and pathogens: Removal, relationships and predictive capabilities in water reclamation facilities. *Water Res.* **2008**, *42*, 4439–4448. [CrossRef] [PubMed]
24. Steele, M.; Odumeru, J. Irrigation Water as Source of Foodborne Pathogens on Fruit and Vegetables. *J. Food Prot.* **2004**, *67*, 2839–2849. [CrossRef]
25. Deviller, G.; Lundy, L.; Despo, F.K. Recommendations to derive quality standards for chemical pollutants in reclaimed water intended for reuse in agricultural irrigation. *Chemosphere* **2020**, *240*, 124911. [CrossRef]
26. Hartemann, P.; Hautemaniere, A. Legionellosis prevention in France. *Bundesgesundheitsblatt* **2011**, *54*, 724–727. [CrossRef]
27. Nogueira, R.; Utecht, K.U.; Exner, M.; Verstraete, W.; Rosenwinkel, K.H. Strategies for the reduction of Legionella in biological treatment systems. *Water Sci. Technol.* **2016**, *74*, 816–823. [CrossRef]
28. U.S. Environmental Protection Agency (USEPA). *Risk Assessment Guidance for Superfund. Human Health Evaluation Manual, Part A*; Office of Emergency and Remedial Response: Washington, DC, USA, 1989.
29. Zhang, J.Z.; Zhang, H.X.; Li, L.W.; Wang, Q.; Yu, J.W.; Chen, Y.S. Microbial community analysis and correlation with 2-methylisoborneol occurrence in landscape lakes of Beijing. *Environ. Res.* **2020**, *183*, 109217. [CrossRef]
30. Chen, Z.C.; Zhang, X.L.; Yang, Y.; Zhou, K.; Wragg, N.; Liu, Y.; Lewis, M.; Liu, C.Q. Fabrication and characterization of 3D complex hydroxyapatite scaffolds with hierarchical porosity of different features for optimal bioactive performance. *Ceram. Int.* **2017**, *43*, 336–344. [CrossRef]
31. Sadeghi, J.; Chaganti, S.R.; Shahraki, A.H.; Heath, D.D. Microbial community and abiotic effects on aquatic bacterial communities in north temperate lakes. *Sci. Total Environ.* **2021**, *781*, 146771. [CrossRef]
32. Becerra-Castro, C.; Gonçalo, M.; Silva, A.M.T.; Manaia, C.M.; Nunes, O.C. Proteobacteria become predominant during regrowth after water disinfection. *Sci. Total Environ.* **2016**, *573*, 313–323. [CrossRef] [PubMed]
33. Newton, R.J.; Jones, S.E.; Eiler, A.; McMahon, K.D.; Bertilsson, S. Guide to the Natural History of Freshwater Lake Bacteria. *Microbiol. Mol. Biol.* **2011**, *75*, 14–49. [CrossRef] [PubMed]
34. Mondol, M.A.; Shin, S.J.; Islam, M.T. Diversity of secondary metabolites from marine bacillus species: Chemistry and biological activity. *Mar. Drugs* **2013**, *11*, 2846–2872. [CrossRef] [PubMed]
35. Zheng, Y.; Yang, J.; Amalfitano, S.; Yu, X.Q.; Liu, L.M. Effects of water stratification and mixing on microbial community structure in a subtropical deep reservoir. *Sci. Rep.* **2014**, *4*, 5821.
36. Woodhouse, J.N.; Ongley, S.E.; Brown, M.V.; Neilan, B.A. Microbial diversity and diazotrophy associated with the freshwater non-heterocyst forming cyanobacterium lyngbya robusta. *J. Appl. Phycol.* **2013**, *25*, 1039–1045. [CrossRef]
37. Sepa. *Environmental Quality Standard for Surface Water (Eqssw), Gb3838–2002*; Inspection and Quarantine of PR China: Beijing, China, 2002. (In Chinese)
38. Jing, S.; Ji, D.F.; Lin, M.; Chen, Y.Q.; Sun, Y.Y.; Huo, L.L.; Zhu, J.C.; Xi, B.D. Developing Surface Water Quality Standards in China. *Resour. Conserv. Recycl.* **2017**, *117*, 294–303.
39. Theodorou, C.M.; Stokes, S.C.; Hegazi, M.S.; Brown, E.G.; Saadai, P. Is pseudomonas infection associated with worse outcomes in pediatric perforated appendicitis? *J. Pediatr. Surg.* **2020**, *56*, 1826–1830. [CrossRef]

40. Dong, P.Y.; Cui, Q.J.; Fang, T.T.; Huang, Y.; Wang, H. Occurrence of antibiotic resistance genes and bacterial pathogens in water and sediment in urban recreational water. *J. Environ. Sci.* **2019**, *77*, 65–74. [CrossRef]
41. Cui, Q.J.; Fang, T.T.; Huang, Y.; Dong, P.Y.; Wang, H. Evaluation of bacterial pathogen diversity, abundance and health risks in urban recreational water by amplicon next-generation sequencing and quantitative PCR. *J. Environ. Sci.* **2017**, *57*, 137–149. [CrossRef]
42. Ranjbar, M.; Behrouz, B.; Norouzi, F.; Gargari, S.L.M. Anti-Pcrv Igy antibodies protect against pseudomonas aeruginosa infection in both acute pneumonia and burn wound models. *Mol. Immunol.* **2019**, *116*, 98–105. [CrossRef]
43. El-Saed, A.; Balkhy, H.H.; Al-Dorzi, H.M.; Khan, R.; Rishu, A.H.; Arabi, Y.M. Acinetobacter is the most common pathogen associated with late-onset and recurrent ventilator-associated pneumonia in an adult intensive care unit in saudi arabia. *Int. J. Infect. Dis.* **2013**, *17*, e696–e701. [CrossRef] [PubMed]
44. Ebringer, A.; Rashid, T.; Wilson, C. Chapter 29—Multiple sclerosis and creutzfeldt-jakob disease are autoimmune diseases probably caused by exposure to the nasal microbe Acinetobacter. In *Infection and Autoimmunity*, 2nd ed.; Shoenfeld, Y., Agmon-Levin, N., Rose, N.R., Eds.; Academic Press: Amsterdam, The Netherlands, 2015; pp. 505–517.
45. Te, S.H.; Tan, B.F.; Thompson, J.R.; Gin, K.Y.H. Relationship of microbiota and cyanobacterial secondary metabolites in planktothricoides-dominated bloom. *Environ. Sci. Technol.* **2017**, *51*, 4199–4209. [CrossRef] [PubMed]
46. Cuzman, O.A.; Rescic, S.; Richter, K.; Wittig, L.; Tiano, P. Sporosarcina pasteurii use in extreme alkaline conditions for recycling solid industrial wastes. *J. Biotechnol.* **2015**, *214*, 49–56. [CrossRef] [PubMed]
47. Wang, H.; Liu, X.; Wang, Y.; Zhang, S.; Zhang, G.; Han, Y.; Li, M.; Liu, L. Spatial and temporal dynamics of microbial community composition and factors influencing the surface water and sediments of urban rivers. *J. Environ. Sci.* **2023**, *124*, 187–197. [CrossRef]
48. Bunse, C.; Bertos-Fortis, M.; Sassenhagen, I.; Sildever, S.; Sjöqvist, C.; Godhe, A.; Gross, S.; Kremp, A.; Lips, I.; Lundholm, N.; et al. Spatio-temporal interdependence of bacteria and phytoplankton during a baltic sea spring bloom. *Front. Microbiol.* **2016**, *7*, 517. [CrossRef]
49. Bleves, S.; Viarre, V.; Salacha, R.; Michel, G.P.F.; Filloux, A.; Voulhoux, R. Protein secretion systems in pseudomonas aeruginosa: A wealth of pathogenic weapons. *Int. J. Med. Microbiol.* **2010**, *300*, 534–543. [CrossRef]
50. Curran, C.S.; Bolig, T.; Torabi-Parizi, P. Mechanisms and targeted therapies for Pseudomonas aeruginosa lung infection. *Am. J. Respir. Crit. Care Med.* **2018**, *197*, 708–727. [CrossRef]
51. Casciaro, B.; d'Angelo, I.; Zhang, X.; Loffredo, M.R.; Conte, G.; Cappiello, F.; Quaglia, F.; Di, Y.P.P.; Ungaro, F.; Mangoni, M.L. Poly (Lactide-Co-Glycolide) Nanoparticles for prolonged therapeutic efficacy of esculentin-1a-derived antimicrobial peptides against pseudomonas aeruginosa lung infection: In vitro and in vivo studies. *Biomacromolecules* **2019**, *20*, 1876–1888. [CrossRef]
52. Savini, V.; Fazii, P.; Favaro, M.; Astolfi, D.; Polilli, E.; Pompilio, A.; Vannucci, M.; D'Amario, C.; Bonaventura, G.D.; Fontana, C.; et al. Tuberculosis-Like pneumonias by the aerobic actinomycetes Rhodococcus, Tsukamurella and Gordonia. *Microbes Infect.* **2012**, *14*, 401–410. [CrossRef]
53. Zhang, H.H.; Wang, Y.; Chen, S.N.; Zhao, Z.F.; Feng, J.; Zhang, Z.H.; Lu, K.Y.; Jia, J.Y. Water Bacterial and Fungal Community Compositions Associated with Urban Lakes, Xi'an, China. *Int. J. Environ. Res. Public Health* **2018**, *15*, 469. [CrossRef]
54. Fang, T.T.; Cui, Q.J.; Huang, Y.; Dong, P.Y.; Wang, H.; Liu, W.-T.; Ye, Q.H. Distribution Comparison and Risk Assessment of Free-Floating and Particle-Attached Bacterial Pathogens in Urban Recreational Water: Implications for Water Quality Management. *Sci. Total Environ.* **2018**, *613–614*, 428–438. [CrossRef] [PubMed]
55. Chattopadhyay, S.; Perkins, S.D.; Shaw, M.; Nichols, T.L. Evaluation of exposure to brevundimonas diminuta and Pseudomonas aeruginosa during showering. *J. Aerosol. Sci.* **2017**, *114*, 77–93. [CrossRef] [PubMed]
56. Montoya, J.E.M.; Moran, M.A.M.; Ardila, J.A.B.; Henao, P.G.C.; Rodriguez, E.E.M.; Meza, G.A.C. Brain abscess by Kocuria Rosea: Case report and literature review. *Interdiscip. Neurosurg.* **2017**, *7*, 59–61. [CrossRef]
57. Moreira, J.S.; Riccetto, A.G.L.; Silva, M.T.N.D.; Vilela, M.M.D.S. Endocarditis by Kocuria Rosea in an immunocompetent child. *Braz. J. Infect. Dis.* **2015**, *19*, 82–84. [CrossRef] [PubMed]
58. Baruah, M.; Lyngdoh, C.; Lyngdoh, W.V.; Talukdar, R. Noncatheter-related bacteraemia due to Chryseobacterium Indologenes in an immunocompetent patient. *Indian J. Med. Microbiol.* **2016**, *34*, 380–381. [CrossRef]
59. Bhuyar, G.; Jain, S.; Shah, H.; Mehta, V.K. Urinary tract infection by Chryseobacterium Indologenes. *Indian J. Med. Microbiol.* **2012**, *30*, 370–372. [CrossRef]
60. Hogenhout, S.A.; Loria, R. Virulence mechanisms of Gram-Positive plant pathogenic bacteria. *Curr. Opin. Plant Biol.* **2008**, *11*, 449–456. [CrossRef]
61. Loria, R.; Kers, J.; Madhumita, J. Evolution of plant pathogenicity in Streptomyces. *Annu. Rev. Phytopathol.* **2006**, *44*, 469–487. [CrossRef]
62. Chen, S.; Zhang, B.; Ning, D.L.; Zhang, Y.; Dai, T.J.; Wu, L.W.; Li, T.L.; Liu, W.; Zhou, J.Z.; Wen, X.H. Seasonal dynamics of the microbial community in two full-scale wastewater treatment plants: Diversity, composition, phylogenetic group based assembly and co-occurrence pattern. *Water Res.* **2021**, *200*, 117295.

International Journal of
Environmental Research and Public Health

MDPI

Article

Numerical Investigations of Urban Pollutant Dispersion and Building Intake Fraction with Various 3D Building Configurations and Tree Plantings

Qingman Li [1,2,†], Jie Liang [1,2], Qun Wang [3,†], Yuntong Chen [1], Hongyu Yang [1,2], Hong Ling [1,2,*], Zhiwen Luo [4] and Jian Hang [1,2]

1 Southern Marine Science and Engineering Guangdong Laboratory (Zhuhai), School of Atmospheric Sciences, Sun Yat-sen University, Zhuhai 519082, China; liqm23@mail2.sysu.edu.cn (Q.L.); liangj58@mail2.sysu.edu.cn (J.L.); chenyt226@mail2.sysu.edu.cn (Y.C.); yanghy46@mail2.sysu.edu.cn (H.Y.); hangj3@mail.sysu.edu.cn (J.H.)
2 Key Laboratory of Tropical Atmosphere-Ocean System, Ministry of Education, Sun Yat-sen University, Zhuhai 519000, China
3 Department of Mechanical Engineering, The University of Hong Kong, Pokfulam Road, Hong Kong SAR, China; qunwang@connect.hku.hk
4 School of Construction Management and Engineering, University of Reading, Whiteknights, Reading RG6 6AH, UK; z.luo@reading.ac.uk
* Correspondence: lingh23@mail.sysu.edu.cn; Tel.: +86-20-84112436
† These authors contributed equally to this work.

Abstract: Rapid urbanisation and rising vehicular emissions aggravate urban air pollution. Outdoor pollutants could diffuse indoors through infiltration or ventilation, leading to residents' exposure. This study performed CFD simulations with a standard k-ε model to investigate the impacts of building configurations and tree planting on airflows, pollutant (CO) dispersion, and personal exposure in 3D urban micro-environments (aspect ratio = H/W = 30 m, building packing density $\lambda_p = \lambda_f = 0.25$) under neutral atmospheric conditions. The numerical models are well validated by wind tunnel data. The impacts of open space, central high-rise building and tree planting (leaf area density LAD= 1 m^2/m^3) with four approaching wind directions (parallel 0° and non-parallel 15°, 30°, 45°) are explored. Building intake fraction <*P_IF*> is adopted for exposure assessment. The change rates of <*P_IF*> demonstrate the impacts of different urban layouts on the traffic exhaust exposure on residents. The results show that open space increases the spatially-averaged velocity ratio (*VR*) for the whole area by 0.40–2.27%. Central high-rise building (2*H*) can increase wind speed by 4.73–23.36% and decrease the CO concentration by 4.39–23.00%. Central open space and high-rise building decrease <*P_IF*> under all four wind directions, by 6.56–16.08% and 9.59–24.70%, respectively. Tree planting reduces wind speed in all cases, raising <*P_IF*> by 14.89–50.19%. This work could provide helpful scientific references for public health and sustainable urban planning.

Keywords: CFD simulation; ventilation; pollutant dispersion; open space; urban tree planting; personal intake fraction

Citation: Li, Q.; Liang, J.; Wang, Q.; Chen, Y.; Yang, H.; Ling, H.; Luo, Z.; Hang, J. Numerical Investigations of Urban Pollutant Dispersion and Building Intake Fraction with Various 3D Building Configurations and Tree Plantings. *Int. J. Environ. Res. Public Health* **2022**, *19*, 3524. https://doi.org/10.3390/ijerph19063524

Academic Editor: Cheng Yan

Received: 3 January 2022
Accepted: 12 March 2022
Published: 16 March 2022

1. Introduction

Rapid urbanisation has aggravated urban environmental problems over the past several decades. The rapidly increasing vehicular emissions in street networks deteriorate urban air quality and have become one of the main pollutant sources in modern cities [1–3]. Urban air pollutant exposure has induced rising risks of respiratory and cardiovascular diseases, or even premature mortality [4,5]. People spend more than 90% of their lifetime indoors, on average. Moreover, outdoor air pollutants can diffuse into the indoor environment by infiltration or ventilation via windows, vents, and so on. The indoor pollutant exposure is closely influenced by the outdoor air quality, especially for buildings with natural ventilation. Therefore, near-road residents usually suffer from much higher air pollutant exposure than those in other regions [4,6–8]. Special attention is required to develop sustainable urban designs to improve urban ventilation and reduce urban residents' exposure [9,10].

The urban canopy layer (UCL) represents the atmospheric layer from the ground to the building rooftops, where most urban residents live. To mitigate the pollutant exposure of residents in the UCL, improving the ventilation and pollutant dilution capacity is one of the major solutions [9,10]. Recently, field observations, computational fluid dynamic (CFD) simulations, and laboratory-scale physical modelling (wind tunnel or water channel experiments) have been widely employed to investigate the ventilation and pollutant dispersion at the street scale (~100 m) or neighbourhood scale (~1 km) [11–18]. Field observation can directly monitor the critical characteristics of air flow and dispersion in real cities, but is restricted by low spatial resolution, uncontrollable boundary conditions, and complicated building configurations [15]. Although laboratory-scale physical modelling techniques can control boundary conditions and building configurations well, and have been widely used to validate numerical models, they have to meet the similarity criteria requirements and the costs are relatively high [13,15,17,18]. Numerical modelling with a high temporal–spatial resolution turns out to be a more efficient and relatively low-cost tool to study the flow features and dispersion characteristics, but sometimes it has challenges in attaining satisfactory validation by experimental data [19–25]. In this work, CFD modelling is applied to simulate the flow field and pollutant dispersion at the street scale.

Key urban morphological parameters include street aspect ratio ($AR = H/W$, where H is the building height, W is the street width) [26,27], tree planting [28–30], the direction of approaching flow [31,32], building packing densities [33,34], building height variation [35,36], special building designs including open space [37–39] and high-rise building [40,41]. Previous studies have investigated their impacts on the urban ventilation and pollutant dispersion. Nevertheless, while most studies focused on the airflow and pollutant dispersion in the street canyon, the integrated impacts of different urban layouts on residents' exposure in three-dimensional (3D) urban models are still rare. Therefore, this study aims at evaluating the synthetic impacts of these urban parameters (urban open space, tree planting and central high-rise building in this work) on ventilation, pollutant dispersion and related human exposure. The work provides a scientific reference and effective methodologies for sustainable urban design and public health.

2. Methodology

2.1. Definition of Crucial Parameters

2.1.1. Velocity Ratio (*VR*)

Velocity ratio (*VR*) is used throughout the work to normalise and quantify the wind environment experienced by pedestrians [11], defined by Equation (1):

$$VR = V_p / V_\delta \tag{1}$$

where V_p is the wind velocity at the pedestrian level (z = 2 m) and V_δ is the wind velocity at the top of the boundary layer. Here, V_δ = 4.34 m/s at z = 300 m [39].

2.1.2. Building Intake Fraction <*P_IF*>

The concept of the building intake fraction <*P_IF*>, derived from the personal intake fraction *P_IF*, is employed to evaluate the impacts of urban layouts on the residents' pollutant exposure. *P_IF* represents the total pollutant inhalation per person, which is widely adopted to quantify the indoor and street-scale (~100 m) vehicular pollutant exposure [42–44].

The intake fraction (IF) for a certain population is defined in Equation (2) [45–48]:

$$IF = \sum_{i}^{N}\sum_{j}^{M} P_i \times Br_{i,j} \times \Delta t_{i,j} \times Ce_j / m \quad (2)$$

where N stands for the total number of population age groups considered in this research; M is the total number of micro-environment types; P_i is the number of the population in the age group i; $Br_{i,j}$ (m^3/s) is the volume-mean breathing rate for individuals of the age group i in the micro-environment group j; $\Delta t_{i,j}$ is the time that group i stays in the micro-environment j; Ce_j (kg/m^3) is the time-averaged concentration C of the certain vehicular pollutant in the micro-environment j; and m (kg) is the total emission of the vehicular pollutant over the research period.

The population in the research is divided into three groups (N = 3) according to Luo et al. [46]. The composition of the target population is: children (<18, i = 1), adults (18–60, i = 2) and elders (>60, i = 3). Chau et al. [49] considered four types of micro-environment (M = 4) in their work, including indoor at home (j = 1), other indoor locations (j = 2), near vehicles (j = 3) and other outdoor locations (j = 4). To simplify the computation, only one micro-environment (j = 1, indoor at home) is considered in this study, and the buildings are assumed to be the residential type with natural ventilation. As Table A1 in the supplement shows, the percentage of children, adults and elders are 21.2%, 63.3% and 15.5%, respectively, in this paper. The breathing rate Br and time percentage spent indoors at home for different age groups is 12.5 and 61.70% (children), 13.8 and 59.50% (adults) and 13.1 and 71.60% (elders), respectively [46,49,50].

IF has been used to express the source-to-intake relationship for vehicular pollutants in realistic street canyons [48]. Since IF would change linearly with the variation of the population, it has been optimised by defining personal intake fraction P_IF in Equation (3) [26,42]. P_IF is independent of population size and density, and it represents the average IF for each person.

$$P\_IF = IF / \sum_{j}^{M} P_i \quad (3)$$

To estimate the influence of urban layouts on personal exposure, P_IF is employed in this work to quantitatively evaluate the pollutant inhalation of residents.

Previous researchers [6,7] found that the ratio of indoor and outdoor pollutant concentration (I/O) was approximately 1 for buildings with natural ventilation. Therefore, to reduce the number of grids and computational resources, the inner space of buildings is not considered in simulations. The pollutant concentrations on building surfaces are adopted as the indoor concentrations in this work, and the vehicle emission is assumed as the only source of the indoor environment [26,43,45]. Building intake fraction $<P\_IF>$ is the spatial mean of P_IF at all building surfaces. Throughout this work, $<P\_IF>$ is used to present the spatially-averaged personal intake exposure for the whole urban area. The change rate of $<P\_IF>$ could represent the varied exposure risks for the indoor residents due to the impacts of different urban layouts.

2.2. Set-Up for Numerical Modelling

CFD simulation has been widely used for urban micro-climate research in recent decades [51–55]. Compared with Reynolds-averaged Navier–Stokes (RANS) approaches, large eddy simulations (LES) are more accurate in simulating and predicting turbulence [54,56–60]. However, LES models need enormous computational resources. Thus, RANS models are still widely applied for turbulence simulation [61–65]. Among the RANS models, the standard k-ε model has remarkable performance in predicting urban airflows and pollutant dispersion [27,33,44,66–68]. In this paper, the Ansys FLUENT 15.0 with standard k-ε model is applied for airflow simulations under isothermal conditions.

2.2.1. CFD Model Description

The governing equations of mass, momentum, turbulent kinetic energy (k) and its dissipation rate (ε) of the employed CFD model are shown in Equations (4)–(7), as follows: the mass conservation equation:

$$\frac{\partial \overline{u_i}}{\partial x_i} = 0 \quad (4)$$

the momentum equation:

$$\overline{u_j}\frac{\partial \overline{u_i}}{\partial x_i} = -\frac{1}{\rho}\frac{\partial \overline{p}}{\partial x_i} + \frac{\partial}{\partial x_j}\left(v\frac{\partial \overline{u_i}}{\partial x_j} - \overline{u_i{}'' u_j{}''}\right) \quad (5)$$

the transport equations of turbulent kinetic energy (k) and its dissipation rate (ε):

$$\overline{u_i}\frac{\partial k}{\partial x_i} = \frac{\partial}{\partial x_i}\left[\left(v + \frac{v_t}{\sigma_k}\right)\frac{\partial k}{\partial x_i}\right] + \frac{1}{\rho}P_k - \varepsilon \quad (6)$$

$$\overline{u_i}\frac{\partial \varepsilon}{\partial x_i} = \frac{\partial}{\partial x_i}\left[\left(v + \frac{v_t}{\sigma_\varepsilon}\right)\frac{\partial \varepsilon}{\partial x_i}\right] + \frac{1}{\rho}C_{\varepsilon 1}\frac{\varepsilon}{k}\left(P_k + C_{\varepsilon 3}G_b\right) - C_{\varepsilon 2}\frac{\varepsilon^2}{k} \quad (7)$$

where $\overline{u}_j$ stands for time-averaged velocity components ($\overline{u}_j = \overline{u}, \overline{v}, \overline{w}$ as j = 1, 2, 3); v is the kinematic viscosity; and v_t is the kinetic eddy viscosity ($v_t = C_\mu \frac{k^2}{\varepsilon}$). The constant C_μ is 0.09. $-\overline{u''{}_i u''{}_j} = v_t\left(\frac{\partial \overline{u_i}}{\partial x_j} + \frac{\partial \overline{u_j}}{\partial x_i}\right) - \frac{2}{3}k\delta_{ij}$ is the Reynolds stress tensor. δ_{ij} is the Kroneker delta. $\delta_{ij} = 1$ when $i = j$ and $\delta_{ij} = 0$ otherwise. $P_k = v_t \times \frac{\partial \overline{u_i}}{\partial x_j}\left(\frac{\partial \overline{u_i}}{\partial x_j} + \frac{\partial \overline{u_j}}{\partial x_i}\right)$ is the turbulence production term.

The SIMPLE scheme is applied for coupling pressure and velocity. The under-relaxation factors for pressure term, momentum term, k and ε terms are 0.3, 0.7, 0.5 and 0.5, respectively. When all the absolute residuals are smaller than 10^{-6}, the iteration is converged.

2.2.2. Model Set-Up and Boundary Conditions

The 3D idealised full-scale UCL model with neutral atmosphere conditions is adopted in this study. The whole UCL model has a 5 × 5 building matrix composed of 25 cubic models ($H = B = W$ = 30 m) with moderate packing density (aspect ratio H/W = 1, building packing density $\lambda_p = \lambda_f = 0.25$). The designed UCL model is an idealised typical urban residential area in miniature, especially relating to the communities in small and medium-sized towns, or the communities in the old town of modern cities. To evaluate the impacts of approaching winds with different directions, θ (the included angle of the approaching wind and axis x) is set as 0°, 15°, 30° and 45° for every scenario. The setup of the simulation domain is shown in Figure 1a,b.

Figure 1a depicts the simulation area of the cases with parallel approaching wind (θ = 0°), with the geometry of 1700 m (x) × 870 m (y) × 300 m (z). The distances from the UCL model to the domain inlet, outlet and lateral boundaries are 6.7H, 41H and 10H, respectively. The symmetry boundary condition is adopted at the domain top and the lateral boundaries, while the domain outlet takes the zero normal gradient boundary condition [39,69,70].

The domain geometry of the cases with non-parallel approaching wind (θ = 15°, 30° and 45°) is 1700 m (x) × 1700 m (y) × 300 m (z) (Figure 1b). In this condition, the distances from the UCL model to the domain inlets and outlets are 6.7H and 41H, respectively. The symmetry boundary condition is only adopted at the domain top.

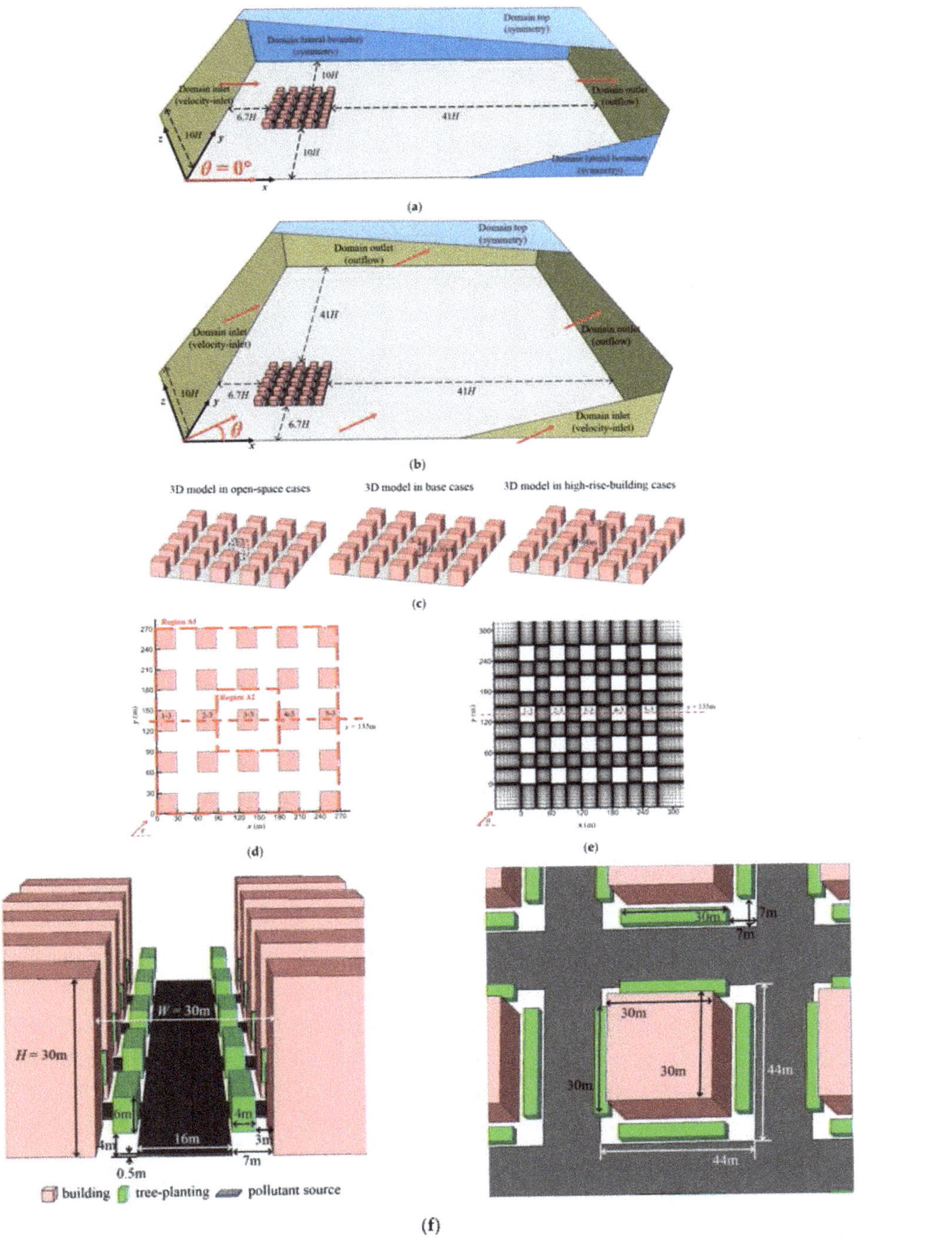

Figure 1. Computational domain of (**a**) Case [Base, 0°] and (**b**) Case [Base, θ] (θ = 15°, 30°, 45°). (**c**) 3D model description of open-space cases, base cases and high-rise-building cases. (**d**) Model description and (**e**) grid arrangements from top view in base cases. (**f**) Setups of building, tree planting and pollutant source.

The boundary condition for the domain inlet is provided by Equations (8)–(10) [27,67,71,72]:

$$U_{in}(z) = U_{ref} \times (z/H)^{0.16} \tag{8}$$

$$k_{in}(z) = u^{*2}/\sqrt{C_\mu} \tag{9}$$

$$\varepsilon_{in}(z) = C_\mu^{\frac{3}{4}} k_{in}^{\frac{3}{2}}/(\kappa_v z) \tag{10}$$

where at the building height (H = 30 m), the reference velocity U_{ref} is 3 m/s. The friction velocity u^* is 0.24 m/s. The von Kármán constant κ_v is 0.41. The empirical constant C_μ is 0.09.

Six building configurations are considered in the current study (Table 1). The UCL model of base cases is a building matrix with 5-row and 5-column blocks (Figure 1c). For the study of the open-space effect on city ventilation and human exposure, we remove the central building (Building 3-3) to obtain the open space (Location 3-3). For high-rise building scenarios, we double the height of Building 3-3. Each type of building configuration is combined with tree-planting and tree-free types. According to the settings above, the cases are named "Case [urban layout, wind direction]", as shown in Table 1.

Table 1. Scenarios tested in this work. Different urban designing and wind direction are considered for the assessment.

Building Arrangement	Vegetation Planning	Wind Direction (θ)	Case Name
Base	Tree-free	0°, 15°, 30°, 45°	[Base, θ]
	With tree	0°, 15°, 30°, 45°	[Base-tree, θ]
Open space	Tree-free	0°, 15°, 30°, 45°	[Open, θ]
	With tree	0°, 15°, 30°, 45°	[Open-tree, θ]
High-rise building	Tree-free	0°, 15°, 30°, 45°	[High, θ]
	With tree	0°, 15°, 30°, 45°	[High-tree, θ]

For the comprehensive data analysis, we present the results both in the entire UCL Region A1 (5 × 5 building matrix) and in the Region A2 (the central area of Region A1, including Building 3-3/Location 3-3 and the surroundings), as shown in Figure 1d. For all tested cases, the minimum size of the hexahedral cells near wall surfaces is 0.2 m. The total number of grids ranges from approximately 5 million and 15 million for the tree-planting and tree-free cases, respectively. Figure 1e illustrates the grid arrangement from the top view and side view. This grid arrangement is sufficient to ensure the requirement recommended by CFD guidelines [69,70].

2.2.3. Description of Pollutant Dispersion Modelling

Scientists have found that traffic emissions have critical negative impacts on respiratory and cardiovascular function [73,74]. The situation in Asian cities might be more severe due to the very high population density and more residents living in close proximity to road traffic compared with those in European cities. To model the dispersion of traffic-related pollutants in the street canyon, reactive gaseous as primary particles [75], inert gas like CO [76], as well as reactive gaseous pollutants such as NO_x and VOCs [77] are adopted in the simulation for investigating the dispersion of traffic-related air pollutants. Unlike various studies focusing on the reactive compounds from traffic such as NO_x, VOCs and particles, we adopted monoxide (CO) as an indicator of the traffic emissions. Although NO_x, VOCs and particles have more significant health impacts than CO, these compounds are more or less chemically or photo-chemically reactive, which means they could not be used as a tracer for the variation of the traffic emissions. To numerically investigate the impacts of various urban layouts on the physical dispersion of the traffic-related pollutants and related human exposure, a stable indicator is needed. As one of the main inert pollutants, CO has been widely used as a tracer of traffic emissions [42,46,48]. This study mainly focuses on the dynamic dispersion of traffic pollutants influenced by different urban layouts. Deposition and chemical reactions are not considered. The CO emission source is settled from $z = 0$ m to 0.5 m, with a width of 16 m and is 7 m away from the kerbside building (marked with dark grey colour in Figure 1f).

The governing equation of time-averaged CO concentration C (kg/m^3) is applied as Equation (11):

$$\overline{u_j}\frac{\partial C}{\partial x_j} - \frac{\partial}{\partial x_j}\left[(D_m + D_t)\frac{\partial C}{\partial x_j}\right] = S \tag{11}$$

where $\overline{u_j}$ is the time-averaged velocity component in the direction of j. D_m and D_t are molecular diffusivity and the turbulent diffusivity of the pollutant. $D_t = \nu_t / S_{Ct}$, while ν_t is the kinematic eddy viscosity, and S_{Ct} is the turbulent Schmidt number. It is a parameter describing an important property of the flow defined as the ratio of the eddy diffusivity of momentum to the eddy diffusivity of mass. Di Bernardino et al. [78] found that S_{Ct} increased with the height above the canopy, with the maxima of about 0.6 in their water-channel experiments and simulations. In our preliminary work, we found that modelling with $S_{Ct} = 0.7$ had the best performance compared with the wind tunnel experiment results from Gromke and Blocken [63]. Thus, $S_{Ct} = 0.7$ is used throughout this work. The CO emission rate S is set as 1.25×10^{-6} kg/m^3/s, derived from a field observation campaign in a real street of Hong Kong [79]. Such an emission source setting has been adopted in CFD simulations for many studies of urban pollutant dispersion [10,27,42,80].

2.2.4. Description of the Vegetation Modelling

Tree planting is simulated as a series of cubic blocks on both sides of the streets over the whole urban area, except the surroundings of Building 3-3 (Figure 1f). Since the size of the tree trunk is much smaller than that of the crown, the impact of the trunk on the airflow is assumed to be negligible, therefore only the crowns are simulated in the tree-planting cases. According to Yang et al. [80], the *y-density* is set to 1, which means that the tree crown is continuous in the y-direction. As shown in Figure 1f, the scale of the crown cubic is designed as 4 m × 6 m × 30 m. The distance between the crown bottom and the ground surface is 4 m, and that between the crown and the adjacent building wall is 3 m.

Differing from solid obstacles such as buildings, the airflow can pass through the tree crown from spaces in between the branches and leaves. Previous studies found that vegetation models with the porous medium for airflow and pollutant dispersion were more consistent with the wind tunnel experimental results than those with a non-porous medium [81–83]. Accordingly, we adopt the porous fluid zones to simulate tree planting, and the governing equations are listed in Equations (12)–(14).

$$S_{\overline{u_i}} = -\rho C_d LAD \overline{u_i} \boldsymbol{U} \tag{12}$$

$$S_k = \rho C_d LAD \left(\beta_p \boldsymbol{U} \overline{u_i}^3 - \beta_d \boldsymbol{U} k\right) \tag{13}$$

$$S_\varepsilon = \rho C_d LAD \frac{\varepsilon}{k}\left(C_{\varepsilon 4}\beta_p \boldsymbol{U}^3 - C_{\varepsilon 5}\beta_d \boldsymbol{U} k\right) \tag{14}$$

where $S_{\overline{u_i}}$, S_k, S_ε are the additional source and sink terms of momentum, turbulent kinetic energy and turbulent dissipation rate for trees, respectively. ρ (kg/m^3) is the air density. C_d is the leaf drag coefficient, ranging from 0.1 to 0.3, which is related to the tree species. In this paper, we adopt the commonly used empirical value of $C_d = 0.2$ to avoid species particularity [81]. *LAD* (m^2/m^3) is the leaf area density, which represents the one-side leaf area per unit volume of the crown [80,84]. It is related to tree species and crown-height variations, and ranges between 0.5 and 2.0 m^2/m^3 for deciduous trees. To simplify the model, we supposed that the trees in the simulation domain were all deciduous trees with a homogeneous crown height, and thus set the *LAD* value to 1 m^2/m^3 for the simulation. $\overline{u_i}$ is the time-averaged velocity component on direction i, and $\boldsymbol{U}$ is the magnitude of the velocity. β_p is the portion of turbulent kinetic energy converted from mean kinetic energy under the influence of drag, and β_d is the dimensionless coefficient of the Kolmogorov cascade. We adopt β_p as 1.0, and β_d as 5.1, according to [81,82], respectively. Both $C_{\varepsilon 4}$ and $C_{\varepsilon 5}$ are empirical constants of 0.9.

2.2.5. Validations for Flow, Dispersion and Vegetation Modelling

The direct validation for the CFD model of real urban areas is difficult due to the very limited field observation data. The uncontrollable boundary condition is another challenge for repetitive experiments [85]. However, the wind tunnel experiment is a credible solution for model validation if the Reynolds number (*Re*) independence is satisfied (*Re* >> 11,000) [86–88]. We have implemented a series of comprehensive validations for the flow, the dispersion and the vegetation model applied throughout this work, based on the published wind tunnel experiment datasets [86,89,90]. Similar validation methods have been employed and proven valid in the literature [43,80,88].

- *Flow validation by wind tunnel tests of cubic arrays*

Figure A1 presents the results of the flow validation. The UCL model (moderate building density) with a 7 × 11 building matrix was used in the employed wind tunnel dataset [86]. The size of each building model is $H = B = W = 15$ cm. The measuring points for the vertical profiles of the stream-wise velocity ($\overline{u}$) and turbulence kinetic energy (k) are set in the centre of each street, named Pi (i = 1–6) (Figure A1a,b). As we described in Section 2.2.2, similar model configurations are set for the case studies: at full scale with a scale ratio of 200:1 ($H = B = W = 30$ m) to the wind tunnel scale. All settings are similar, except the length from the urban boundary to the domain outlet. Referring to the reference velocity (U_{ref} = 3 m/s) and the model geometry (H = 0.15 m or 30 m), *Re* is approximately 3×10^4 and 6×10^6 at the wind tunnel scale and full scale, satisfying the requirements for Reynolds number independence.

Figure A1c–j tests the grid independence (with a minimum grid size of 0.4 m, 0.2 m and 0.1 m) and the performance of different turbulence models (standard k-ε, RNG k-ε and realisable k-ε models) with standard wall function. The results illustrate that differences generated by the mesh setting are negligible. Thus, the moderate grid size (0.2 m for minima) is applied for all cases to save computational resources. Furthermore, the depicted vertical profiles of $\overline{u}$ and k verify that modelling using the standard k-ε model has better agreement with the wind tunnel data than those using the RNG k-ε and realisable k-ε models. Important statistics are summarised in Table A2, including the normalised mean square error (NMSE), fractional bias (FB) and correlation coefficient (R). The results denote that the standard k-ε model is employed in the study throughout this work.

- *Pollutant dispersion validation by wind tunnel tests without tree models*

Wind tunnel experiment data of inert gas dispersion [89] is employed in our work for the validation of pollutant dispersion. The configurations of the experiment are illustrated in Figure A2a,b. The UCL model consists of a 3 × 3 model matrix, with each prism size being $B_x \times B_y \times H$ = 27.6 cm × 18.4 cm × 8 cm. Inert gas C_2H_6 is used as the tracer, emitting from the line source (L = 18.8 cm, d_x = 0.5 cm) settled in the UCL model area. Similar model settings at full scale ($B_x \times B_y \times H$ = 138 m × 92 m × 40 m) are configured for the simulation to validate the dispersion. Fitting vertical profiles of monitored $\overline{u}$, k and ε in the wind tunnel experiments [89] are set for the domain inlet. The standard k-ε model and standard wall function are adopted in the simulation. Since the tracer gas concentration provided by the wind tunnel experiment is in a non-dimension form, the normalised concentration K [89] is derived referring to Equation (15), in convenience for comparing the experimental data and simulation results.

$$K = C \cdot H \cdot U_{ref} / E \cdot d_x \tag{15}$$

where C is the inert gas concentration, and the emission rate E is 0.01 m/s. Important statistics are summarised in Table A3, including the NMSE, FB and R. The good agreement between the results of the wind tunnel experiment and the CFD simulation confirm that the selected turbulence model and wall function is appropriate for evaluating the pollutant dispersion in our work.

- *Pollutant dispersion validation by wind tunnel tests with tree models*

The validation for vegetation modelling in this paper is performed on the basis of the wind tunnel experiment conducted by Gromke and Ruck [90]. The configurations and boundary conditions are set as in Figure A3a. The 2D street canyon is constructed by two parallel building models, with the same sizes for both in the wind tunnel experiment and the CFD simulation ($L \times W \times H$ = 1.2 m × 0.12 m × 0.12 m). The standard k-ε model and standard wall function coupled with the porous crown model (details in Section 2.2.4) are adopted in the simulation. The vertical profiles of $\overline{u}$, k and ε for the domain inlet are provided by Gromke and Ruck [90]. U_{ref} = 4.7 m/s is applied; thus, the reference Re is 38,630 >> 11,000. The normalised concentration K of the inert gas (SF_6) is used to compare the results of the wind tunnel experiment and the simulation. Vertical profiles of K are presented in Figure A3b,c. Important statistics are summarised in Table A4, including the NMSE, FB and R. In general, the results satisfy the recommended criteria [91,92], except that of y/H = 2 at the leeward wall. Nevertheless, as we focus on the pollutant dispersion in the 2D street canyon, the results of y/H = 0 (central region of the canyon with fully developed turbulence) are more representative of the pollutant distribution features. Good agreements of these statistics at y/H = 0 also confirm the modelling accuracy and reliability of the simulations. These results verify that the porous crown model with the standard k-ε model and standard wall function has good performance, and is suitable for studying the tree-planting effects in this work.

3. Results

3.1. Impacts of Building Configurations and Tree Planting on Flow Pattern

3.1.1. Impact of Open Space and High-Rise Building on Airflow

Figure 2a,c,e presents the streamlines and velocity ratio (VR) at the pedestrian level (z = 2 m) of tree-free cases under the parallel approaching wind, named Case [Base, 0°], Case [Open, 0°] and Case [High, 0°]. It shows the impacts of open space and high-rise buildings on the ambient airflows. Both building configurations are found to apparently change the structure of the vortex and airflow, especially within Region A2. VR in the downstream regions of Case [Open, 0°] is slightly strengthened compared with Case [Base, 0°]. However, the mean wind speed of Case [Base, 0°] and Case [Open, 0°] in the whole region are at the same level. New vortices are generated around Region A2 with the addition of the open area. Comparing to the base cases, VR value increases significant0ly with the addition of high-rise building (Case [High, 0°]), especially in Region A2. The flow structure is changed in the whole urban area, and new vortices are found near the leeward wall or lateral wall of most buildings.

Figure 3 displays a detailed flow field in Region A2 at z = 2 m. As the axis of symmetry of the airflow field is the axis of y =135 m, Figure 3 only shows the streamlines in half of the zone. The urban design of open space decreases the wind speed and complicates the recirculation region (Figure 3c). On the contrary, the wind speed in Region A2 is strongly enhanced by the high-rise building (Figure 3e). The VR in Region A2 increases obviously compared with the base case, with a maximum value increase of 0.20. Both building layouts could deform the structure of the wind field around the building. Small vortices are formed near the windward side of the open space and the leeward side of the high-rise building.

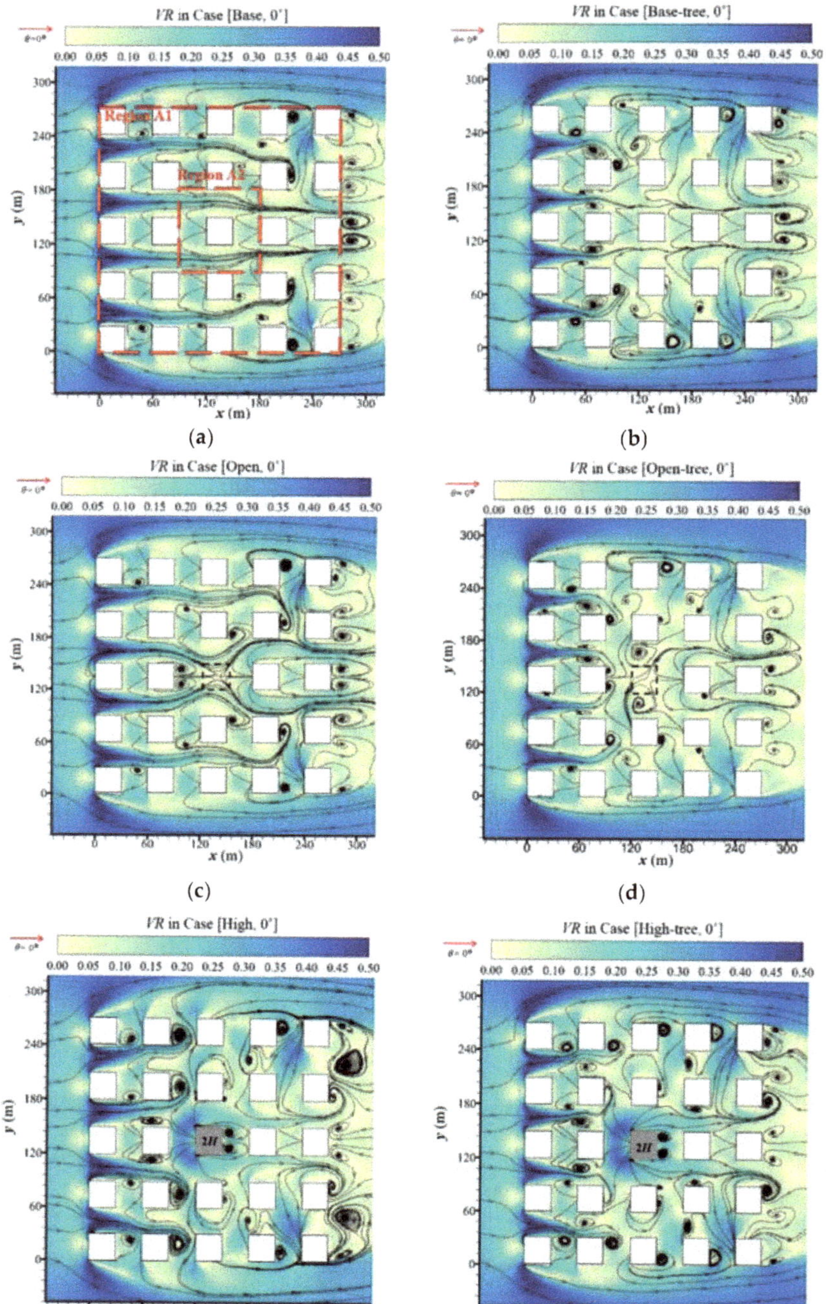

Figure 2. Streamline and velocity ratio (*VR*) at z = 2m in (**a**) Case [Base, 0°], (**b**) Case [Base-tree, 0°], (**c**) Case [Open, 0°], (**d**) Case [Open-tree, 0°], (**e**) Case [High, 0°] and (**f**) Case [High-tree, 0°].

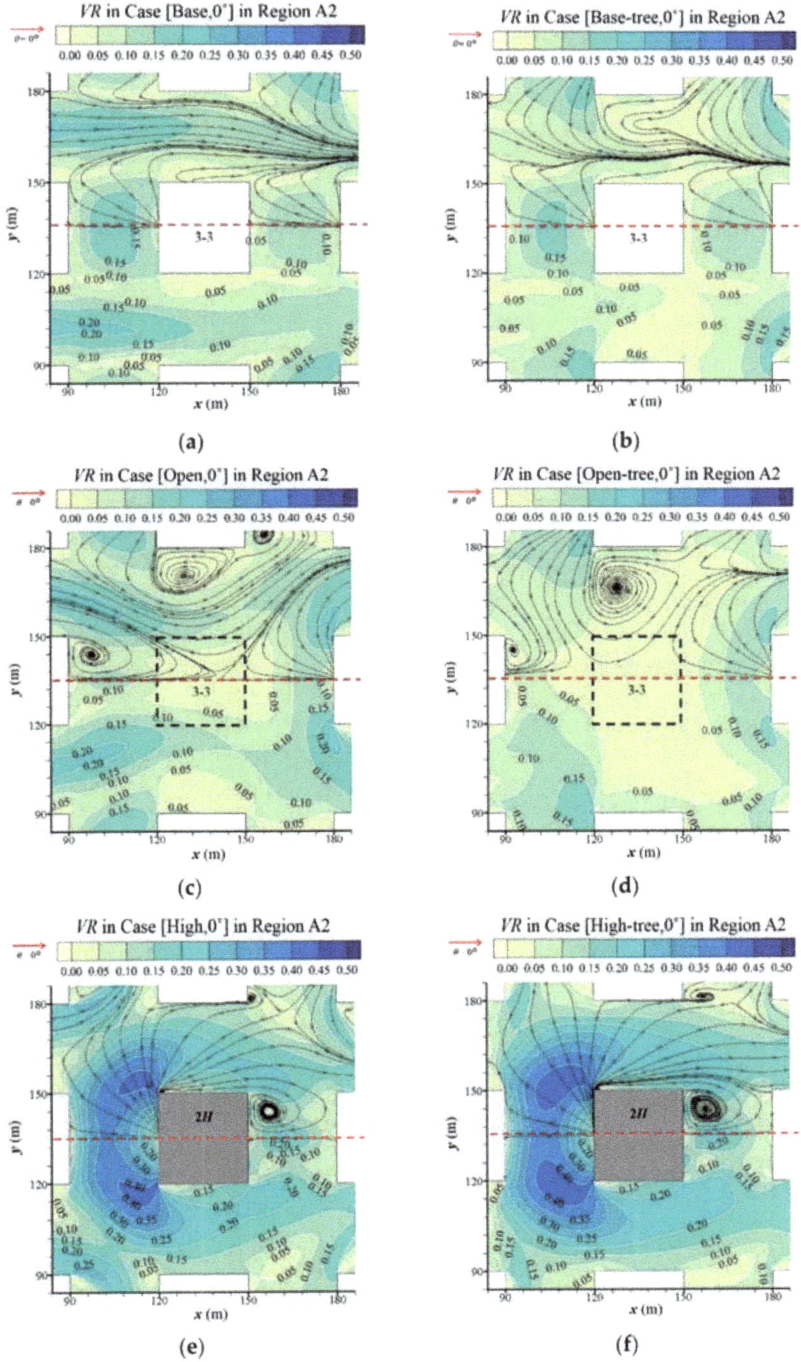

Figure 3. Streamline and velocity ratio (*VR*) at z = 2m in Region A2 in (**a**) Case [Base, 0°], (**b**) Case [Base-tree, 0°], (**c**) Case [Open, 0°], (**d**) Case [Open-tree, 0°], (**e**) Case [High, 0°] and (**f**) Case [High-tree, 0°].

To determine the impact of building configurations on the airflow and the vortices' structure, Figure 4a,c,e depicts the airflow and streamlines at the central *x*-*z* plane (y = 135 m) with the parallel approaching wind of Case [Base, 0°], Case [Open, 0°] and Case [High, 0°]. The designs of the open space and high-rise buildings have a significant influence on the geometry and the structures of the vortices in the street canyons, leading to obvious changes in the flow field. With the open space (Figure 4c), the vortex on the right side of Location 3-3 extends and occupies the space where Building 3-3 was. The vortex on the left side of Location 3-3 is compressed and the centre of the vortex rises up. Meanwhile, weakened airflows are observed compared with base case (Figure 4a). On the contrary, with the high-rise building (Figure 4e), the wind speed is significantly enhanced, especially near the windward side of the high-rise building. Compared with the base case, the former vortex on the left side of Building 3-3 is concentrated to one-half of the original size and the right vortex has disappeared.

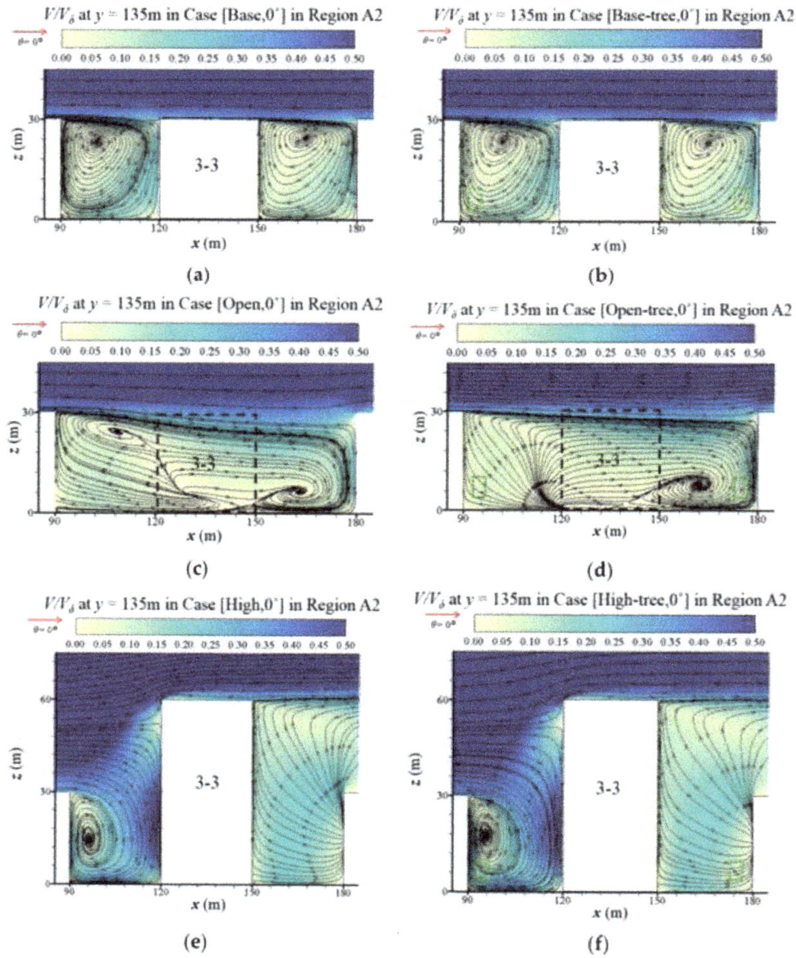

Figure 4. Normalised velocity (V/V_δ, V_δ = 4.34 m/s) at vertical plane (y = 135m) in Region A2 in (**a**) Case [Base, 0°], (**b**) Case [Base-tree, 0°], (**c**) Case [Open, 0°], (**d**) Case [Open-tree, 0°], (**e**) Case [High, 0°] and (**f**) Case [High-tree, 0°].

3.1.2. Influence of Tree Planting on Airflow

Figure 2b,d,f presents the flow field in the domain with tree planting. Whether for base cases (Figure 2a,b), open-space cases (Figure 2c,d) or high-rise building cases (Figure 2e,f), tree planting slightly changes the flow field in the entire domain (Region A1). Small vortices are generated over the whole domain, and the continuity of the flow is obstructed. However, the overall wind speed in Region A1 remains a similar value with the addition of vegetation under a parallel approaching wind.

This finding is further confirmed in Region A2, as presented in Figure 3. The wind vector around trees and street corners becomes denser and more complex. Tree planting slightly reduces the pedestrian-level wind speed in the central area of the base cases (Figure 3a,b) and open-space cases (Figure 3c,d). However, for high-rise building cases (Figure 3e,f), the VR value in Region A2 slightly increases with tree planting.

The vertical profiles of the flow field in Region A2 at y = 135 m are shown in Figure 4. We find that tree planting has a slight influence on the vertical airflow for all three building configurations. Compared with Case [Open, 0°], the two-vortices structure of the flow field is destroyed (Figure 4c,d). In the base cases (Figure 4a,b) and high-rise building cases (Figure 4e,f), the centre of the vortex on the left side of Building 3-3 rises up.

3.1.3. Quantitative Analysis of Impact of Different Urban Layouts on Velocity Field

To quantify the impacts of different urban layouts on the flow field, Figure 5a,b and Table A5 summarise the spatial mean *VR* at z = 2 m of Region A1 ($<VR>_{A1}$) and A2 ($<VR>_{A2}$). Cases with different approaching wind directions (θ = 0°, 15°, 30° and 45°) are discussed as well. For Case [Base, θ], both 2 m $<VR>_{A1}$ and $<VR>_{A2}$ with the approaching wind direction of θ = 0° are much lower than those of other wind directions (θ = 15°, 30° or 45°). The maximum $<VR>_{A1}$ (0.21) and $<VR>_{A2}$ (0.16) both appear at θ = 45°.

For Case [Open, θ], the designed open space leads to a slight increase of 2 m *VR* on the spatial mean in Region A1 (Figure 5a) by 0.40–2.27%. However, in Region A2 (Figure 5b), the 2 m *VR* values in the open-space cases are decreased in comparison with those in base cases under the non-parallel approaching wind (θ = 15°, 30° and 45°), by 8.40–12.06%. Both the 2 m $<VR>_{A1}$ and $<VR>_{A2}$ of the open-space cases increase with the rising θ, with a maximum value of 0.21 and 0.14 at θ = 45°.

For Case [High, θ], the *VR* at z = 2 m has been significantly enhanced in both Region A1 and A2 compared with Case [Base, θ]. The ambient airflow of the high-rise building ($<VR>_{A2}$) is enhanced by 52.78–119.05%, more strongly than that in the whole building matrix ($<VR>_{A1}$) by 4.73–23.36%.

The results in Figure 5a,b and Table A5 show that tree planting reduces the *VR* in both Region A1 and A2 for different building configurations with all four approaching wind directions (θ = 0°, 15°, 30°, 45°), except Case [Base-tree, 0°] and Case [High-tree, 0°]. Tree planting significantly decreases the urban wind speed at z = 2 m ($<VR>_{A1}$) on the basis of either open space or high-rise building designs, by 4.63–14.99% or 2.04–16.68%, respectively. The $<VR>_{A1}$ with non-parallel approaching wind directions (θ = 15°, 30°, 45°) is reduced more in comparison with parallel approaching wind (θ = 0°). Taking Case [High-tree, θ] as an example, $<VR>_{A1}$ is decreased by 6.87–16.68% compared with Case [High, θ] when $\theta \neq 0°$, more than the 2.04% when θ = 0°. These results are also consistent with the results presented in Figures 2–5. Contrary to the reductive effect of most tree-planting cases, tree planting increases the $<VR>_{A1}$ of Case [Base-tree, 0°] and $<VR>_{A2}$ of Case [High-tree, 0°] by 6.27% and 8.98%, respectively. This phenomenon still needs more discussion in the ongoing work.

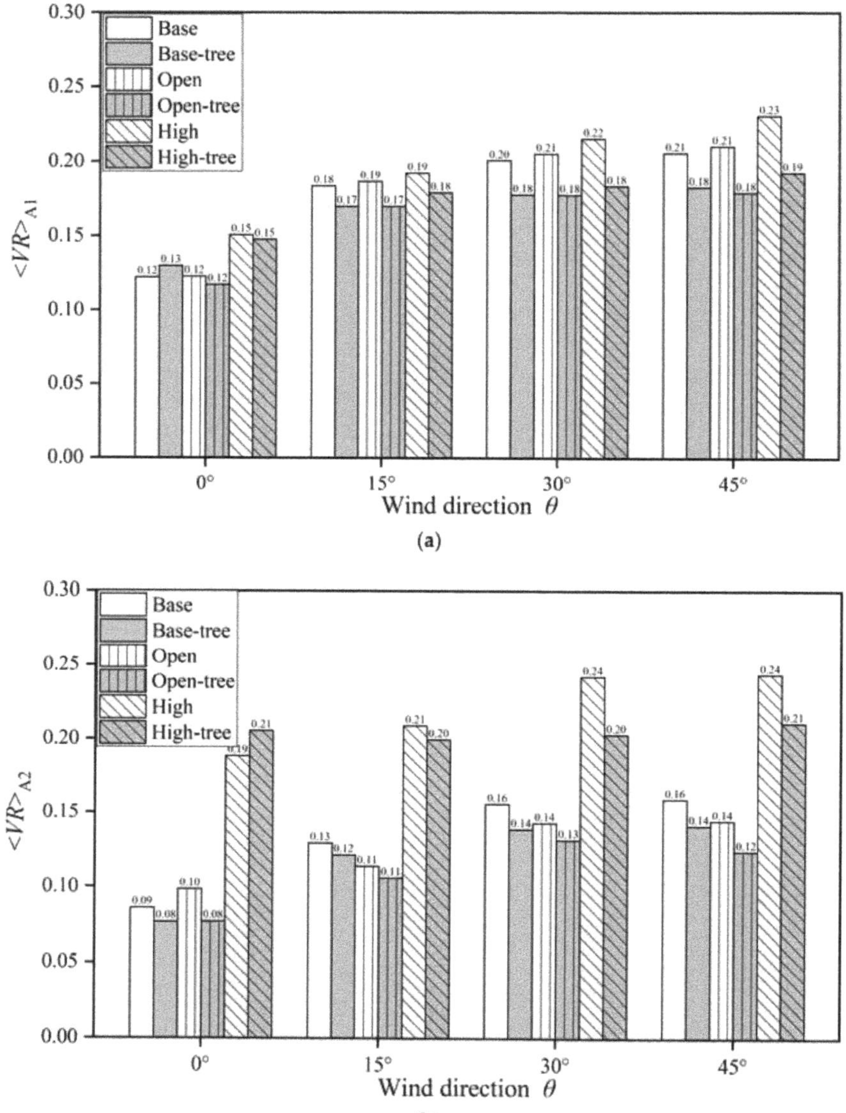

Figure 5. Spatially-averaged *VR* in different scenarios at *z* = 2 m in (**a**) Region A1 and (**b**) Region A2.

3.2. *Impacts of Building Configurations and Tree Planting on Pollutant Dispersion*

3.2.1. Influence of Open Space and High-Rise Building on Pollutant Dispersion

To investigate the impacts of building configurations on pollutant diffusion, Figure 6a,c,e presents the distributions of CO concentration (*C*) at the pedestrian level (*z* = 2 m) in tree-free cases. Overall, the level of *C* in the three tree-free cases is similar, but the regions of high *C* are affected by different building configurations. In general, open space enhances the CO accumulation on the leeward side of the buildings in the central and downstream areas. Nevertheless, the high-rise building enhances the CO accumulation in the central and upstream area significantly. However, CO in the central area is diluted to a quite low level due to the strongly strengthened wind velocity surrounding the high-rise building.

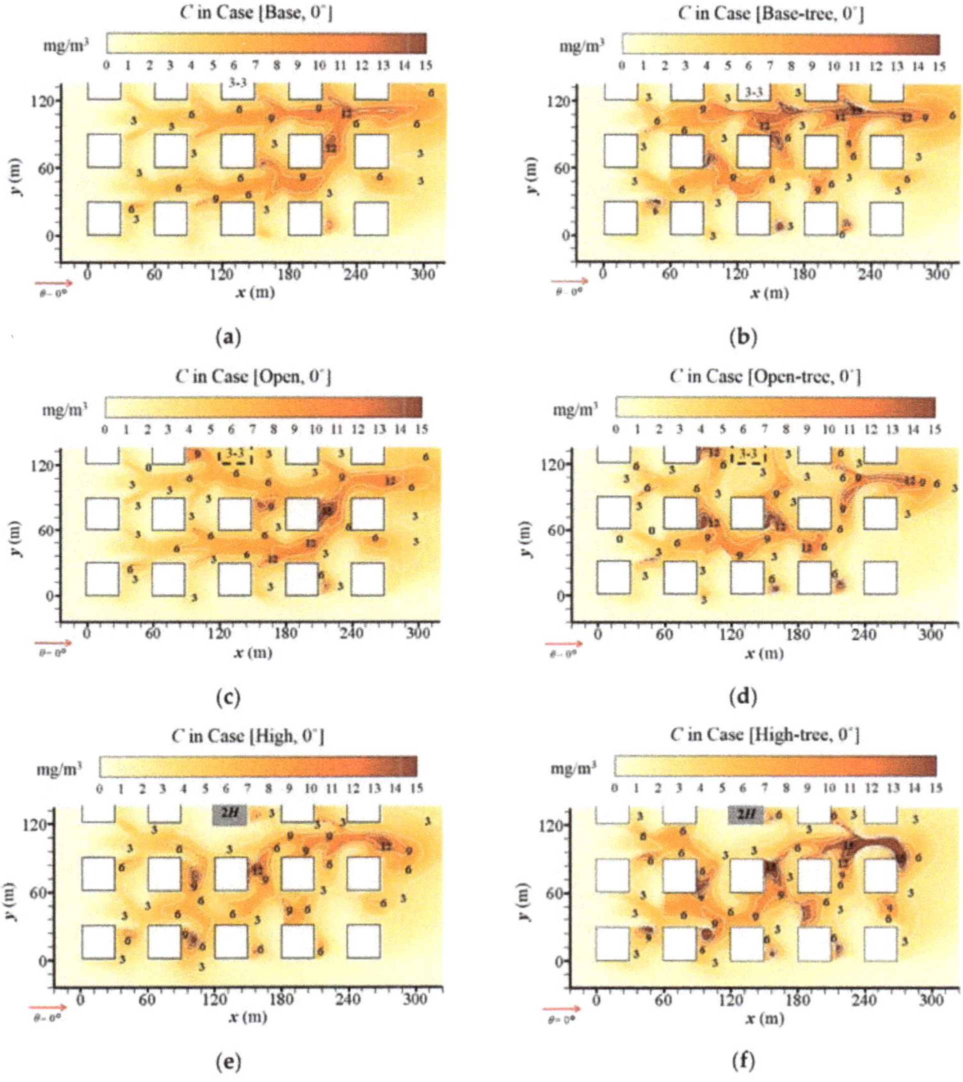

Figure 6. CO concentration (C) at z = 2 m in (**a**) Case [Base, 0°], (**b**) Case [Base-tree, 0°], (**c**) Case [Open, 0°], (**d**) Case [Open-tree, 0°], (**e**) Case [High, 0°] and (**f**) Case [High-tree, 0°].

To better understand how building configurations affect the pollutant dispersion in the central region (Region A2), Figure 7 shows the detailed vertical distribution of C at the x-z plane (y = 135 m). Comparing Figure 7c with Figure 7a, the low wind speed weakens the dilution and leads to high C levels in the open area. Particularly at the near-ground level of the upwind area, C is higher than 13 mg/m^3, while C in the same area of the base case is about 5 mg/m^3. In contrast, the strong airflow in the upwind of the high-rise building evidently decreases the C (Figure 7e). The near-ground C is decreased to about 2 mg/m^3. Additionally, the CO distribution in the upwind of the high-rise building is reduced to a very limited vertical range, while that in the downwind is expanded due to the existence of the high-rise building.

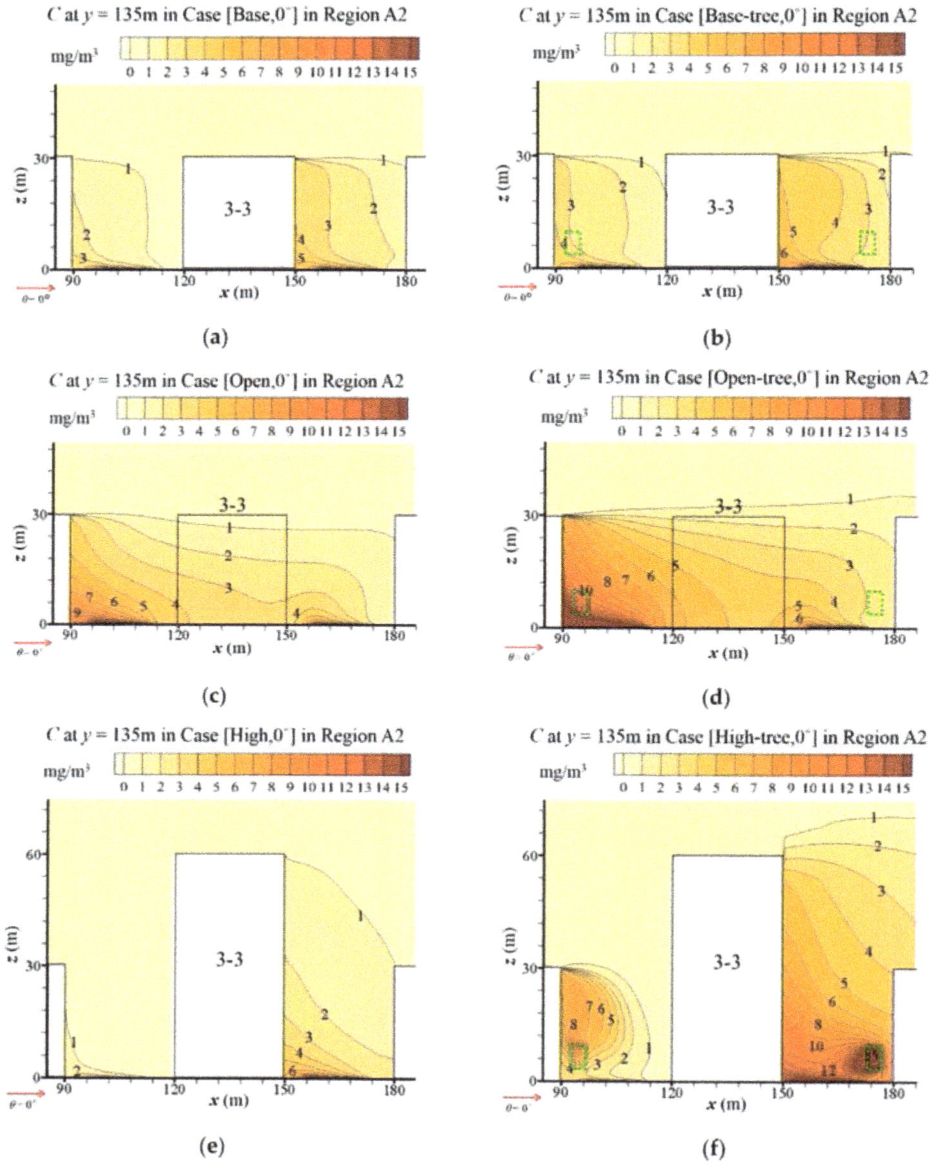

Figure 7. Vertical profile of C in Region A2 at y = 135 m: (**a**) Case [Base, 0°], (**b**) Case [Base-tree, 0°], (**c**) Case [Open, 0°], (**d**) Case [Open-tree, 0°], (**e**) Case [High, 0°] and (**f**) Case [High-tree, 0°].

3.2.2. Influence of Tree Planting on Pollutant Dispersion

Figure 6b,d,f illustrates the *C* distribution in the urban area with tree planting, coupled with the basic design, open-space and high-rise-building design, respectively. Comparing them with Figure 6a,c,e, a significant increase of *C* is found in the whole urban area under the tree-planting design, no matter which type of building configuration is considered. The CO dispersion is significantly weakened by trees in the whole domain, and new hotspots with high *C* appear.

Detailed vertical distributions of *C* surrounding Building 3-3 in the central region (Region A2) with different building configurations are illustrated in Figure 7b,d,f. Compared with Figure 7a,c,e, tree planting evidently increases the near-ground *C* on the leeward side of all buildings. Particularly in Case [Open-tree, 0°], the near-ground *C* in the upwind of the open space increases to higher than 15 mg/m^3. Trees around the high-rise building also have a significant influence on CO dispersion (Figure 7f). An area with a high *C* appears at the upper layer of the building wall on the upwind of Building 3-3, corresponding to the vortex of the flow field in Figure 4. At the downwind of Building 3-3, the ground-level *C* is higher than 15 mg/m^3. Moreover, another hotspot with *C* higher than 15 mg/m^3 appears in the area of the tree crown.

3.2.3. Quantitative Analysis for Impact of Building Configurations and Tree Planting on Pollutant Dispersion

Figure 8a,b and Table A6 summarise the mean CO concentration (*C*) at the pedestrian level (z = 2 m) in Region A1 ($<CO>_{A1}$) and A2 ($<CO>_{A2}$). The impacts of three building configurations, tree planting and wind directions (θ = 0°, 15°, 30° and 45°) are quantitatively assessed.

For the base cases, $<CO>_{A1}$ slightly rises from 4.14 mg/m^3 to 5.24 mg/m^3 with the θ varying from 0° to 45°. $<CO>_{A2}$, with the range of 4.87–7.11 mg/m^3, is higher than $<CO>_{A1}$ with the same direction of approaching flows. Both $<CO>_{A1}$ and $<CO>_{A2}$ decrease in the cases with open space and high-rise buildings (Figure 8a,b). For the open-space cases, $<CO>_{A1}$ and $<CO>_{A2}$ decrease rapidly with all four wind directions compared with the base cases, by 7.83–20.54% and 0.08–24.43%, respectively. For the high-rise building cases, the decrement ranges from 4.39% to 23.00% for $<CO>_{A1}$, and ranges from 43.88% to 47.40% for $<CO>_{A2}$, in comparison with base cases. The wind direction influences *C* more significantly for the area around the high-rise building (Region A2) than for the whole domain (Region A1).

For cases with tree planting, the results show that both the $<CO>_{A1}$ and $<CO>_{A2}$ evidently increase in all conditions with increasing rates of 2.84–31.88% and 2.85–35.46%, respectively. Taking θ = 0° as an example, $<CO>_{A1}$ increases by 20.19%, 22.44% and 12.61% in Case [Base-tree, 0°], Case [Open-tree, 0°] and Case [High-tree, 0°], compared with the tree-free cases. For the base cases, the largest increasing ratios of $<CO>_{A1}$ and $<CO>_{A2}$ both appear when θ = 0°. With tree planting, the CO concentration at the pedestrian level and in the central area ($<CO>_{A2}$) are higher than in the entire urban area ($<CO>_{A1}$) for both the base cases and open-space cases. The $<CO>_{A2}$ is particularly high in open-space cases with tree planting, with the values ranging from 6.59 mg/m^3 to 7.89 mg/m^3. For the high-building cases with tree planting, both $<CO>_{A1}$ and $<CO>_{A2}$ increase by 12.61–26.38% and 6.10–21.19%, respectively, compared with the tree-free cases. Regardless of the building configuration, the tree-planting design obviously weakens the dilution and dispersion capacity of pollutants and remarkably increases the CO concentration.

3.3. *Impacts of Building Configurations and Tree Planting on <P_IF>*

As illustrated in Section 2.1.2, we use <*P_IF*> to quantify the influence of urban layouts on personal exposure in UCL. As the buildings are assumed to be a residential type with natural ventilation, the pollutant concentration (C) on the building surfaces is adopted as the indoor concentration due to I/O $\approx$ 1 [6,7].

Figure 9a–f plots *C* on the building surfaces in different cases when θ = 0°. The *C* distribution is apparently influenced by different building configurations and tree planting, especially in the central area. Figure 10 and Table A7 summarise the parameter <*P_IF*> to specify and quantify the CO exposure under scenarios with base conditions, open space, high-rise building and tree planting. Four wind directions (θ = 0°, 15°, 30°, 45°) are also considered in the evaluation.

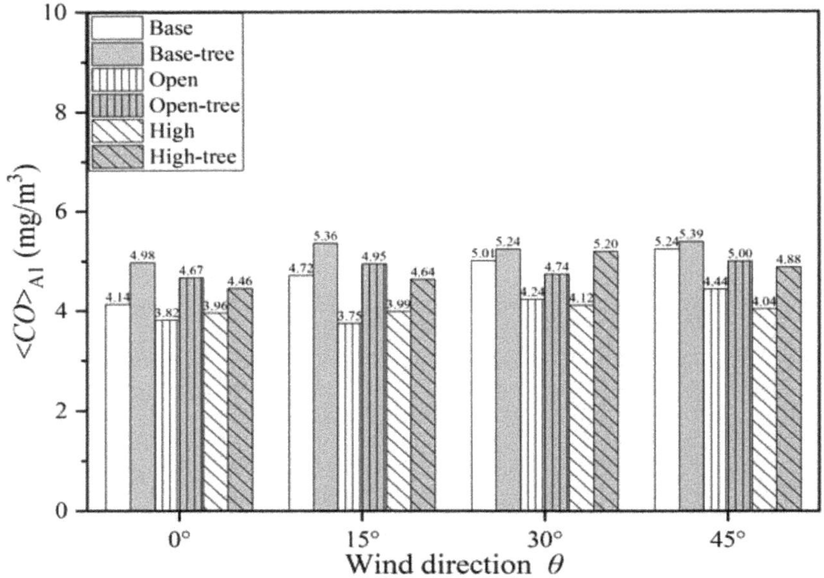

(a)

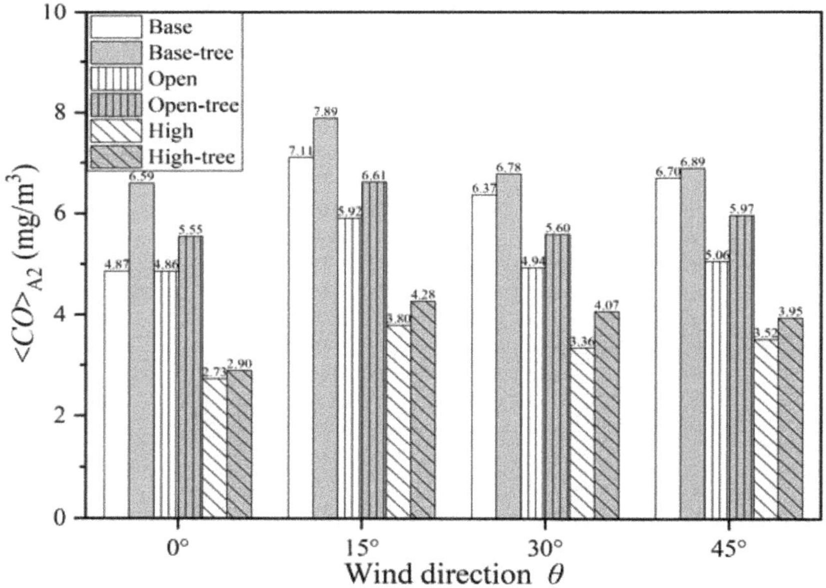

(b)

Figure 8. Spatially-averaged CO concentration in different scenarios at z = 2 m in (**a**) Region A1 and (**b**) Region A2.

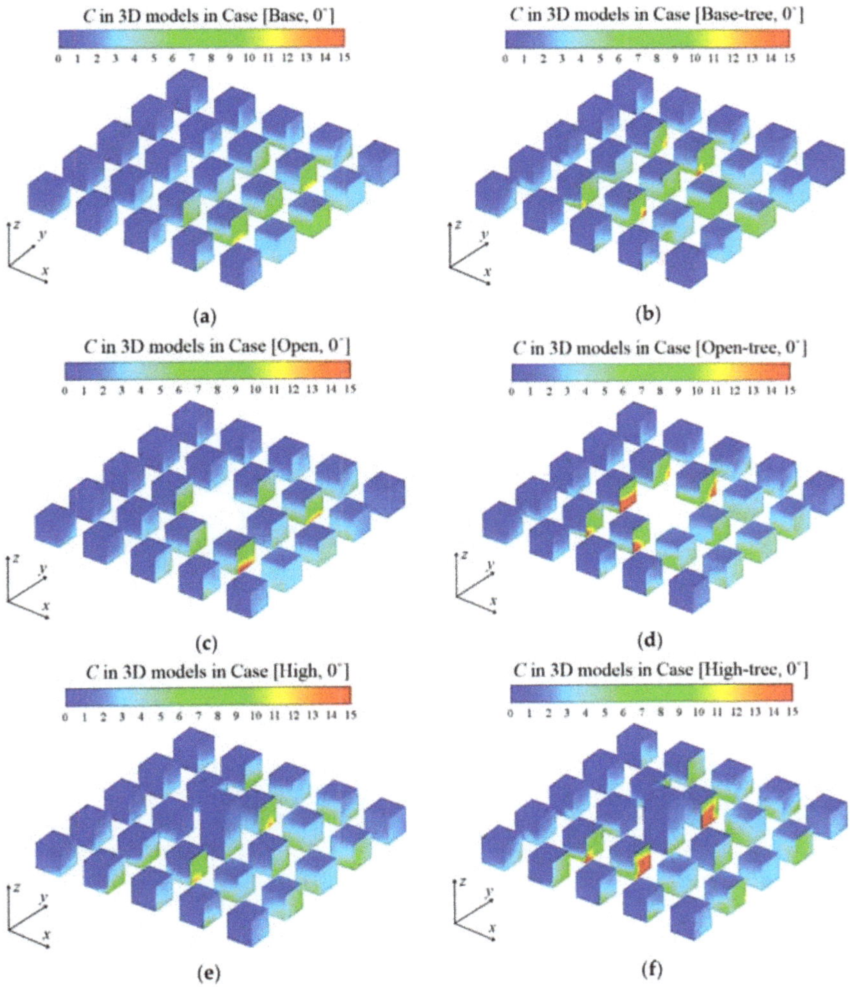

Figure 9. CO concentration (C) at building walls in 3D models: (**a**) Case [Base, 0°], (**b**) Case [Base-tree, 0°], (**c**) Case [Open, 0°], (**d**) Case [Open-tree, 0°], (**e**) Case [High, 0°] and (**f**) Case [High-tree, 0°].

As displayed in Figure 6a,c,e, the open space and high-rise building change the C distribution on the building surfaces. Both layouts make CO accumulate in the centre of the building matrix, while regions of high C in the base cases are in the downstream area of the approaching flow (Figure 9a–f). Comparing Figure 6b,d,f with Figure 6a,c,e, tree planting significantly increases the C on building surfaces, especially of the central 3 × 3 building matrix.

Similar results can also be found in Figure 10 and Table A7. For tree-free cases, both open space and high-rise building could decrease <*P_IF*> with the wind from all four directions by 6.56–16.08% and 9.59–24.70%, respectively. For different wind directions, the maximum <*P_IF*> of the tree-free cases always appears when θ = 45°, with <*P_IF*> = 2.53 ppm ([Base, 45°]), 2.19 ppm ([Open, 45°]) and 1.90 ppm ([High, 45°]), respectively. For cases with tree planting, the personal exposure in all building configurations (~2.05–2.90 ppm) is significantly increased compared with those of the tree-free cases (~1.54–2.53 ppm). The increasing ratio of <*P_IF*> ranges from 14.89% to 50.19% compared with tree-free cases. The maximum <*P_IF*> among all cases is 2.90 ppm, appearing in Case [Base-tree, 45°], while the maximum increasing ratio of <*P_IF*> is 50.19%, in Case [High-tree, 15°].

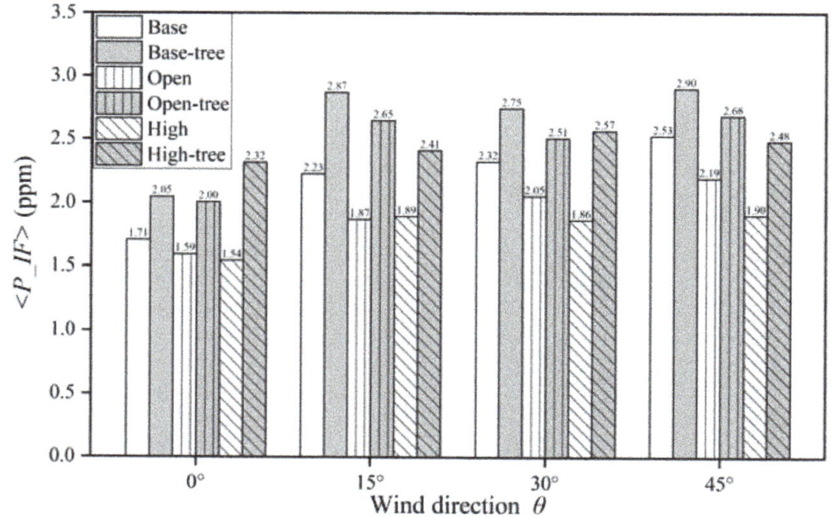

Figure 10. Building intake fraction <*P_IF*> in all cases.

3.4. Velocity, CO Concentration and <P_IF> in Surrounding Area of A2 (Region A1−A2)

Tables A8 and A9 summarise *VR* and CO concentrations in the surrounding area of A2 (Region A1−A2) in terms of spatial mean. In most of the scenarios, the building configurations and tree-planting plans have similar impacts as those in Region A2. However, the opposite effects exist in Region A1−A2 with certain conditions, especially for Case [Open-tree]. The *VR* values are increased in this region with all four wind directions, while the *VR* values in Region A2 are restrained for Case [Open-tree]. Meanwhile, the CO concentration of Case [Open-tree] in Region A1−A2 is decreased accordingly. The changes of the flow and dispersion are probably induced by the channelling effect owing to the narrowed street between the boundary and the open space. The phenomenon still needs more discussion in the ongoing work. Furthermore, the CO concentration in Region A1−A2 of Case [Open] is increased contrarily to that of Region A2, although most of the *VR* in Region A2 of Case [Open] is decreased. However, the open space improves the dilution conditions in Region A2. Thus, open space has the opposite impact on CO dispersion in the central area (Region A2) compared with the surrounding area (Region A1−A2). Moreover, since the variation of <*P_IF*> is closely related to CO concentration, the <*P_IF*> of Case [Open] will increase and that of Case [Open-tree] will decrease in Region A1−A2. It is also the opposite of that in Region A2.

4. Discussion

As critical determinants for urban ventilation and pollutant dispersion, the impacts of tree-planting plans and varied aspect ratios of 2D street canyons have been investigated in previous studies through both field experiments [93] and numerical simulations [80]. Chen et al. investigated the effect of different tree-planting parameters on the urban thermal and wind environment by conducting scaled outdoor field experiments [93]. Tree planting was found to reduce the pedestrian-level wind velocity in street canyons with all investigated *AR* values. The decreasing rate ranged 29–70%. Although the experiments are conducted in 2D idealised street canyon models, we also find that tree planting has a restraining effect on the urban wind in our work as well. Yang et al. [80] evaluated the integrated impact of tree planting and various *AR* values in a full-scale street canyon by CFD modelling (standard k-ε model) with the same emission settings as ours, and concluded that tree planting can lead to the reduction of velocity by various magnitude and an increase in CO exposure. In the canyon with *AR* = 1, the tree-induced CO increment is almost 70% (from 9.63 mg/m^3 to 16.30 mg/m^3). However, in 2D street canyon models, only the condition with perpendicular approaching wind to the street axis is considered, which corresponds to the worst ventilation situation, since only air exchange across the street roof contributes to pollutant removal. In this work with a 3D urban canopy, the ventilation can be better than in 2D models and is closer to that of the real urban community. With our 3D UCL model, the largest decreasing rate contributed by the tree planting is about 22% in spatial mean (Case [Open-tree, 0°] vs Case [Open, 0°]). Meanwhile, the tree-induced CO increment ranges 2.84–35.46% in this work (Table A6). We can conclude that even with the same *AR* (*AR* = 1) and tree-planting plan (tree planting on both sides of the street), the natural ventilation and dispersion conditions in the 3D building matrix are much better than that in 2D street canyons.

Using dimensional variables to evaluate the variance of the residents' exposure owing to the varied impactors in different cities has huge challenges, because the emission strengths of different sources are not the same and may even be at different orders. Meanwhile, the size of the target population in different studies may vary significantly. As mentioned in Section 2.1.2, the variable *IF* has been used to express the source-to-intake relationship for vehicular pollutants in realistic street canyons [48], but this would be strongly affected by the population size and the spatial scale. For example, Habilomatis and Chaloulakou [45] found that the *IF* of vehicular ultrafine particles is 371 ppm in a 2D street canyon in the central area of Athens. The *IF* at the city scale is relatively small. Marshall et al. [94] reported that the *IF* of particles in US cities ranges from 1 to 10 ppm and the *IF* of CO is 270 ppm in Hong Kong, with a huge population size [46]. The *IF* of particles at the regional scale are reported to range from 0.12 to 25 ppm in the entire United States [95].

Consequently, the variable <P_IF> is derived and applied for the exposure assessment in this work. This normalised exposure index <P_IF> is more suitable for evaluating and comparing the exposure risks in different areas, since the influence caused by different orders of the population size and pollutant emission rates is avoided. Hang et al. found the <P_IF> of CO in the tree-free idealised 2D street canyon (*AR* = 1) was 5.21 ppm [42]. Yang et al. found the <P_IF> of CO in tree-planted 2D street canyons (*AR* = 1, *LAD* = 1) were 5.60 and 5.58 ppm, and the values raised with increased *AR*. When *AR* is raised to 5, the <P_IF> is an order of magnitude larger than that with *AR* = 0.5, 1 and 3. Comparing these works in 2D street canyons, the <P_IF> of CO in the 3D UCL model ranges from 1.54 to 2.87 with various tree-planting plans, building configurations and wind directions. The values in all scenarios are much lower than those in 2D street canyons.

According to the comparison above, an important suggestion for the urban designer is to avoid building 2D street canyons either too deep or too long in urban districts. If it cannot be avoided, more leakages and a wider roadway could improve the ventilation conditions in 2D street canyons. Moreover, if the street canyon is designed longer than 8H [17], the axis of the street should be approximately parallel to the prevailing wind direction.

To simplify the calculation process, we adopt idealised 3D UCL models in this paper. The building models are all assumed to be residential-type and are highly simplified with the same configuration (in a 5 × 5 building array) for the case study. The trees are treated as cubes of porous media. Neutral atmospheric conditions are adopted, and inert gas (CO) is considered as the tracer pollutant from the traffic emissions. Only four wind directions (θ = 0°, 15°, 30°, 45°) are considered in this research. Nevertheless, the real urban environment is affected by various parameters. Thus, it is worth mentioning that the results may be significantly different if urban morphologies, atmospheric conditions or other parameters are changed.

The impacts of urban morphological parameters in realistic urban areas are much more complicated than in such an idealised model. The study of different building coating plans and the direct radiation effect of the aerosol within urban canopy, as well as their impacts on the urban thermal environment and human outdoor thermal comfort (Figures 3 and 4), is being implemented now. In future work, more kinds of realistic factors and conditions will be carefully considered and evaluated, including non-neutral atmospheric conditions and radiation impacts, the chemical reactions and composition of air pollutants, and more complicated urban morphological arrangements. Furthermore, the different tree species and the pollutant deposition on trees will also be considered in the ongoing work. CFD simulations coupling turbulence and radiation models will be validated by our scaled outdoor experiments (H = 1.2 m), as reported by Chen et al. [96,97]. These works will be adopted in numerical studies for full-scale realistic or idealised urban models. Our work is a step-by-step approximation of the real urban situation using the idealised model, and we are constantly improving our work on the way towards approaching the final target.

5. Conclusions

This paper is novel in that it numerically investigates the integrated impacts of open space, high-rise buildings and tree planting on urban airflow, pollutant dispersion and related human exposure in 3D idealised UCL models (5-row and 5-column, aspect ratio $H/W = 1$, building plan area fraction λ_p = frontal area aspect ratio $\lambda_f = 0.25$) under neutral atmospheric conditions. Four approaching wind directions (parallel 0° and non-parallel 15°, 30°, 45°) are considered. The computational fluid dynamics (CFD) simulations with the standard k-ε model are well validated by the wind tunnel data from the literature. The personal intake fraction *P_IF* and its spatially-averaged value for the entire UCL building surfaces <*P_IF*> are adopted to quantify the pollutant exposure on residents.

The CFD simulation results show that open space, high-rise building and tree planting all have strong effects on the flow structure, pollutant dispersion and residents' exposure. Some meaningful findings are concluded as follows:

(1) Without tree planting, in contrast to the general 5 × 5 uniform-height building cluster ($H = B = W$ = 30m), open space (the central building is removed) increases the spatially-averaged velocity ratio (*VR*) for the whole urban area under all four approaching wind directions (0°, 15°, 30° and 45°) by 0.40–2.27%. Designing the central building to be taller (2*H*) than the surroundings (*H*) can increase the *VR* for the entire urban area by 4.73–23.36%. In particular, the mean wind speed at the pedestrian level (z = 2 m) in the area around the high-rise building is significantly increased by 52.78–119.05%. However, tree planting significantly decreases the urban wind speed at z = 2 m on the basis of either open space or high-rise building designs, by 4.63–14.99% or 2.04–16.68%, respectively.

(2) Pollutant dispersion is determined by urban airflow characteristics. CO is released near the ground as a surrogate of traffic emissions. Without tree planting, both open space and central high-rise building would decrease the mean C at the pedestrian level for the whole urban area by 7.83–20.54% (0.32–0.97 mg/m^3) and 4.39–23.00% (0.18–1.2 mg/m^3) separately. This decreasing effect on C is significantly stronger for the high-rise building in the central area, by 43.88–47.40% (2.14–3.18 mg/m^3). On the contrary, urban tree planting evidently weakens the pollutant dilution in all scenarios, with the increasing rate of 2.84–31.88% (0.15–1.2 mg/m^3) for C at the pedestrian level in the entire urban area.

(3) The traffic-related CO exposure on residents in kerbside buildings is evaluated by <*P_IF*>. For the tree-free scenarios, both open space and high-rise buildings could decrease <*P_IF*> with the wind from all four directions by 6.56–16.08% and 9.59–24.70%, respectively. In contrast, tree planting obviously increases personal exposure in all scenarios by 14.89–50.19%. The <*P_IF*> of the tree-free cases ranges from 1.54 to 2.53 ppm, while <*P_IF*> ranges from 2.05 to 2.90 ppm in cases with tree planting.

This work provides a practical and efficient method to investigate the impacts of synthetic urban layouts on urban ventilation and pollutant dispersion. This work also extends the application of the CFD methodology to the assessment of exposure, and consequently connects to the area of public health. The method is applicable for further study coupling with more kinds of urban configurations under various atmospheric conditions. The results can provide helpful references for urban designers developing the sustainability of the city.

Author Contributions: Conceptualization, J.H. and Z.L.; data curation, Q.L., J.L. and Y.C.; simulation, Q.L., J.L., Y.C. and H.Y.; validation, H.Y.; writing—original draft preparation, H.L., Q.L., J.L. and Q.W., writing—review and editing, H.L., J.H., Q.W., Q.L. and J.L. All authors have read and agreed to the published version of the manuscript.

Funding: This work was supported by the National Natural Science Foundation of China (NSFC, No. 42175094, 41805102 and 41875015), as well as the Special Fund for Science and Technology Innovation Strategy of Guangdong Province (International cooperation) (China, No 2019A050510021). The support from the UK GCRF Rapid Response Grant on 'Transmission of SARS-CoV-2 virus in crowded indoor environment' and the Innovation Group Project of the Southern Marine Science and Engineering Guangdong Laboratory (Zhuhai) (No. 311020001) are also gratefully acknowledged.

Conflicts of Interest: The authors declare no conflict of interest.

Nomenclature

AR, H/W	aspect ratio
B, H, W	building width, building height and street width (m)
Br	volume-mean breathing rate (m^3/s)
C	time-averaged pollutant (CO) concentration (kg/m^3)
C_d	leaf drag coefficient
$<CO>_{A1}$, $<CO>_{A2}$	the spatial mean CO concentration at z = 2 m in Region A1 and in Region A2 (mg/m^3)
D_m, D_t	molecular and turbulent diffusivity of the pollutant (m^2/s)
I/O	indoor/outdoor pollutant concentration ratio
IF	intake fraction
k, ε	turbulent kinetic energy (m^2/s^2) and its dissipation rate (m^2/s^3)
LAD	leaf area density (m^2/m^3)
m	total pollutant emission over the considered period (kg)
M, N	total number of micro-environment types and population age groups
P	number of the population
P_IF, $<P\_IF>$	personal intake fraction (ppm), building intake fraction (ppm)
Re	Reynolds number
S	realistic CO emission rate ($kg/m^3/s$)
Sc_t	turbulent Schmidt number
$S_{\overline{u_i}}$, S_k, S_ε	additional source and sink terms of momentum, k and ε ($kg/m/s^3$)
t	time (s)
$\overline{u}_j$	time-averaged velocity component (m/s) on stream-wise ($\overline{u}$), span-wise ($\overline{v}$, lateral) and vertical ($\overline{w}$) directions, as j = 1, 2, 3
U	velocity magnitude (m/s)
$U_{in}(z)$	velocity profiles used at CFD domain inlet (m/s)
U_{ref}, u^*	reference velocity at building height and friction velocity (m/s)
v, v_t	kinematic viscosity and kinetic eddy viscosity (m^2/s)
V_p, V_δ	wind velocity at pedestrian-level and at the top of boundary layer (m/s)
VR	velocity ratio
$<VR>_{A1}$, $<VR>_{A2}$	the spatial mean velocity ratio at z = 2 m in Region A1 and in Region A2 (mg/m^3)

x_j	spatial coordinates (m) on stream-wise (x), span-wise (y) and vertical (z) directions, as $j = 1, 2, 3$
β_d	portion of turbulent kinetic energy converted from mean kinetic energy under the influence of drag
β_p	dimensionless coefficient of the Kolmogorov cascade
θ	wind direction (°)
κ_v	von Kármán constant
λ_p, λ_f	building plan area density, frontal area density
ρ	air density (kg/m^3)

Appendix A

1. Factors for exposure assessment

Table A1. Population composition and related factors for exposure assessment.

Item	Juveniles	Adults	Elderly
Percentage of total population	21.2%	63.3%	15.5%
Breathing rate Br indoors at home (m^3/day)	12.5	13.8	13.1
Time spent indoors at home (percentage)	61.70%	59.50%	71.60%

2. Validation studies for the flow, dispersion and vegetation modelling

   2.1. CFD validation of flow modelling

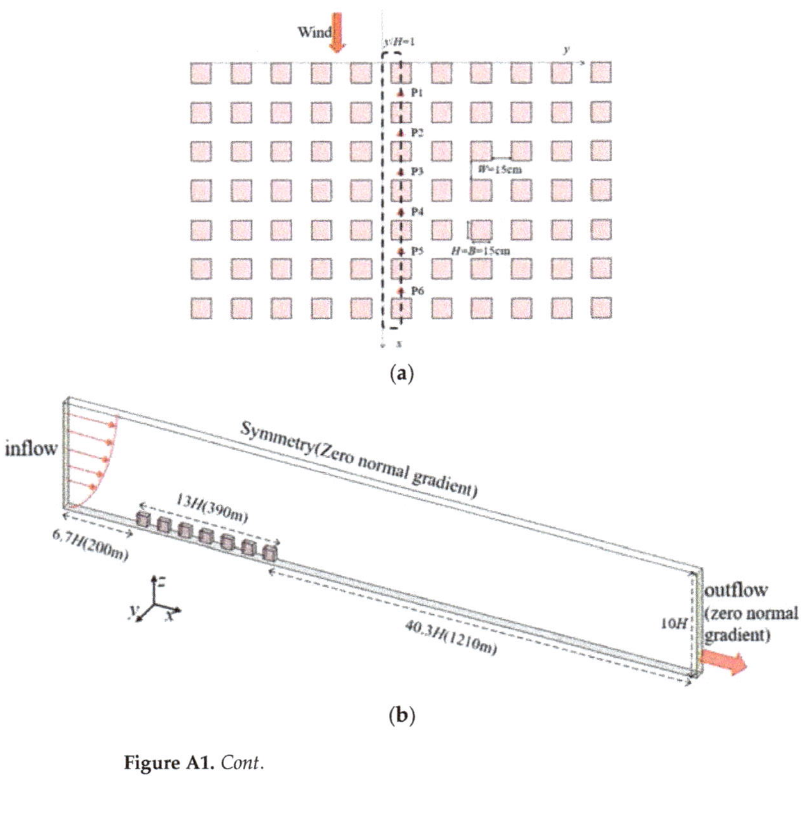

(a)

(b)

Figure A1. *Cont.*

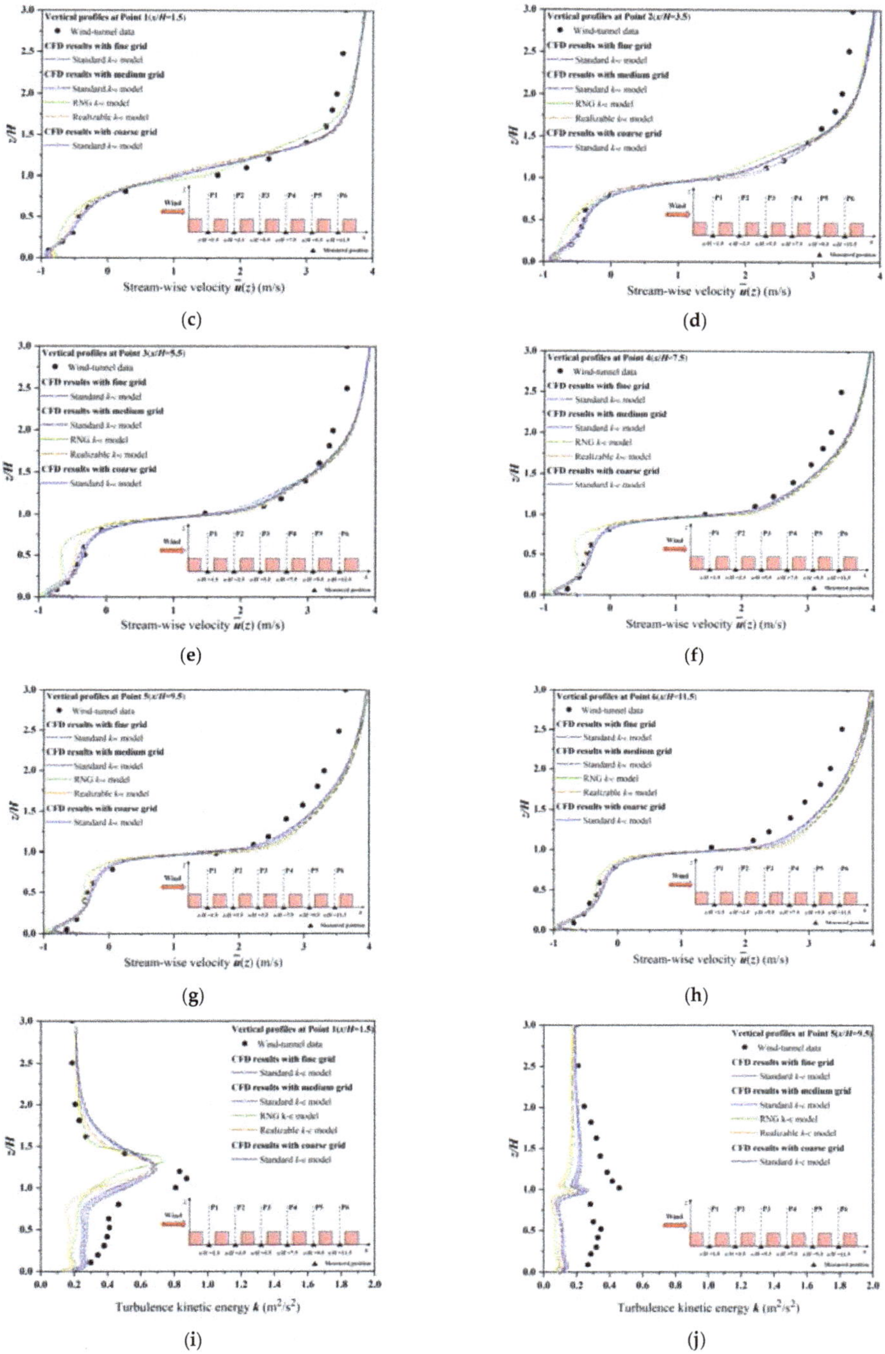

Figure A1. Flow validation by wind tunnel data. (**a**) Geometry of the UCL model in wind tunnel dataset. (**b**) Setting of the computational domain and boundary conditions. (**c**–**h**) Vertical profiles of monitored and modelled stream-wise velocity ($\overline{u}$) at Point P1–P6. (**i**,**j**) Vertical profiles of monitored and modelled turbulence kinetic energy (k) at Point P1 and P5.

Table A2. Statistical analysis between wind tunnel data and CFD simulation results—flow validation.

Variable (Position)	**Grid Size**	**Turbulence Model**	**NMSE** *	**FB** **	**R** ***
Criteria			**≤1.5**	**−0.3–0.3**	**→ 1.0**
Stream-wise velocity at P1	Fine grid	STD	0.054	0.084	0.990
	Medium grid	STD	0.049	0.087	0.990
		RNG	0.107	0.018	0.996
		RKE	0.115	0.094	0.986
	Coarse grid	STD	0.038	0.081	0.991
Stream-wise velocity at P2	Fine grid	STD	0.007	0.002	0.999
	Medium grid	STD	0.009	0.002	0.997
		RNG	0.045	−0.090	0.994
		RKE	0.016	−0.083	0.998
	Coarse grid	STD	0.011	0.013	0.997
Stream-wise velocity at P3	Fine grid	STD	0.012	−0.063	0.996
	Medium grid	STD	0.005	−0.024	0.998
		RNG	0.093	−0.179	0.996
		RKE	0.022	−0.100	0.997
	Coarse grid	STD	0.005	−0.013	0.997
Stream-wise velocity at P4	Fine grid	STD	0.014	−0.057	0.999
	Medium grid	STD	0.014	−0.034	1.000
		RNG	0.053	−0.175	0.998
		RKE	0.014	−0.102	0.999
	Coarse grid	STD	0.015	−0.023	0.999
Stream-wise velocity at P5	Fine grid	STD	0.019	−0.051	0.998
	Medium grid	STD	0.021	−0.017	0.998
		RNG	0.033	−0.130	0.998
		RKE	0.016	−0.073	0.997
	Coarse grid	STD	0.022	−0.009	0.997
Stream-wise velocity at P6	Fine grid	STD	0.037	−0.104	0.995
	Medium grid	STD	0.037	−0.058	0.997
		RNG	0.039	−0.147	0.993
		RKE	0.039	−0.106	0.993
	Coarse grid	STD	0.035	−0.048	0.998
TKE at P1	Fine grid	STD	0.137	0.177	0.820
	Medium grid	STD	0.115	0.150	0.842
		RNG	0.275	0.343	0.662
		RKE	0.361	0.366	0.727
	Coarse grid	STD	0.091	0.115	0.867
TKE at P5	Fine grid	STD	0.610	0.615	0.076
	Medium grid	STD	0.621	0.615	0.135
		RNG	1.299	0.837	−0.105
		RKE	1.096	0.802	−0.378
	Coarse grid	STD	0.526	0.575	0.255

* normalised mean square error: $NMSE = \frac{\overline{(Obs-Sim)^2}}{\overline{Obs}*\overline{Sim}}$. ** fractional bias-$FB = \frac{\overline{Obs}-\overline{Sim}}{0.5*(\overline{Obs}+\overline{Sim})}$. *** correlation coefficient–$R = \frac{\sum(Obs-\overline{Obs})(Sim-\overline{Sim})}{\sqrt{\sum(Obs-\overline{Obs})^2\sum(Sim-\overline{Sim})^2}}$.

2.2. CFD validation of pollutant dispersion without tree models

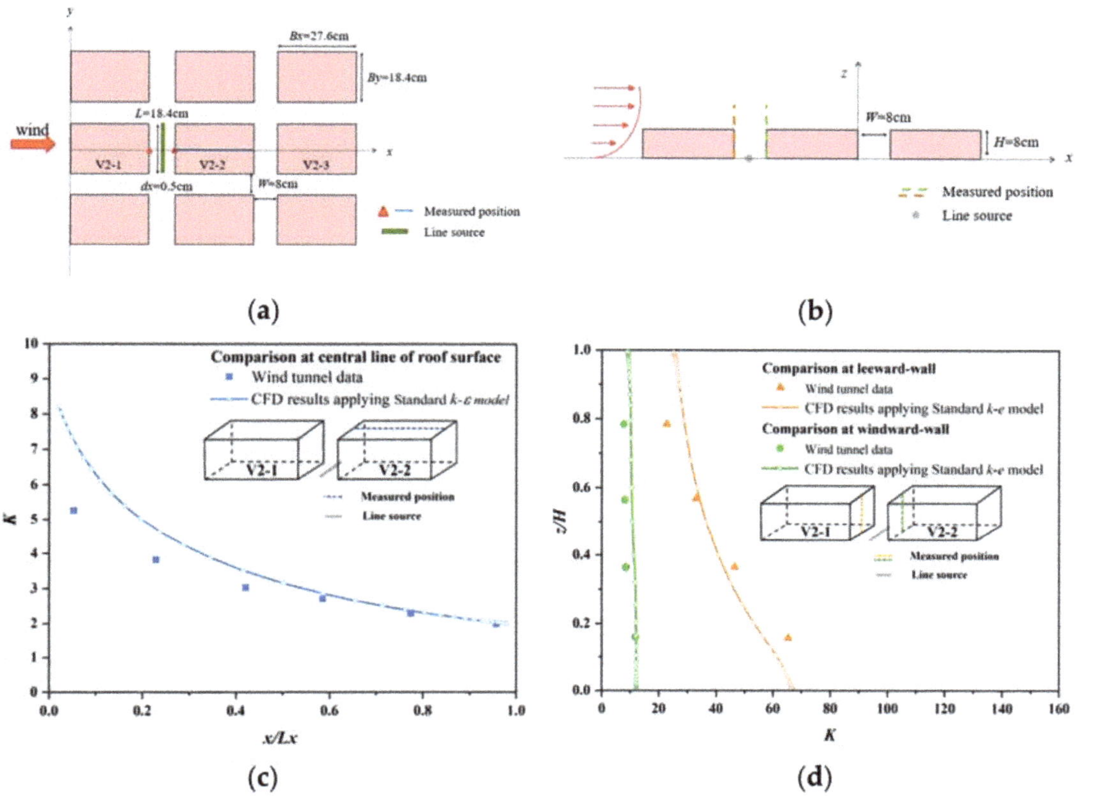

Figure A2. Validation for pollutant dispersion. (**a**) Configurations of the wind tunnel experiment with top view. (**b**) Configurations of the wind tunnel experiment with lateral view. (**c**) Vertical profiles of normalized inert gas concentration *K* at the roof top of the model. (**d**) Vertical profiles of *K* at the leeward and windward wall.

Table A3. Statistical analysis between wind tunnel data and CFD simulation results—dispersion modelling.

Variable (Position) Criteria	NMSE * ≤1.5	FB ** −0.3–0.3	R *** → 1.0
Leeward wall	0.021	−0.012	0.997
Windward wall	0.064	−0.223	0.855
Central line	0.029	−0.13	0.998

* normalised mean square error: $NMSE = \frac{\overline{(Obs-Sim)^2}}{\overline{Obs}*\overline{Sim}}$. ** fractional bias-$FB = \frac{\overline{Obs}-\overline{Sim}}{0.5*(\overline{Obs}+\overline{Sim})}$. *** correlation coefficient-$R = \frac{\Sigma(Obs-\overline{Obs})(Sim-\overline{Sim})}{\sqrt{\Sigma(Obs-\overline{Obs})^2\Sigma(Sim-\overline{Sim})^2}}$.

2.3. CFD validation of pollutant dispersion with tree models

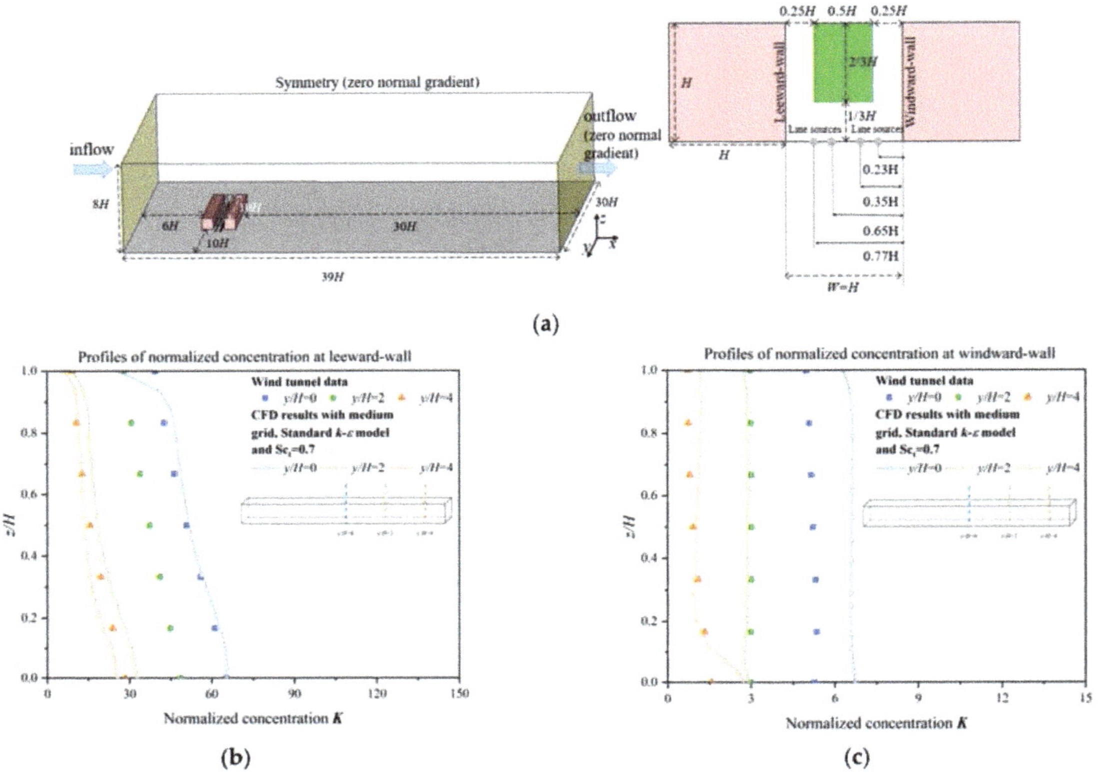

Figure A3. Validation for the vegetation modelling. (**a**) Configurations of the wind tunnel experiment with vegetation model. (**b**) Vertical profiles of K at the leeward wall. (**c**) Vertical profiles of K at the windward wall.

Table A4. Statistical analysis between wind tunnel data and CFD simulation results—vegetation modelling.

Variable (Position) Criteria		NMSE * ≤1.5	FB ** −0.3–0.3	R *** → 1.0
Leeward side	$y/H = 0$	0.159	0.011	0.918
	$y/H = 2$	1.306	0.594	0.987
	$y/H = 4$	0.108	0.128	0.974
Windward side	$y/H = 0$	0.055	−0.233	0.630
	$y/H = 2$	0.007	0.049	0.784
	$y/H = 4$	0.221	−0.256	0.700

* normalised mean square error: $NMSE = \frac{\overline{(Obs-Sim)^2}}{\overline{Obs}*\overline{Sim}}$. ** fractional bias-$FB = \frac{\overline{Obs}-\overline{Sim}}{0.5*(\overline{Obs}+\overline{Sim})}$. *** correlation coefficient–$R = \frac{\Sigma(Obs-\overline{Obs})(Sim-\overline{Sim})}{\sqrt{\Sigma(Obs-\overline{Obs})^2\,\Sigma(Sim-\overline{Sim})^2}}$.

3. Quantitative investigations of the flow, pollutant concentration and traffic-related exposure

Table A5. Spatial mean velocity ratio (*VR*) at z = 2 m in Region A1 ($<VR>_{A1}$) and Region A2 ($<VR>_{A2}$) of each scenario and the change rate.

Case	$<VR>_{A1}$				$<VR>_{A2}$			
	0°	15°	30°	45°	0°	15°	30°	45°
Case [Base]	0.12	0.18	0.20	0.21	0.09	0.13	0.16	0.16
Case [Open]	0.12 +0.40%	0.19 +1.70%	0.21 +2.27%	0.21 +2.22%	0.10 +14.53%	0.11 −12.06%	0.14 −8.40%	0.14 −9.29%
Case [High]	0.15 +23.36%	0.19 +4.73%	0.22 +7.30%	0.23 +12.07%	0.19 +119.05%	0.21 +61.30%	0.24 +55.06%	0.24 +52.78%
Case [Base-tree]	0.13 +6.27%	0.17 −7.51%	0.18 −11.59%	0.18 −11.41%	0.08 −10.76%	0.12 −6.22%	0.14 −11.08%	0.14 −11.43%
Case [Open-tree]	0.12 −4.63%	0.17 −9.01%	0.18 −13.65%	0.18 −14.99%	0.08 −21.92%	0.11 −6.91%	0.13 −7.84%	0.12 −14.18%
Case [High-tree]	0.15 −2.04%	0.18 −6.87%	0.18 −14.93%	0.19 −16.68%	0.21 +8.98%	0.20 −4.42%	0.20 −16.11%	0.21 −13.44%

The percentage number in each cell denotes the change rate of each case. Case [Base] is the comparison reference for Case [Open], Case [High] and Case [Base-tree]. The change rate of Case [Open-tree] refers to Case [Open], and that of Case [High-tree] refers to Case [High].

Table A6. Spatial mean CO concentration (*C*) at *z* = 2 m in Region A1 ($<CO>_{A1}$) and Region A2 ($<CO>_{A2}$) of each scenario and the change rate.

Case	$<CO>_{A1}$ (mg/m^3)				$<CO>_{A2}$ (mg/m^3)			
	0°	15°	30°	45°	0°	15°	30°	45°
Case [Base]	4.14	4.72	5.01	5.24	4.87	7.11	6.37	6.70
Case [Open]	3.82 −7.83%	3.75 −20.54%	4.24 −15.31%	4.44 −15.27%	4.86 −0.08%	5.92 −16.82%	4.94 −22.48%	5.06 −24.43%
Case [High]	3.96 −4.39%	3.99 −15.47%	4.12 −17.78%	4.04 −23.00%	2.73 −43.88%	3.80 −46.58%	3.36 −47.24%	3.52 −47.40%
Case [Base-tree]	4.98 +20.19%	5.36 +13.55%	5.24 +4.73%	5.39 +2.84%	6.59 +35.46%	7.89 +10.93%	6.78 +6.47%	6.89 +2.85%
Case [Open-tree]	4.67 +22.44%	4.95 +31.88%	4.74 +11.80%	5.00 +12.64%	5.55 +14.09%	6.61 +11.76%	5.60 +13.35%	5.97 +17.93%
Case [High-tree]	4.46 +12.61%	4.64 +16.31%	5.20 +26.38%	4.88 +20.88%	2.90 +6.10%	4.28 +12.57%	4.07 +21.19%	3.95 +12.04%

The percentage data denote the change rate of each case in contrast to Case [Base]. The change rate of Case [Open-tree] refers to Case [Open], and that of Case [High-tree] refers to Case [High].

Table A7. Building intake fraction $<P\_IF>$ and the change rate in different cases.

Case	$<P\_IF>$ (ppm)			
	0°	15°	30°	45°
Case [Base]	1.71	2.23	2.32	2.53
Case [Open]	1.59 −6.56%	1.87 −16.08%	2.05 −11.65%	2.19 −13.19%
Case [High]	1.54 −9.59%	1.89 −15.00%	1.86 −19.74%	1.90 −24.70%

Table A7. *Cont.*

Case	<P_IF> (ppm) 0°	15°	30°	45°
Case [Base-tree]	2.05 +19.94%	2.87 +28.92%	2.75 +18.24%	2.90 +14.89%
Case [Open-tree]	2.00 +25.67%	2.65 +41.62%	2.51 +22.22%	2.68 +22.43%
Case [High-tree]	2.32 +50.19%	2.41 +27.28%	2.57 +37.63%	2.48 +30.51%

The percentage number in each cell denotes the rate of change of each case. Case [Base] is the comparison reference for Case [Open], Case [High] and Case [Base-tree]. The change rate of Case [Open-tree] refers to Case [Open], and that of Case [High-tree] refers to Case [High].

Table A8. Spatial mean velocity ratio (*VR*) at z = 2 m in the Region A1−A2 ($<VR>_{A1-A2}$) of each scenario and the change rate.

Case Name	$<VR>_{A1-A2}$ 0°	15°	30°	45°
Case [Base]	0.55	0.83	0.90	0.92
Case [Open]	0.59 +7.71%	0.76 −7.61%	0.79 −11.64%	0.81 −11.41%
Case [High]	0.55 −7.90%	0.85 +11.35%	0.93 +16.88%	0.95 +16.60%
Case [Base-tree]	0.53 −2.94%	0.77 −9.16%	0.79 −14.14%	0.81 −15.06%
Case [Open-tree]	0.63 +19.69%	0.83 +6.92%	0.92 +15.91%	0.99 +23.36%
Case [High-tree]	0.61 −3.82%	0.77 −7.20%	0.79 −14.76%	0.82 −17.11%

The percentage number in each cell denotes the rate of change of each case. Case [Base] is the comparison reference for Case [Open], Case [High] and Case [Base-tree]. The change rate of Case [Open-tree] refers to Case [Open], and that of Case [High-tree] refers to Case [High].

Table A9. Spatial mean CO concentration (*C*) at z = 2 m in the Region A1−A2 ($<CO>_{A1-A2}$) of each scenario and the change rate.

Case Name	$<CO>_{A1-A2}$ 0°	15°	30°	45°
Case [Base]	4.05	4.42	4.84	5.06
Case [Open]	4.78 +17.89%	5.04 +14.07%	5.05 +4.45%	5.20 +2.83%
Case [High]	3.69 −22.81%	3.48 −31.00%	4.15 −17.79%	4.36 −16.13%
Case [Base-tree]	4.56 +23.82%	4.74 +36.15%	4.63 +11.57%	4.88 +11.87%
Case [Open-tree]	4.11 −9.88%	4.01 −15.29%	4.21 −9.13%	4.10 −16.01%
Case [High-tree]	4.65 +13.15%	4.69 +16.75%	5.34 +26.89%	5.00 +21.83%

The percentage number in each cell denotes the rate of change of each case. Case [Base] is the comparison reference for Case [Open], Case [High] and Case [Base-tree]. The change rate of Case [Open-tree] refers to Case [Open], and that of Case [High-tree] refers to Case [High].

References

1. Chan, C.K.; Yao, X. Air pollution in mega cities in China. *Atmos. Environ.* **2008**, *42*, 1–42. [CrossRef]
2. Fenger, J. Urban air quality. *Atmos. Environ.* **1999**, *33*, 4877–4900. [CrossRef]
3. Pu, Y.; Yang, C. Estimating urban roadside emissions with an atmospheric dispersion model based on in-field measurements. *Environ. Pollut.* **2014**, *192*, 300–307. [CrossRef] [PubMed]
4. Ji, W.; Zhao, B. Estimating mortality derived from indoor exposure to particles of outdoor origin. *PLoS ONE* **2015**, *10*, e0124238. [CrossRef] [PubMed]
5. Peters, A.; Pope III, C.A. Cardiopulmonary mortality and air pollution. *Lancet* **2002**, *360*, 1184–1185. [CrossRef]
6. Chen, C.; Zhao, B.; Zhou, W.; Jiang, X.; Tan, Z. A methodology for predicting particle penetration factor through cracks of windows and doors for actual engineering application. *Build. Environ.* **2012**, *47*, 339–348. [CrossRef]
7. Quang, T.N.; He, C.; Morawska, L.; Knibbs, L.D.; Falk, M. Vertical particle concentration profiles around urban office buildings. *Atmos. Chem. Phys. Discuss.* **2012**, *12*, 5017–5030. [CrossRef]
8. Zaeh, S.E.; Koehler, K.; Eakin, M.N.; Wohn, C.; Diibor, I.; Eckmann, T.; Wu, T.D.; Clemons-Erby, D.; Gummerson, C.E.; Green, T.; et al. Indoor air quality prior to and following school building renovation in a mid-Atlantic school district. *Int. J. Environ. Res. Public Health* **2021**, *18*, 12149. [CrossRef]
9. Bady, M.; Kato, S.; Huang, H. Towards the application of indoor ventilation efficiency indices to evaluate the air quality of urban areas. *Build. Environ.* **2008**, *43*, 1991–2004. [CrossRef]
10. Ng, W.-Y.; Chau, C.-K. A modeling investigation of the impact of street and building configurations on personal air pollutant exposure in isolated deep urban canyons. *Sci. Total Environ.* **2014**, *468*, 429–448. [CrossRef]
11. Ng, E. Policies and technical guidelines for urban planning of high-density cities-air ventilation assessment (ava) of Hong Kong. *Build. Environ.* **2009**, *44*, 1478–1488. [CrossRef] [PubMed]
12. Chang, C.H.; Meroney, R.N. Concentration and flow distributions in urban street canyons: Wind tunnel and computational data. *J. Wind. Eng. Ind. Aerodyn.* **2003**, *91*, 1141–1154. [CrossRef]
13. Chew, L.W.; Aliabadi, A.A.; Norford, L.K. Flows across high aspect ratio street canyons: Reynolds number independence revisited. *Environ. Fluid Mech.* **2018**, *18*, 1275–1291. [CrossRef]
14. Grimmond, C.; Roth, M.; Oke, T.R.; Au, Y.; Best, M.; Betts, R.; Carmichael, G.; Cleugh, H.; Dabberdt, W.; Emmanuel, R. Climate and more sustainable cities: Climate information for improved planning and management of cities (producers/capabilities perspective). *Procedia Environ. Sci.* **2010**, *1*, 247–274. [CrossRef]
15. Li, X.-X.; Liu, C.-H.; Leung, D.Y.C.; Lam, K.M. Recent progress in CFD modelling of wind field and pollutant transport in street canyons. *Atmos. Environ.* **2006**, *40*, 5640–5658. [CrossRef]
16. Peng, Y.L.; Buccolieri, R.; Gao, Z.; Ding, W. Indices employed for the assessment of "urban outdoor ventilation"—A review. *Atmos. Environ.* **2020**, *223*, 117211. [CrossRef]
17. Tominaga, Y.; Stathopoulos, T. CFD simulation of near-field pollutant dispersion in the urban environment: A review of current modeling techniques. *Atmos. Environ.* **2013**, *79*, 716–730. [CrossRef]
18. Vardoulakis, S.; Fisher, B.E.; Pericleous, K.; Gonzalez-Flesca, N. Modelling air quality in street canyons: A review. *Atmos. Environ.* **2003**, *37*, 155–182. [CrossRef]
19. Antoniou, N.; Montazeri, H.; Wigo, H.; Neophytou, M.K.A.; Blocken, B.; Sandberg, M. CFD and wind-tunnel analysis of outdoor ventilation in a real compact heterogeneous urban area: Evaluation using "air delay". *Build. Environ.* **2017**, *126*, 355–372. [CrossRef]
20. Blocken, B.; Stathopoulos, T.; Carmeliet, J. CFD simulation of the atmospheric boundary layer: Wall function problems. *Atmos. Environ.* **2007**, *41*, 238–252. [CrossRef]
21. Gromke, C. A vegetation modeling concept for building and environmental aerodynamics wind tunnel tests and its application in pollutant dispersion studies. *Environ. Pollut.* **2011**, *159*, 2094–2099. [CrossRef] [PubMed]
22. Liu, J.L.; Zhang, X.L.; Niu, J.L.; Tse, K.T. Pedestrian-level wind and gust around buildings with a 'lift-up' design: Assessment of influence from surrounding buildings by adopting LES. *Build. Simul.* **2019**, *12*, 1107–1118. [CrossRef]
23. Santiago, J.L.; Martilli, A.; Martin, F. CFD simulation of airflow over a regular array of cubes. Part I: Three-dimensional simulation of the flow and validation with wind-tunnel measurements. *Bound.-Layer Meteorol.* **2007**, *122*, 609–634. [CrossRef]
24. Yang, H.Y.; Lam, C.K.C.; Lin, Y.Y.; Chen, L.; Mattsson, M.; Sandberg, M.; Hayati, A.; Claesson, L.; Hang, J. Numerical investigations of re-independence and influence of wall heating on flow characteristics and ventilation in full-scale 2d street canyons. *Build. Environ.* **2021**, *189*, 107510. [CrossRef]
25. Zhang, M.; Gao, Z.; Guo, X.; Shen, J. Ventilation and pollutant concentration for the pedestrian zone, the near-wall zone, and the canopy layer at urban intersections. *Int. J. Environ. Res. Public Health* **2021**, *18*, 11080. [CrossRef]
26. He, L.; Hang, J.; Wang, X.; Lin, B.; Li, X.; Lan, G. Numerical investigations of flow and passive pollutant exposure in high-rise deep street canyons with various street aspect ratios and viaduct settings. *Sci. Total Environ.* **2017**, *584*, 189–206. [CrossRef]
27. Hang, J.; Xian, Z.; Wang, D.; Mak, C.M.; Wang, B.; Fan, Y. The impacts of viaduct settings and street aspect ratios on personal intake fraction in three-dimensional urban-like geometries. *Build. Environ.* **2018**, *143*, 138–162. [CrossRef]
28. Duarte, D.H.S.; Shinzato, P.; Gusson, C.D.; Alves, C.A. The impact of vegetation on urban microclimate to counterbalance built density in a subtropical changing climate. *Urban Clim.* **2015**, *14*, 224–239. [CrossRef]

29. Moradpour, M.; Hosseini, V. An investigation into the effects of green space on air quality of an urban area using CFD modelling. *Urban Clim.* **2020**, *34*, 100686. [CrossRef]
30. Yang, X.; Peng, L.L.H.; Chen, Y.; Yao, L.; Wang, Q. Air humidity characteristics of local climate zones: A three-year observational study in Nanjing. *Build. Environ.* **2020**, *171*, 106661. [CrossRef]
31. Gromke, C.; Ruck, B. Pollutant concentrations in street canyons of different aspect ratio with avenues of trees for various wind directions. *Bound.-Layer Meteorol.* **2012**, *144*, 41–64. [CrossRef]
32. Hang, J.; Chen, L.; Lin, Y.Y.; Buccolieri, R.; Lin, B.R. The impact of semi-open settings on ventilation in idealized building arrays. *Urban Clim.* **2018**, *25*, 196–217. [CrossRef]
33. Chen, L.; Hang, J.; Sandberg, M.; Claesson, L.; Di Sabatino, S.; Wigo, H. The impacts of building height variations and building packing densities on flow adjustment and city breathability in idealized urban models. *Build. Environ.* **2017**, *118*, 344–361. [CrossRef]
34. Ramponi, R.; Blocken, B.; de Coo, L.B.; Janssen, W.D. CFD simulation of outdoor ventilation of generic urban configurations with different urban densities and equal and unequal street widths. *Build. Environ.* **2015**, *92*, 152–166. [CrossRef]
35. Gu, Z.-L.; Zhang, Y.-W.; Cheng, Y.; Lee, S.-C. Effect of uneven building layout on air flow and pollutant dispersion in non-uniform street canyons. *Build. Environ.* **2011**, *46*, 2657–2665. [CrossRef]
36. Hang, J.; Li, Y.; Sandberg, M.; Buccolieri, R.; Di Sabatino, S. The influence of building height variability on pollutant dispersion and pedestrian ventilation in idealized high-rise urban areas. *Build. Environ.* **2012**, *56*, 346–360. [CrossRef]
37. Chen, L.; Wen, Y.; Zhang, L.; Xiang, W.-N. Studies of thermal comfort and space use in an urban park square in cool and cold seasons in shanghai. *Build. Environ.* **2015**, *94*, 644–653. [CrossRef]
38. He, Y.Y.; Tablada, A.; Wong, N.H. Effects of non-uniform and orthogonal breezeway networks on pedestrian ventilation in Singapore's high-density urban environments. *Urban Clim.* **2018**, *24*, 460–484. [CrossRef]
39. Sha, C.; Wang, X.; Lin, Y.; Fan, Y.; Chen, X.; Hang, J. The impact of urban open space and 'lift-up' building design on building intake fraction and daily pollutant exposure in idealized urban models. *Sci. Total Environ.* **2018**, *633*, 1314–1328. [CrossRef]
40. Cheung, J.O.P.; Liu, C.-H. CFD simulations of natural ventilation behaviour in high-rise buildings in regular and staggered arrangements at various spacings. *Energy Build.* **2011**, *43*, 1149–1158. [CrossRef]
41. Zhang, Y.; Kwok, K.C.; Liu, X.-P.; Niu, J.-L. Characteristics of air pollutant dispersion around a high-rise building. *Environ. Pollut.* **2015**, *204*, 280–288. [CrossRef]
42. Hang, J.; Luo, Z.; Wang, X.; He, L.; Wang, B.; Zhu, W. The influence of street layouts and viaduct settings on daily carbon monoxide exposure and intake fraction in idealized urban canyons. *Environ. Pollut.* **2017**, *220*, 72–86. [CrossRef]
43. Lin, Y.; Chen, G.; Chen, T.; Luo, Z.; Yuan, C.; Gao, P.; Hang, J. The influence of advertisement boards, street and source layouts on co dispersion and building intake fraction in three-dimensional urban-like models. *Build. Environ.* **2019**, *150*, 297–321. [CrossRef]
44. Zhang, K.; Chen, G.; Wang, X.; Liu, S.; Mak, C.M.; Fan, Y.; Hang, J. Numerical evaluations of urban design technique to reduce vehicular personal intake fraction in deep street canyons. *Sci. Total Environ.* **2019**, *653*, 968–994. [CrossRef]
45. Habilomatis, G.; Chaloulakou, A. A CFD modeling study in an urban street canyon for ultrafine particles and population exposure: The intake fraction approach. *Sci. Total Environ.* **2015**, *530*, 227–232. [CrossRef]
46. Luo, Z.; Li, Y.; Nazaroff, W.W. Intake fraction of nonreactive motor vehicle exhaust in Hong Kong. *Atmos. Environ.* **2010**, *44*, 1913–1918. [CrossRef]
47. Nazaroff, W.W. Inhalation intake fraction of pollutants from episodic indoor emissions. *Build. Environ.* **2008**, *43*, 269–277. [CrossRef]
48. Zhou, Y.; Levy, J.I. The impact of urban street canyons on population exposure to traffic-related primary pollutants. *Atmos. Environ.* **2008**, *42*, 3087–3098. [CrossRef]
49. Chau, C.K.; Tu, E.Y.; Chan, D.W.T.; Burnett, C.J. Estimating the total exposure to air pollutants for different population age groups in hong kong. *Environ. Int.* **2002**, *27*, 617–630. [CrossRef]
50. Allan, M.; Richardson, G.M.; Jones-Otazo, H. Probability density functions describing 24-h inhalation rates for use in human health risk assessments: An update and comparison. *Hum. Ecol. Risk Assess.* **2008**, *14*, 372–391. [CrossRef]
51. Blocken, B. Computational fluid dynamics for urban physics: Importance, scales, possibilities, limitations and ten tips and tricks towards accurate and reliable simulations. *Build. Environ.* **2015**, *91*, 219–245. [CrossRef]
52. Gao, Z.L.; Bresson, R.; Qu, Y.F.; Milliez, M.; de Munck, C.; Carissimo, B. High resolution unsteady RANS simulation of wind, thermal effects and pollution dispersion for studying urban renewal scenarios in a neighborhood of Toulouse. *Urban Clim.* **2018**, *23*, 114–130. [CrossRef]
53. Li, X.-X.; Britter, R.E.; Norford, L.K. Transport processes in and above two-dimensional urban street canyons under different stratification conditions: Results from numerical simulation. *Environ. Fluid Mech.* **2015**, *15*, 399–417. [CrossRef]
54. Wang, W.; Ng, E. Air ventilation assessment under unstable atmospheric stratification-a comparative study for Hong Kong. *Build. Environ.* **2018**, *130*, 1–13. [CrossRef]
55. Zhang, Y.; Gu, Z.; Yu, C.W. Review on numerical simulation of airflow and pollutant dispersion in urban street canyons under natural background wind condition. *Aerosol Air Qual. Res.* **2018**, *18*, 780–789. [CrossRef]
56. Blocken, B. LES over RANS in building simulation for outdoor and indoor applications: A foregone conclusion? *Build. Simul.* **2018**, *11*, 821–870. [CrossRef]

57. Meroney, R.N. Ten questions concerning hybrid computational/physical model simulation of wind flow in the built environment. *Build. Environ.* **2016**, *96*, 12–21. [CrossRef]
58. Zhong, J.; Cai, X.-M.; Bloss, W.J. Modelling the dispersion and transport of reactive pollutants in a deep urban street canyon: Using large-eddy simulation. *Environ. Pollut.* **2015**, *200*, 42–52. [CrossRef]
59. Fuka, V.; Xie, Z.-T.; Castro, I.P.; Hayden, P.; Carpentieri, M.; Robins, A.G. Scalar fluxes near a tall building in an aligned array of rectangular buildings. *Bound.-Layer Meteorol.* **2018**, *167*, 53–76. [CrossRef]
60. McMullan, W.A.; Angelino, M. The effect of tree planting on traffic pollutant dispersion in an urban street canyon using large eddy simulation with a recycling and rescaling inflow generation method. *J. Wind. Eng. Ind. Aerodyn.* **2022**, *221*, 104877. [CrossRef]
61. Cui, D.J.; Mak, C.M.; Kwok, K.C.S.; Ai, Z.T. CFD simulation of the effect of an upstream building on the inter-unit dispersion in a multi-story building in two wind directions. *J. Wind. Eng. Ind. Aerodyn.* **2016**, *150*, 31–41. [CrossRef]
62. Cui, D.; Hu, G.; Ai, Z.; Du, Y.; Mak, C.M.; Kwok, K. Particle image velocimetry measurement and CFD simulation of pedestrian level wind environment around u-type street canyon. *Build. Environ.* **2019**, *154*, 239–251. [CrossRef]
63. Gromke, C.; Blocken, B. Influence of avenue-trees on air quality at the urban neighborhood scale. Part I: Quality assurance studies and turbulent Schmidt number analysis for RANS CFD simulations. *Environ. Pollut.* **2015**, *196*, 214–223. [CrossRef]
64. Panagiotou, I.; Neophytou, M.K.A.; Hamlyn, D.; Britter, R.E. City breathability as quantified by the exchange velocity and its spatial variation in real inhomogeneous urban geometries: An example from central London urban area. *Sci. Total Environ.* **2013**, *442*, 466–477. [CrossRef] [PubMed]
65. Wang, W.; Xu, Y.; Ng, E.; Raasch, S. Evaluation of satellite-derived building height extraction by CFD simulations: A case study of neighborhood-scale ventilation in Hong Kong. *Landsc. Urban Plan.* **2018**, *170*, 90–102. [CrossRef]
66. Ashie, Y.; Kono, T. Urban-scale CFD analysis in support of a climate-sensitive design for the Tokyo bay area. *Int. J. Clim.* **2011**, *31*, 174–188. [CrossRef]
67. Hang, J.; Wang, Q.; Chen, X.; Sandberg, M.; Zhu, W.; Buccolieri, R.; Di Sabatino, S. City breathability in medium density urban-like geometries evaluated through the pollutant transport rate and the net escape velocity. *Build. Environ.* **2015**, *94*, 166–182. [CrossRef]
68. Zhang, K.; Chen, G.W.; Zhang, Y.; Liu, S.H.; Wang, X.M.; Wang, B.M.; Hang, J. Integrated impacts of turbulent mixing and nox-o-3 photochemistry on reactive pollutant dispersion and intake fraction in shallow and deep street canyons. *Sci. Total Environ.* **2020**, *712*, 135553. [CrossRef]
69. Franke, J.; Hellsten, A.; Schluenzen, K.H.; Carissimo, B. The cost 732 best practice guideline for CFD simulation of flows in the urban environment: A summary. *Int. J. Environ. Pollut.* **2011**, *44*, 419–427. [CrossRef]
70. Tominaga, Y.; Mochida, A.; Yoshie, R.; Kataoka, H.; Nozu, T.; Yoshikawa, M.; Shirasawa, T. Aij guidelines for practical applications of CFD to pedestrian wind environment around buildings. *J. Wind. Eng. Ind. Aerodyn.* **2008**, *96*, 1749–1761. [CrossRef]
71. Lien, F.S.; Yee, E. Numerical modelling of the turbulent flow developing within and over a 3-d building array, part I: A high-resolution reynolds-averaged navier-stokes approach. *Bound.-Layer Meteorol.* **2004**, *112*, 427–466. [CrossRef]
72. Lin, M.; Hang, J.; Li, Y.; Luo, Z.; Sandberg, M. Quantitative ventilation assessments of idealized urban canopy layers with various urban layouts and the same building packing density. *Build. Environ.* **2014**, *79*, 152–167. [CrossRef]
73. Lu, Y.; Lin, S.; Fatmi, Z.; Malashock, D.; Hussain, M.M.; Siddique, A.; Carpenter, D.O.; Lin, Z.; Khwaja, H.A. Assessing the association between fine particulate matter (pm2.5) constituents and cardiovascular diseases in a mega-city of Pakistan. *Environ. Pollut.* **2019**, *252*, 1412–1422. [CrossRef]
74. Sinharay, R.; Gong, J.; Barratt, B.; Ohman-Strickland, P.; Ernst, S.; Kelly, F.; Zhang, J.; Collins, P.; Cullinan, P.; Chung, K.F. Respiratory and cardiovascular responses to walking down a traffic-polluted road compared with walking in a traffic-free area in participants aged 60 years and older with chronic lung or heart disease and age-matched healthy controls: A randomised, crossover study. *Lancet* **2018**, *391*, 339–349.
75. Blocken, B.; Vervoort, R.; van Hooff, T. Reduction of outdoor particulate matter concentrations by local removal in semi-enclosed parking garages: A preliminary case study for Eindhoven city center. *J. Wind. Eng. Ind. Aerodyn.* **2016**, *159*, 80–98. [CrossRef]
76. Lin, L.; Hang, J.; Wang, X.; Wang, X.; Fan, S.; Fan, Q.; Liu, Y. Integrated effects of street layouts and wall heating on vehicular pollutant dispersion and their reentry toward downstream canyons. *Aerosol Air Qual. Res.* **2016**, *16*, 3142–3163. [CrossRef]
77. Kwak, K.-H.; Baik, J.-J.; Lee, K.-Y. Dispersion and photochemical evolution of reactive pollutants in street canyons. *Atmos. Environ.* **2013**, *70*, 98–107. [CrossRef]
78. Di Bernardino, A.; Monti, P.; Leuzzi, G.; Querzoli, G. Turbulent Schmidt number measurements over three-dimensional cubic arrays. *Bound.-Layer Meteorol.* **2020**, *174*, 231–250. [CrossRef]
79. Xia, L.P.; Shao, Y.P. Modelling of traffic flow and air pollution emission with application to Hong Kong island. *Environ. Model. Softw.* **2005**, *20*, 1175–1188. [CrossRef]
80. Yang, H.; Chen, T.; Lin, Y.; Buccolieri, R.; Mattsson, M.; Zhang, M.; Hang, J.; Wang, Q. Integrated impacts of tree planting and street aspect ratios on co dispersion and personal exposure in full-scale street canyons. *Build. Environ.* **2020**, *169*, 106529. [CrossRef]
81. Buccolieri, R.; Santiago, J.-L.; Rivas, E.; Sanchez, B. Review on urban tree modelling CFD simulations: Aerodynamic, deposition and thermal effects. *Urban For. Urban Green.* **2018**, *31*, 212–220. [CrossRef]
82. Katul, G.G.; Mahrt, L.; Poggi, D.; Sanz, C. One- and two-equation models for canopy turbulence. *Bound.-Layer Meteorol.* **2004**, *113*, 81–109. [CrossRef]
83. Sanz, C. A note on k-epsilon modelling of vegetation canopy air-flows. *Bound.-Layer Meteorol.* **2003**, *108*, 191–197. [CrossRef]

84. Abhijith, K.V.; Kumar, P.; Gallagher, J.; McNabola, A.; Baldauf, R.; Pilla, F.; Broderick, B.; Di Sabatino, S.; Pulvirenti, B. Air pollution abatement performances of green infrastructure in open road and built-up street canyon environments—A review. *Atmos. Environ.* **2017**, *162*, 71–86. [CrossRef]
85. Toparlar, Y.; Blocken, B.; Maiheu, B.; van Heijst, G.J.F. A review on the CFD analysis of urban microclimate. *Renew. Sustain. Energy Rev.* **2017**, *80*, 1613–1640. [CrossRef]
86. Brown, M.J.; Lawson, R.E.; Decroix, D.S.; Lee, R.L. Comparison of centerline velocity measurements obtained around 2d and 3d building arrays in a wind tunnel. *Int. Soc. Environ. Hydraul. Conf.* **2001**, *5*, 495.
87. Snyder, W.H. Similarity criteria for the application of fluid models to the study of air pollution meteorology. *Bound.-Layer Meteorol.* **1972**, *3*, 113–134. [CrossRef]
88. Zhang, Y.; Yang, X.; Yang, H.Y.; Zhang, K.; Wang, X.M.; Luo, Z.W.; Hang, J.; Zhou, S.Z. Numerical investigations of reactive pollutant dispersion and personal exposure in 3d urban-like models. *Build. Environ.* **2020**, *169*, 106569. [CrossRef]
89. Chang, C.H.; Meroney, R.N. Numerical and physical modeling of bluff body flow and dispersion in urban street canyons. *J. Wind. Eng. Ind. Aerodyn.* **2001**, *89*, 1325–1334. [CrossRef]
90. Gromke, C.; Ruck, B. Influence of trees on the dispersion of pollutants in an urban street canyon-experimental investigation of the flow and concentration field. *Atmos. Environ.* **2007**, *41*, 3287–3302. [CrossRef]
91. Chang, J.C.; Hanna, S.R. Air quality model performance evaluation. *Meteorol. Atmos. Phys.* **2004**, *87*, 167–196. [CrossRef]
92. Di Sabatino, S.; Buccolieri, R.; Olesen, H.R.; Ketzel, M.; Berkowicz, R.; Franke, J.; Schatzmann, M.; Schlunzen, K.H.; Leitl, B.; Britter, R.; et al. Cost 732 in practice: The must model evaluation exercise. *Int. J. Environ. Pollut.* **2011**, *44*, 403–418. [CrossRef]
93. Chen, T.H.; Yang, H.Y.; Chen, G.W.; Lam, C.K.C.; Hang, J.; Wang, X.M.; Liu, Y.L.; Ling, H. Integrated impacts of tree planting and aspect ratios on thermal environment in street canyons by scaled outdoor experiments. *Sci. Total Environ.* **2021**, *764*, 142920. [CrossRef]
94. Marshall, J.D.; Teoh, S.-K.; Nazaroff, W.W. Intake fraction of nonreactive vehicle emissions in us urban areas. *Atmos. Environ.* **2005**, *39*, 1363–1371. [CrossRef]
95. Greco, S.L.; Wilson, A.M.; Spengler, J.D.; Levy, J.I. Spatial patterns of mobile source particulate matter emissions-to-exposure relationships across the united states. *Atmos. Environ.* **2007**, *41*, 1011–1025. [CrossRef]
96. Chen, G.W.; Wang, D.Y.; Wang, Q.; Li, Y.G.; Wang, X.M.; Hang, J.; Gao, P.; Ou, C.Y.; Wang, K. Scaled outdoor experimental studies of urban thermal environment in street canyon models with various aspect ratios and thermal storage. *Sci. Total. Environ.* **2020**, *726*, 138147. [CrossRef]
97. Chen, G.W.; Yang, X.; Yang, H.Y.; Hang, J.; Lin, Y.Y.; Wang, X.M.; Wang, Q.; Liu, Y.L. The influence of aspect ratios and solar heating on flow and ventilation in 2d street canyons by scaled outdoor experiments. *Build. Environ.* **2020**, *185*, 107159. [CrossRef]

International Journal of
Environmental Research and Public Health

MDPI

Article

Long-Term Exposure to Phenanthrene Induced Gene Expressions and Enzyme Activities of *Cyprinus carpio* below the Safe Concentration

Xin Kang [1], Dongpeng Li [1], Xiaoxiang Zhao [1], Yanfeng Lv [1], Xi Chen [2], Xinshan Song [1], Xiangyu Liu [3], Chengrong Chen [3] and Xin Cao [1,*]

[1] Textile Pollution Controlling Engineering Center of Ministry of Environmental Protection, College of Environmental Science and Engineering, Donghua University, Shanghai 201620, China; cathy3449@163.com (X.K.); lddest@163.com (D.L.); zxx@dhu.edu.cn (X.Z.); lvyf_up@outlook.com (Y.L.); newmountain@163.com (X.S.)

[2] Agricultural Genomics Institute at Shenzhen, Chinese Academy of Agricultural Sciences, Shenzhen 518124, China; chenxi02@caas.cn

[3] Australian Rivers Institute, School of Environment and Science, Griffith University, Brisbane, QLD 4111, Australia; xiangyu.liu2@griffithuni.edu.au (X.L.); c.chen@griffith.edu.au (C.C.)

* Correspondence: caoxin@dhu.edu.cn; Tel.: +86-21-6779-2550

Abstract: Phenanthrene (PHE) is a typical compound biomagnified in the food chain which endangers human health and generally accumulates from marine life. It has been listed as one of the 16 priority PAHs evaluated in toxicology. In order to evaluate the changes of *CYP1A GST* mRNA expression and EROD GST enzyme activity in carp exposed to lower than safe concentrations of PHE. Long-term exposure of carp to PHE at lower than safe concentrations for up to 25 days. The mRNA expression level and cytochrome P450 (CYP1A/EROD (7-Ethoxylesorufin O-deethylase)) and glutathione S-transferase (GST) activity were measured in carp liver and brain tissue. The results showed that PHE stress induced low-concentration induction and high-concentration inhibition of *CYP1A* expression and EROD enzyme activity in the liver and brain of carp. In both two organs, GST enzyme activity was also induced. However, the expression of *GST* mRNA was first induced and then inhibited, after the 15th day. These results indicate that long-term exposure to PHE at lower than safe concentrations still poses a potential threat to carp's oxidase system and gene expression.

Keywords: *Cyprinus carpio*; phenanthrene ecotoxicology; cytochrome P4501A; 7-ethoxylesorufin O-deethylase; glutathione S-transferase

Citation: Kang, X.; Li, D.; Zhao, X.; Lv, Y.; Chen, X.; Song, X.; Liu, X.; Chen, C.; Cao, X. Long-Term Exposure to Phenanthrene Induced Gene Expressions and Enzyme Activities of *Cyprinus carpio* below the Safe Concentration. *Int. J. Environ. Res. Public Health* **2022**, *19*, 2129. https://doi.org/10.3390/ijerph19042129

Academic Editors: Cheng Yan and Paul B. Tchounwou

Received: 7 December 2021
Accepted: 26 January 2022
Published: 14 February 2022

1. Introduction

Polycyclic aromatic hydrocarbons (PAHs) are a widespread pollutant which derived from the incomplete combustion of various fossil fuels and hydrocarbons, such as coal and petroleum. Due to the stable chemical properties and harmful effects, PAHs have become persistent organic pollutants (POPs) and threaten environment and even human beings. Especially, PAHs are known to induce stress and affect marine organisms health [1]. According to the combined structure of benzene rings, PAHs can be divided into two categories: fused ring type and non-fused ring type. Phenanthrene (PHE, $C_{14}H_{10}$) is a representative of the fused ring PAHs [2]. It is a low-molecular-weight PAH-composed of three fused benzene rings with bay-area and K-area, and its carcinogenicity is closely related to K-area [3]. Due to the small molecular weight of PHE, it is more easily bio-amplified than other PAHs and produces toxic effects [4] like acute lethal effect on aquatic organisms at individual level [5] which further limits individual development [6], reproduction and other behaviors [7,8]. At the levels of cells, tissues, and organs, PHE can cause histopathological changes in organisms [9,10]. Carp (*Cyprinus carpio*), as a traditional freshwater economic fish in China, has been widely used in the ecotoxicity tests of environmental pollutants [11,12]. However, the study of carp genes expressions and enzymes activities after low dose PHE exposure were seldom reported. Therefore, it is necessary to explore the molecular response of carp-specific genes and characteristic metabolic enzymes in PHE pollution stress.

Current studies on low-molecular-weight PAHs in China have mainly focused on their sources and distributions in rivers and lakes [13,14]. Concentrations of PAHs in the waters of Hangzhou, Qiantang River, and Daya Bay reached 989~9663 ng/L, 70.3~1844.4 ng/L and 10984~19445 ng/L, respectively. The effects of high concentrations of PAHs on aquatic organisms were extensively explored [15,16], but the effects of pollutants below safe concentrations on aquatic organisms were still elusive. In the study, the acute toxicity of PHE to carp was tested and the 96-h lethal concentration and safe concentration of PHE to carp were determined. In the experiment, carp were exposed to the PHE at concentrations below safe levels to explore the changes in the activities of characteristic enzymes and characteristic genes expression involved in carp metabolism.

Cytochrome P450 (CYP) enzymes were reported as the biomarkers to estimate the influences of persistent organic pollutants (POPs) on various aquatic organisms [17–19]. PAHs, which are known agonists for controlling aromatic hydrocarbon receptors [20,21], could induce the expressions of Phase I and Phase II metabolism enzymes including cytochrome P4501A (CYP1A) and glutathione S-transferase (GST).

Although there have been toxicological studies on PHE stress on aquatic organisms, most of the tested organisms are zebrafish, medaka, tilapia, etc. Moreover, most of the exposure concentrations of phenanthrene are concentrated in high or low concentrations. There are not many articles about carp characteristic gene expression and characteristic enzyme activity changes in the environment of carp long-term exposure to lower than safe concentration of PHE [22–24]. Carp is a common freshwater fish in China that can be eaten by humans. It is very important to explore the mechanism of carp being enriched at a safe concentration. This study provides a more complete explanation of the mechanism of action of PHE in carp at lower than safe concentrations.

In this study, the acute toxicity test of PHE was firstly performed on carp, and the PHE concentration group was set for subacute toxicity test according to the measured 96h-LC50, and the safe concentration was calculated to be 1.12 mg/L. Although the toxicity test of PHE to aquatic organisms in natural water has been studied [25], this experiment is higher than that of PHE in natural water but lower than the safe concentration of carp. Long-term exposures were conducted to explore the chronic effects of PHE on carp. The study aimed to explore the activity changes of Phase I and Phase II metabolic characteristic enzymes (cytochrome P450 enzyme (EROD) and glutathione S-transferase (GST) in carp under the exposure to PHE below safe concentration in different organs and determine the changes in the expression levels of characteristic genes of *CYP1A* and *GST*, which controlled the activities of EROD and GST. In addition, the correlations between the expressions of characteristic enzymes (EROD, GST) from metabolic process and the expressions of characteristic genes (*CYP1A*, *GST*) under the exposure to PHE at or below safe concentration were further discussed. We illustrated the responses of carp metabolizing characteristic genes and characteristic enzymes to low concentrations of PHE.

2. Materials and Methods

2.1. Main Reagents

Phenanthrene (PHE, 95%) was bought from Beijing Braun Technology Ltd. Diethy pyrocarbonate (DEPC), Spin Column Animal Total RNA Purification Kit, and M-MuLV First Strand cDNA Synthesis Kit were obtained from Sangon Biotech (Shanghai, China). Taq Plus PCR MasterMix was bought from Tiangen Biotech (Beijing, China). Agarose was bought from Aladdin. TAE loading buffer and DNA loading buffer were configured according to general methods [26]. Fish 7-ethoxyresorufin-o-deethylase (EROD) kit and fish glutathione S transferase (GST) kit came from Jianglai Biotechnology (Shanghai, China). Fish ELISA kits of the enzymes of EROD and GST were bought from Jianglai Biotechnology (Shanghai,

China). Albumin from bovine serum (BSA) was bought from Jiancheng Bioengineering (Nanjing, China).

2.2. Fish and Treatment

Carp (*Cyprinus carpio*, body weight (9.0~11.0 g) and body length (8.0 ± 10.0 cm)) were obtained from a fish farm in Songjiang, Shanghai, China. The experimental carp were domesticated in seven groups of 10-L glass tanks under natural conditions for 2 weeks, each group for 10 fish. All tanks were supplied with continuous aeration to maintain nearly saturated dissolved oxygen. Dechlorinated tap water was used at a temperature of 20 ± 2 °C, pH 6.8~7.3, with light intensity of 100 Lux for 10 h/d for 25 d and fed twice a day and clean up metabolites on time to ensure that the mortality rate during domestication is less than 1%, and 4 d before the start of the experiment, ensure that there was no fish death. Three treatment groups were exposed to waterborne PHE in acetone under the concentrations of 0.1, 0.5, and 1.0 mg/L (the final content of acetone was less than 0.5%) In addition, the 0 mg/L PHE as the negative control (dechlorinated water) was established. The parallel groups of each concentration were set in other three 10-L tanks at 20.0 ± 2 °C. After the exposure for 1 days, 5 days, 15 days and 25 days (We define 1–5 days as "early period"), the brain and liver of fish were dissected, placed in 1.5-mL microcentrifuge tubes, frozen in liquid nitrogen immediately, and then stored at −80 °C until further analysis. A part of one sample was ground fully in a vessel with liquid nitrogen to analyze the expressions of relevant genes. Another part of the sample was ground fully with a glass homogenate device with PBS to detect the activities of related enzymes. All samples are homogenized before separation.

2.3. Acute Toxicity Test

The PHE dose was set to 5, 10, 20, 30, 40, 50 mg/L (the final content of acetone was less than 0.5%) using the aquatic toxicity test method, and each group was 10 pieces of fish. The pre-test first showed that the carp did not die in 96 h. The highest dose and the lowest dose for all deaths in 24 h. During the test, no feed was fed, semi-static exposure was performed, and the water was changed 50% every day. The activity status of the carp was observed over time and the number of dead fish was counted in time.

The acute toxicity experiment was conducted according to the SECP-Part 12: Fish Acute Toxicity Test (GB/T 31270.12-2014). PHE in eight tanks was set according to equidistant logarithmic concentrations at 5.01, 6.31, 7.94, 10.00, 12.59, 15.85, 19.95, and 25.12 mg/L and there were 10 fish in each tank. Three replicates were arranged for each concentration. In the acute test, after 96 h, the mortality was recorded. The carp were not fed during the test period and water was replaced once a day.

Dead fish were identified according to the following method. Carp were transferred to clear water for 30 s and the fishtail was touched with a glass rod. If there was no visible response, carp were considered to be dead.

The safe concentration (SC) is expressed as follows:

$$(SC) = \frac{(LC50(96\text{h}))}{10} \quad (1)$$

2.4. RNA Extraction and Analysis

All samples from the experimental fish were firstly ground into fine powder under liquid nitrogen. Total RNA was extracted with Trizol Reagent (Sangon, China) according to the manufacturer's instructions. Then the OD (Optical density) values of samples at 260 nm and 280 nm were measured. The RNA sample purity was calculated with Equation (2). Finally, based on the calculation results, the integrity of RNA was detected by 1.5% agarose gel electrophoresis (Figure S1).

The purity of RNA is expressed as:

$$\omega = \frac{A_{260}}{A_{280}} \quad (2)$$

If ω is between 1.8 and 2.0, the RNA sample is qualified; if ω is less than 1.8, the sample is contaminated; if ω is larger than 2.0, the RNA sample is degraded.

2.5. Gene Primer Design and Semi-Quantitative RT-PCR (Sq-RT-PCR) Analysis

After RNA extraction, the RNA was reversely transcribed with the cDNA Synthesis Kit (Sangon, China) according to the manufacturer's protocol (Figures S2 and S3). Then, the mRNA levels were expressed as the ratios relative to the transcription level of the β-action. Primer Premier 5.0 was used to design the primers according to two ends of the target gene based on the sequence of carp *CYP1A* (accession: AB048939.1) and *GST* (accession: LC071505.1) (Table 1).

Table 1. Semi-quantitative RT-PCR (Sq-RT-PCR)primers of *Cyprinus carpio* genes.

Gene	GeneBank Number	Primer Sequence	Product Size [11]
β-actin	AF0570 40	F: CCATCTACGAGGGTTACGCC R: AATGCCAGGGTACATGGTGG	551 bp
CYP1A	AB048939.1	F: CTGAGCCTGACCGCTATGAG R: CCGCTTCCTACGATCTTCCC	503 bp
GSTs	LC071505.1	F: CCGCTTCCTACGATCTTCCC	541 bp

We performed Sq-RT-PCR to check the changes in *GST* and *CYP1A* mRNA levels in the carp liver and your brain. For Sq-RT-PCR amplification, each reaction included 10 μL Taq Plus PCR MasterMix, 0.4 μL 10 μmol·L^{-1} specific primer F, 0.4 μL 10 μmol·L^{-1} specific primer R and 1 μL cDNA. (all are the configuration in the kit instructions). The reaction conditions are as follows: 94 °C/5 min; 35 cycles of 94 °C/30 s, 55 °C/30 s, 72 °C/30 s; and 72 °C/10 min. We used the β-actin gene as a reference sample for the relative expression levels between normalizations. The product was subject to agarose gel electrophoresis to detect the amplified product, using the UVitec Cambridge scanning system and the software ImageQuant TL (GE Healthcare). All mRNA data represent normalization for any difference in reverse transcriptase efficiency relative to β-actin.

2.6. EROD and GST Activities

The total protein content was determined with the method of Bradford protein assay [27]. EROD activity was determined with the EROD Synthesis Kit (Jianglai, China) according to the manufacturer's protocol. The samples of the brain and liver were homogenized in an ice-cold homogenization buffer (0.125 M $Na_2HPO_4 \cdot 12H_2O$, 0.125 M KH_2PO_4, 0.05 mM Na_2EDTA, pH = 7.7) and the supernatant was prepared by centrifugation (10,000 rpm, 10 min). The sample mixture (50 μL) was firstly placed in an enzyme-labeled plate and then 100 μL of HRP-labeled enzyme-labeled antibody was added for the reaction at 37 °C for 10 min. After the supernatant was poured, the PBS solution was added and products were washed five times. Reactions were stopped by adding 50 μL of 2 M H_2SO_4 into each tube and the supernatant was determined at OD_{450}. The GST activities were determined with GST Synthesis Kit (Jianglaibio, China) according to the manufacturer's protocol. Other steps were similar to those of the EROD kit.

2.7. Statistical Analysis

In the statistical analysis, the data were expressed as mean value ± standard deviation (SD). In SPSS 20.0 software (IBM software, Inc., Chicago, IL, USA), one-way ANOVA was carried out. Duncan's method was used to make the multiple comparisons of the means and $p < 0.05$ indicates a significant difference. All figures were drawn by Excel. A correlation analysis was performed to evaluate the relationship between EROD activity and *CYP1A* mRNA expression, as well as the correlation between GST activity and *GST* mRNA expression.

The Gel imaging analysis system (GIS) can quantify the optical density of DNA bands into a spectrum, and automatically integrate it to calculate the response value. The size of the response value can indirectly reflect the content of DNA. Therefore, the amount of mRNA expression from DNA can be obtained by the inverse calculation.

3. Results and Discussion

3.1. Median Lethal Concentration (LC50) and Safe Concentration of Carp

According to the experimental procedure of Section 2.3, the acute toxicity dose gradient pre-experiment was first carried out, and the lowest and highest doses of carp death at 24 h and 96 h were recorded: the highest dose of PHE that did not cause death in 96 h was 5 mg/L, the lowest dose of all death in 24 h was 40 mg/L, and when the dose was 20 mg/L, the 96-h mortality rate of carp was about 90%. At 30 mg/L, all carps died within 96 h (Data not shown). Therefore, the acute toxicity dose gradient exposure is set in the range of 5–25 mg/L.

The lethal concentration of PHE for carp are shown in Table 2. With the increase in the concentration of PHE, the mortality of carp also increased, displaying a significant dose-toxicity effect. Furthermore, the regression equations between the probability unit of carp mortality and the logarithmic concentration of PHE after 24-h (Equation (3)) and 96-h (Equation (4)) cultivation were respectively obtained as follows:

$$Y = 3.970\,2X + 0.2275,\ r^2 = 0.9388 \tag{3}$$

$$Y = 0.479\,8X + 0.30,\ r^2 = 0.9229 \tag{4}$$

(Y: Probability unit of mortality; X: logarithmic concentration).

Table 2. LC50 of *Cyprinus carpio* after 24-h and 96-h exposure to PHE.

Variable	Lethal Concentration							
Concentration/(mg·L^{-1})	5.01	6.31	7.94	10.00	12.59	15.85	19.95	25.12
Log concentration	0.7	0.8	0.9	1.0	1.1	1.2	1.3	1.4
24 h mortality rate/%	0	0	20	20	30	60	70	70
Unit of probability	3.04	3.04	4.16	4.16	4.48	5.25	5.52	5.52
96 h mortality rate/%	10	20	20	30	40	70	90	100
Unit of probability	3.72	4.16	4.16	4.48	4.75	5.52	6.28	6.96

The carp's LC50 value of PHE after 24 h and 96 h were respectively determined as 15.926 mg/L (95% confidence interval: 14.675–17.282 mg/L) and 11.198 mg/L (95% confidence interval: 9.950–12.604 mg/L). According to the classification standard for acute toxicity of fish [8], PHE is highly toxic to carp. According to Equation (1), the safe concentration of PHE to carp was calculated to be 1.12 mg/L.

3.2. Gene Extraction Efficiency

DNA amplification samples were subjected to preset denaturation temperature, cycle parameters, and final repair extension on a gradient PCR instrument. The agarose gel electrophoresis result of 5 μL of samples is shown in Figure S4. The three genes were loaded and electrophoresed simultaneously. From left to right, they were respectively *CYP1A*, *GST*, and *β-ACTIN*, a commonly used internal reference gene. According to the mobility of each DNA and the indication of the marker, the length of DNA fragments in the PCR sample was between 500 bp and 550 bp. Compared with the β-*ACTIN* gene, the *GST* gene fragment was shorter and had a higher electrophoretic mobility, followed by

CYP1A. The target genes were specifically amplified and showed good integrity. Therefore, the samples could be used for the semi-quantitative analysis, recovery, and sequencing for further gene comparison.

The sequencing and splicing results of *CYP1A* and *GST* genes and alignment results obtained from the gene bank (NCBI, National Center for Biotechnology Information) are shown in Table 3. The lengths of the two amplified genes (*CYP1A* and *GST*) were respectively 503 bp and 541 bp. Similarly, the amplified gene had a high similarity of greater than 99% to the target gene in the gene bank. Therefore, the entire RNA extraction, reverse transcription and PCR operations could be qualified.

Table 3. Alignment results of sequenced results of amplified target genes [a].

Genes	Gene Length/bp	Accession No. [b]	Gene Similarity/%
CYP1A	503	AB048939.1	99.7
GST	541	LC071505.1	99.5

[a] The splicing results and peak images of the sequencing results are shown in the supplementary information.
[b] The serial number is the gene number in GeneBank (NCBI).

3.3. *CYP1A* mRNA Expression and EROD Activity after the Exposure to PHE

The *CYP1A* mRNA expression in carp tissues are showed in Figure 1. We examined the effect of 0 mg/L PHE in the brain and liver as negative control. Pearson correlation indicated the significant correlation between *CYP1A* mRNA concentration and time in both brain and liver. *CYP1A* mRNA expression levels in the liver showed a significant increase compared with those in the control group during 25-day exposure to PHE (Figure 1a) and the induction trend gradually became more significant during the exposure. *CYP1A* mRNA levels were induced significantly by 0.50 mg/L PHE during the 25-day exposure. In the brain, *CYP1A* mRNA levels were significant induced compared with those in the control, but the induction effect was not significantly enhanced with the increase in exposure time or doping concentration (Figure 1b).

To detect whether the change in *CYP1A* mRNA brought about the change of related metabolic characteristic enzymes, the enzyme activity of EROD was tested (Figure 1) We examined the effect of 0 mg/L PHE in the brain and liver as negative control. In the liver, the EROD activity did not change significantly in the initial stage, but all experimental groups resulted in a significant EROD induction in the liver in the 15th day and 25th day (Figure 1c). In the brain, the activity of EROD in the early period of PHE exposure (Day 1, 5 and 15) was significantly induced compared with the control assays, ($p < 0.05$). However, on Day 25, the induction was not significant (Figure 1d). The activity induction trend of the EROD enzyme in the liver and brain is consistent with the mRNA expression trend of *CYP1A*. It shows that long-term exposure of carp to a safe concentration of PHE will stimulate the expression of *CYP1A* mRNA in the liver of carp and the increase in mRNA expression of *CYP1A* reflects an increase in protein functionality [28].

It widely reported that when an organism is exposed to pollutants such as heavy metals and persistent organic pollutants (POPs), the activity of antioxidant enzymes will be affected [29]. Cytochrome P450 enzyme (CYP450) is widely present in organisms and is an important metabolic enzyme involved in the conversion of many heterologous compounds [30]. CYP1A is a member of the cytochrome P450 family. CYP1A enzyme activity is considered to be an indicator of fish resistance to PAH.

Previous studies have shown that mild oxidative stress can induce the activity of antioxidant enzymes such as SOD and GST [31]. However, when oxidative stress reaches a certain level, the activity of antioxidant enzymes will be inhibited [32]. In this study, the expression of *CYP1A* mRNA and the activity of EROD enzyme in the liver and brain increased with time and concentration. Moreover, the level of induction in the liver is more pronounced than in the brain. The liver is a key site of detoxification and an important target organ of PAHs [33]. EROD is the important enzyme assisting in the metabolism of toxic compounds. If pollutants enter the body, the organism starts stress reactions.

The differential responses of transcription and expression of characteristic genes can be used as early warning parameters to measure the degree of environmental pollution. The change in *CYP1A* mRNA expression in the liver and brain of carp was the consequence of the stimulation of the signaling pathway. If the concentration of PHE was too high, it might cause damage and inhibit the gene expression and enzyme activity [20]. Similar to previous reports, after intraperitoneal injection of TCDD in goldfish, it was also found that the expression of *CYP1A* mRNA was induced in various tissues, and the induction effect was most significant in liver tissues. It was reported that after the exposure of medaka to pentachlorobiphenyl, CYP1B1 and CYP1C1 mRNA expressions were induced due to the acceleration of Phase I metabolic response [34]. If a pollutant entered the carp body, the organism could defend itself by stimulating the production of Phase I- and Phase II-related enzymes degrading external compounds. However, if excessive external substances existed, the body defense mechanism became destroyed.

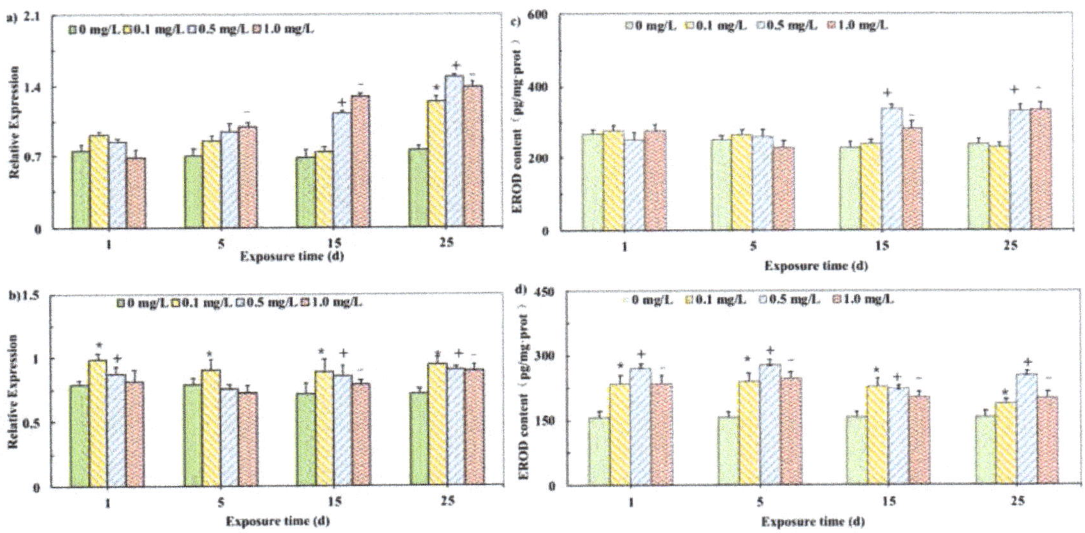

Figure 1. Relative mRNA expression of *CYP1A* in liver (**a**) and brain (**b**) of carp exposed to PHE. Effects of different PHE concentrations on EROD activities in carp (**c**) in the liver. (**d**) in the brain (The final expression results are based on the content of characteristic proteins in the mass of each tissue unit of the carp.). Expression was quantified by Sq-RT-PCR. Values are mean ± SD of three replicates. Significant differences, $p < 0.05$. (*) are under 0.1 mg/L versus negative control of 1 d; (+) are under 0.5 mg/L versus negative control of 1 d; (−) are under 1.0 mg/L versus negative control of 1 d, which is based on Duncan's method.

A similar phenomenon was observed in the liver of tilapia (*Oreochromis niloticus*). The short-term exposure to a low concentration of PHE induced EROD activities in the liver of tilapia *(Oreochromis niloticus)*, whereas the long-term exposure to a high concentration of PHE inhibited EROD activities [35]. It was also found that several PAHs with lower molecular weights caused the differential expressions of P450 enzymes [36]. It was also reported that a low concentration of PHE activated EROD in young *Sparus aurata*, but inhibited it under high concentration [37]. In the brain and liver of carp, the activity of EROD showed an obvious correlation with the expression level of *CYP1A*. Many compounds could induce EROD activity of fishes and the induction of EROD activity might be impeded as chemicals were competitively bound to the structure of AhR or *CYP1A* [38].

3.4. GST mRNA Expression and GST Activity after the Exposure to PHE

The *GST* mRNA expressions in carp tissues are shown in Figure 2. In the liver, the expression of *GST* mRNA was always induced when carp was exposed to 0.1 mg/L PHE and the induction effect was weakened in the 25th day. However, the exposure to 0.5 and 1.0 mg/L PHE showed a significant inhibition effect on *GST* mRNA expression from the first day to the 25th day (Figure 2a). In the brain, the experimental groups of different concentrations showed the significant induction in the first day and the induction effect was reduced in the 5th day. During the PHE exposure period, the expression of *GST* mRNA showed a significant inhibitory effect in the 15th day and the 25th day (Figure 2b).

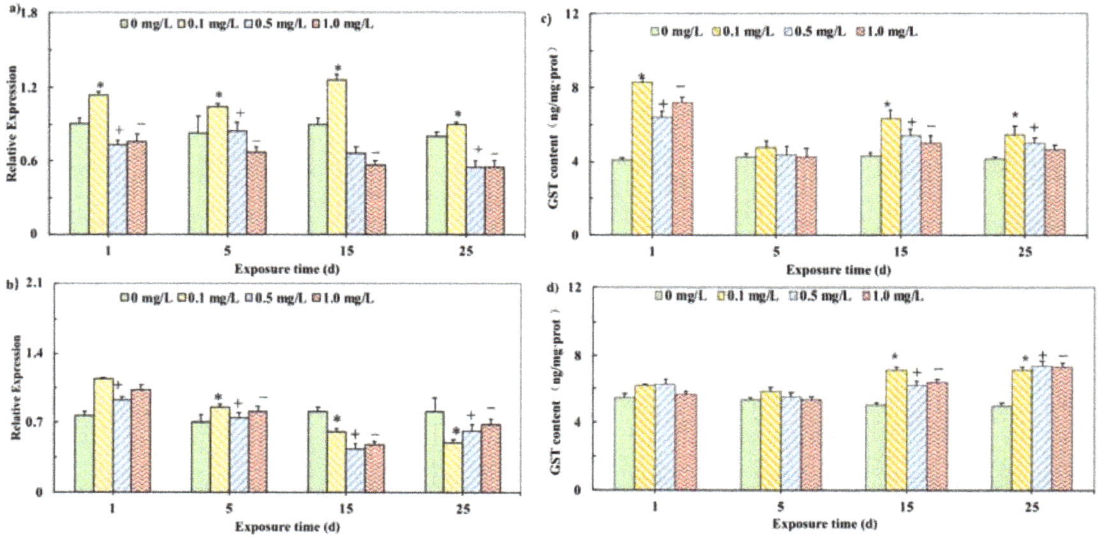

Figure 2. Relative mRNA expression of *GST* in liver (**a**) and brain (**b**) of carp exposed to PHE. Effects of different PHE concentrations on GST activities in the liver (**c**). and the brains (**d**) of carp (The final expression results are based on the content of characteristic proteins in the mass of each tissue unit of the carp.) Expression was quantified by Sq-RT-PCR. Values are mean ± SD of three replicates. Significant differences, $p < 0.05$ [26]. (*) are under 0.1 mg/L versus negative control of 1 d; (+) are under 0.5 mg/L versus negative control of 1 d; (−) are under 1.0 mg/L versus negative control of 1 d, which is based on Duncan's method.

In order to detect whether the change in *GST* mRNA level brought about the change in related characteristic metabolic enzymes, the enzyme activity of GST was tested (Figure 2). In the liver, the GST activities of all experimental groups were significantly stimulated after induction ($p < 0.05$), but the induction effect was gradually weakened (Figure 2c). There was no obvious change in the GST activity at the beginning of the experiment (from the first day to the 5th day) in the brain. However, during the PHE exposure period, the GST activities in the 15th and 25th days had a significant induction effect ($p < 0.05$) (Figure 2d).

GST can catalyze the binding of exogenous compounds such as insecticides, herbicides, and antibiotics to glutathione, and avoids damaging other cell biomolecules, thus playing an important role in combating oxidative stress [39]. In this study, the GST enzyme activity of the liver and brain were both induced by exposure to lower than safe concentrations of phenanthrene. This is similar to the previous report: The reports found that five kinds of PAHs induced the GST activity in the carp liver [40]. Yin et al. also reported that the GST activity in the tissues of catfish was significantly increased after the exposure to PHE [41]. Nahrgang et al. reported that benzo (a) pyrene induced the GST activity in cod liver [20].

However, Olinga et al. found that the short-term exposure to PHE inhibited GST activity in the kidney of *Liza aurata* [42].

However, the expression of mRNA in the liver and brain of carp is not synchronized with the changes in GST enzyme activity. mRNA expression in the liver was inhibited at concentrations of 0.5 and 1.0 mg/L. mRNA expression in the brain is first induced and then suppressed over time. This phenomenon of asynchrony between enzyme activity and gene expression also exists in previous reports: Sun et al. reported that after yellow catfish was exposed to Hg^{+}, the related *SOD* and *GPX1* mRNA expression levels increased significantly, but the SOD enzyme and GPX1 enzyme activities did not change much [43]. The analysis of the cause of this phenomenon may be due to the fact that antioxidant enzymes such as GST are encoded by multiple subtype genes. However, semi-quantitative genetic testing only analyzes one subtype gene, so changes in mRNA levels do not necessarily occur at the enzyme level. On the other hand, there is a time lag effect in the process of gene transcription, translation and protein modification, so the enzyme activity cannot reflect the mRNA level in time [44]. In the study, the expression of *GST* mRNA in carp liver and brain showed the early induction and subsequent inhibition effects, indicating that after the Phase I metabolic reactions, the biological activity of GST in the tissues of carp gradually became unbalanced during the exposure.

3.5. Correlation Analysis

The correlation analysis was carried out to test the expression levels of characteristic enzymes and characteristic mRNA (Figure 3). The correlation coefficient indicates that EROD activity in the liver showed a positive correlation with *CYP1A* mRNA level (r = 0.602, $p < 0.01$) (Figure 3a). In the brain, the correlation between EROD activity and *CYP1A* mRNA level was less significant than that in the liver (r = 0.508, $p < 0.01$) (Figure 3b). However, in the liver or brain, the correlation between GST activity and *GST* mRNA expression was relatively poor (liver, r = 0.395, $p < 0.05$; brain, r = 0.293, $p < 0.05$) (Figure 3c,d).

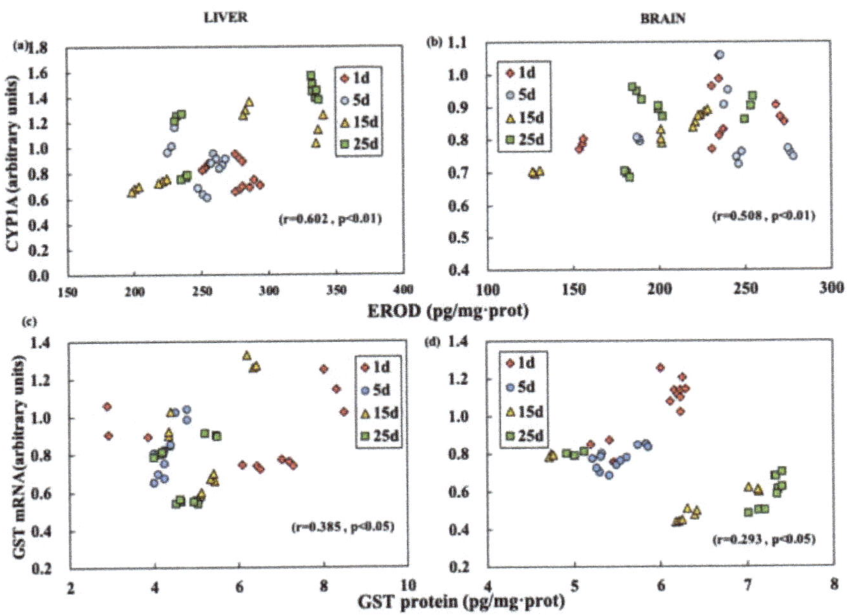

Figure 3. Correlation of EROD and *CYP1A* mRNA (**a**,**b**), GST and *GST* mRNA (**c**,**d**) values in brain (**b**,**d**) and liver (**a**,**c**) at different exposure time.

In the study, the activity of EROD in the liver was more highly correlated with the activity of *CYP1A* mRNA expression. The liver is an important metabolism organ of heterogeneous organisms and the blood mediates the relationship between the brain and other target organs and accelerates the reaction between the characteristic liver enzymes and its mRNA expression. Similar experimental results had been reported in previous studies [7,45]. After the brain and muscle were exposed to soluble components of crude oil, the changes in AchE activity were not obvious compared to other organs [46]. In addition, the toxicity induced by CYP1A indicated the chemical exposure and preferential effects in various biological tissues [38].

The cytochrome P450 system showed a highly correlation than the GST system in both the liver and brain. The cytochrome P450 enzyme system played an important role in the heterogeneous biotransformation in Phase I metabolism of fish and other aquatic animals. The activity of ethoxysalolin O-deethylase (EROD) and the *CYP1A* mRNA level seemed to be the most sensitive catalytic probe. The induction response of the cytochrome P450 system had been determined in many studies [47,48]. The GST is a representative enzyme in Phase II metabolism and play an important role of catalyzing or reducing oxidative substances in the metabolic body. In most cases, only the modest changes in total GST activity were reported [49]. Similar studies reported that *GSTα* mRNA expression in all tissues showed no significant difference compared with *CYP1A* and ABC efflux transporters after the exposure to benzo(*a*)pyrene [45].

In addition, there was a significant decrease in *GST* mRNA in the brain after 15 days of PHE exposure. The higher GST activity and the lower *GST* mRNA expression in the 15th day might be interpreted as follows. From the first day to the 5th day, the activity of GST increased, thus increasing the clearance rate of PHE and reducing the overall mRNA response and *GST* synthesis. As a result, the mRNA expression level was reduced (Figure 2). However, the same phenomenon was not observed in the liver, indicating the differences in the heterogeneous metabolism and PHE-induced enzymes between the brain and liver [20].

3.6. Mechanism

The possible mechanism of PHE exposed to carp is shown in Figure 4. When PHE entered the carp body, the characteristic enzymes of the I-phase reaction were activated to remove pollutants, which resulted in an increase in the expression of *CYP1A* gene. As the body's defense oxidation system proceeds to the II-phase reaction, the expression of GST and *GST* genes is stimulated.

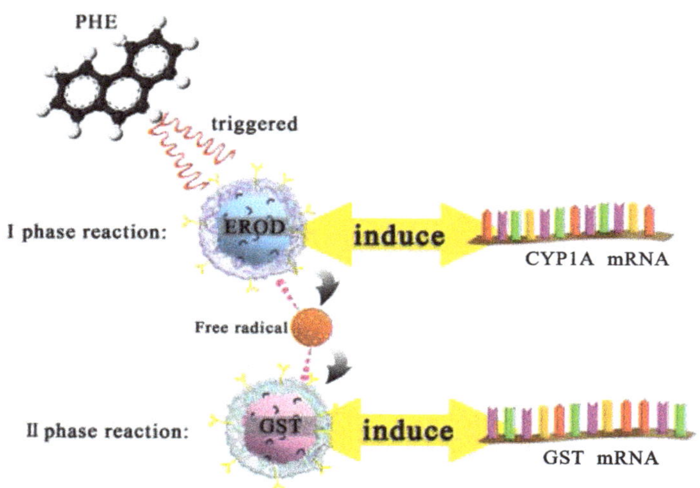

Figure 4. The metabolic mechanism of PHE exposed to carp.

4. Conclusions

In conclusion, our study used carp as the test organism to perform acute toxicity tests and calculate safe concentrations. Exposure of carp to sub-safe concentrations of PHE for 25 days resulted in changes in oxidative stress. The expression of EROD and GST enzymes in the liver was stimulated from day 1 of exposure to induce the expression of *CYP1A* and *GST* genes. The expression activities of EROD and GST enzymes in the brain were stimulated, but the expression of *GST* gene was delayed. Correlation analysis also illustrates this phenomenon. These results suggest that long-term exposure of carp to sub-safe concentrations still affects its oxidase system and gene expression. The regulation of carp's own molecular level under the stress of exogenous pollutants was revealed.

Supplementary Materials: The following supporting information can be downloaded at: https://www.mdpi.com/article/10.3390/ijerph19042129/s1, Figure S1: Electrophoresis diagram of RNA sample (Lanes 1 and 2 indicate the electrophoresis of liver and brain RNA of carp domesticated in clean water, and lanes 3 and 4 are the extraction of total RNA from tissues after PHE stress. Figure S2: GST amplified gene sequencing stitching results. Figure S3: CYP1A amplified gene sequencing stitching results. Figure S4: Electrophoresis diagram of each gene sample. The expression levels of target genes CYP1A and GST were detected by genomic PCR analysis, and β-ACTIN was used as a template control. From top to bottom are the cDNA gel electrophoresis diagrams of PHE stress concentrations of 0, 0.1, 0.5, 1.0 mg/L.

Author Contributions: X.K. developed the methodology and conducted a research and investigation process. X.K. and X.C. (Xin Cao) wrote the first draft of the article. Y.L. and D.L. carried out experiments and had helped the experiment testing and data collection. X.Z. had oversight and leadership responsibility for the research activity planning. X.S. gave a contribution to conceptualization. X.C. (Xin Cao) and X.C. (Xi Chen) revised and edited the draft. X.L. and C.C. reviewed the manuscript. All authors have read and agreed to the published version of the manuscript.

Funding: This research was funded by the National Natural Science Foundation of China grant number No. 51909034 and 52170152; the National Key Research and Development Project grant number No. 2019YFC0408603 and 2019YFC0408604; the Research Project of Shanghai Science and Technology Commission grant number No. 21DZ1202402; the Shanghai Sailing Program grant number No. 19YF1401900. And The APC was funded by the National Natural Science Foundation of China grant number No. 51909034.

Institutional Review Board Statement: This is to certify that the fish that were used in this experiment are in accordance with 'REGULATIONS FOR THE ADMINISTRATION OF AFFAIRS CONCERNING EXPERIMENTAL ANIMALS' which is approved by the State Council on October 31, 1988 and promulgated by Decree No. 2 of the State Science and Technology Commission on November 14, 1988. The fish were caught by permission of the Shanghai Ministry of Environmental Protection. They were handled in such a way to minimize stress and discomfort.

Informed Consent Statement: Not applicable.

Data Availability Statement: Data available on request due to restrictions eg privacy or ethical The data presented in this study are available on request from the corresponding author. The data are not publicly available due to confidentiality agreement.

Acknowledgments: The authors acknowledge the financial support from the National Natural Science Foundation of China (Grant No. 51909034 and 52170152); the National Key Research and Development Project (Grant No. 2019YFC0408603 and 2019YFC0408604); the Research Project of Shanghai Science and Technology Commission (Grant No. 21DZ1202402); and the Shanghai Sailing Program (Grant No. 19YF1401900).

Conflicts of Interest: The authors declare that they have no known competing financial interest or personal relationships that could have appeared to influence the work reported in this paper.

Abbreviations

AhR	aromatic hydrocarbon receptors
BSA	Albumin from bovine serum
CYP1A	cytochrome P4501A
DEPC	Diethy pyrocarbonate
EROD	7-Ethoxylesorufin O-deethylase
GST	glutathione S-transferase
PAHs	polycyclic aromatic hydrocarbons
PBS	Phosphate buffer solution
PHE	Phenanthrene
POPs	persistent organic pollutants
Sq-RT-PCR	semi-quantitative reverse transcription and polymerase Chain reaction

References

1. Carmen, S.; Helmut, S. Cytochrome P4501A (CYP1A) in teleostean fishes. A review of immunohistochemical studies. *Sci. Total Environ.* **2000**, *247*, 313–332. [CrossRef]
2. Wda, B.; Min, L.; Ye, L.; Jza, B.; Xwa, B.; Jing, Y.; Yha, B.; Dza, B.; Dga, B.; Aqa, B. Cross-interface transfer of polycyclic aromatic hydrocarbons (PAHs) in a shallow urban lake in Shanghai, China based on the fugacity model—ScienceDirect. *Sci. Total Environ.* **2020**, *736*, 139369. [CrossRef]
3. Harvey, R.G.; Harvey, R.G. *Polycyclic Aromatic Hydrocarbons: Chemistry and Carcinogenicity*; Cambridge University Press: Cambridge, UK, 1991. [CrossRef]
4. Maskaoui, K.; Zhou, J.L.; Hong, H.S.; Zhang, Z.L. Contamination by polycyclic aromatic hydrocarbons in the Jiulong River Estuary and Western Xiamen Sea, China. *Environ. Pollut.* **2002**, *118*, 109–122. [CrossRef]
5. Zheng, Y.; Li, Y.; Yue, Z.; Samreen; Wang, J. Teratogenic effects of environmentally relevant concentrations of phenanthrene on the early development of marine medaka (*Oryzia melastigma*). *Chemosphere* **2020**, *254*, 126900. [CrossRef]
6. Lotufo, G.R.; Fleeger, J.W. Effects of sediment-associated phenanthrene on survival, development and reproduction of two species of meiobenthic copepods. *Mar. Ecol.-Prog. Ser.* **1997**, *151*, 91–102. [CrossRef]
7. Nam, T.H.; Kim, L.; Jeon, H.J.; Kim, K.; Ok, Y.S.; Choi, S.D.; Lee, S.E. Biomarkers indicate mixture toxicities of fluorene and phenanthrene with endosulfan toward earthworm (*Eisenia fetida*). *Environ. Geochem. Health* **2017**, *39*, 307–317. [CrossRef]
8. Sobanska, M.; Scholz, S.; Nyman, A.M.; Cesnaitis, R.; Alonso, S.G.; Kluever, N.; Kuehne, R.; Tyle, H.; De, K.J.; Dang, Z. Applicability of the fish embryo acute toxicity (FET) test (OECD 236) in the regulatory context of Registration, Evaluation, Authorisation, and Restriction of Chemicals (REACH). *Environ. Toxicol. Chem.* **2018**, *37*, 657–670. [CrossRef]
9. Horng, C.Y.; Lin, H.C.; Lee, W. A Reproductive Toxicology Study of Phenanthrene in Medaka (*Oryzias latipes*). *Arch. Environ. Contam. Toxicol.* **2010**, *58*, 131–139. [CrossRef]
10. Sun, S.; Wang, H.; Fu, B.; Zhang, H.; Xu, J. Non-bioavailability of extracellular 1-hydroxy-2-naphthoic acid restricts the mineralization of phenanthrene by *Rhodococcus* sp. WB9. *Sci. Total Environ.* **2019**, *704*, 135331. [CrossRef]
11. Woo, S.J.; Kim, N.Y.; Kim, S.H.; Ahn, S.J.; Seo, J.S.; Jung, S.H.; Cho, M.Y.; Chung, J.K. Toxicological effects of trichlorfon on hematological and biochemical parameters in *Cyprinus carpio* L. following thermal stress. *Comp. Biochem. Physiol. C-Toxicol. Pharmacol* **2018**, *209*, 18–27. [CrossRef]
12. Sidika, S.; Giang, P.T.; Burkina, V.; Zamaratskaia, G. The effects of sewage treatment plant effluents on hepatic and intestinal biomarkers in common carp (*Cyprinus carpio*). *Sci. Total Environ.* **2018**, *635*, 1160–1169. [CrossRef]
13. Huang, Y.; Sui, Q.; Lyu, S.; Wang, J.; Yu, G. Tracking emission sources of PAHs in a region with pollution-intensive industries, Taihu Basin: From potential pollution sources to surface water. *Environ. Pollut.* **2020**, *264*, 114674. [CrossRef] [PubMed]
14. Yang, T.; Diao, X.; Cheng, H.; Wang, H.; Chen, C.M. Comparative study of polycyclic aromatic hydrocarbons (PAHs) and heavy metals (HMs) in corals, sediments and seawater from coral reefs of Hainan, China. *Environ. Pollut.* **2020**, *264*, 114719. [CrossRef] [PubMed]
15. Mager, E.M.; Pasparakis, C.; Stieglitz, J.D.; Hoenig, R.; Morris, J.M.; Benetti, D.D.; Grosell, M. Combined effects of hypoxia or elevated temperature and Deepwater Horizon crude oil exposure on juvenile mahi-mahi swimming performance. *Mar. Environ. Res.* **2018**, *139*, 129–135. [CrossRef]
16. Wilson, H.K.; Buckeridge, S.A.; Yau, W.; Howerth, E.W.; Gato, W.E. Investigating the toxic effects of 2-aminoanthracene ingestion in pregnant Sprague Dawley dams. *J. Environ. Sci. Health Part. B-Pestic. Contam. Agric. Wastes* **2018**, *53*, 283–289. [CrossRef]
17. Mortensen, A.S.; Tolfsen, C.C.; Arukwe, A. Gene Expression Patterns in Estrogen (Nonylphenol) and Aryl Hydrocarbon Receptor Agonists (PCB-77) Interaction Using Rainbow Trout (*Oncorhynchus mykiss*) Primary Hepatocyte Culture. *J. Toxicol. Env. Health Part. A* **2006**, *69*, 1–19. [CrossRef]
18. Kais, B.; Ottermanns, R.; Scheller, F.; Braunbeck, T. Modification and quantification of in vivo EROD live-imaging with zebrafish (*Danio rerio*) embryos to detect both induction and inhibition of CYP1A. *Sci. Total Environ.* **2017**, *615*, 330–347. [CrossRef]
19. Roy, M.A.; Sant, K.E.; Venezia, O.L.; Shipman, A.B.; Timme-Laragy, A.R. The emerging contaminant 3,3′-dichlorobiphenyl (PCB-11) impedes Ahr activation and Cyp1a activity to modify embryotoxicity of Ahr ligands in the zebrafish embryo model (*Danio rerio*). *Environ. Pollut.* **2019**, *254*, 113027. [CrossRef]

20. Nahrgang, J.; Camus, L.; Gonzalez, P.; Goksoyr, A.; Christiansen, J.S.; Hop, H. PAH biomarker responses in polar cod (*Boreogadus saida*) exposed to benzo(a)pyrene. *Aquat. Toxicol.* **2009**, *94*, 309–319. [CrossRef]
21. Van der Oost, R.; Beyer, J.; Vermeulen, N.P.E. Fish bioaccumulation and biomarkers in environmental risk assessment: A review. *Environ. Toxicol. Pharmacol* **2003**, *13*, 57–149. [CrossRef]
22. Mai, Y.; Peng, S.; Li, H.; Lai, Z. Histological, biochemical and transcriptomic analyses reveal liver damage in zebrafish (*Danio rerio*) exposed to phenanthrene. *Comp. Biochem. Physiol. C-Toxicol. Pharmacol.* **2019**, *225*, 108582. [CrossRef] [PubMed]
23. Mu, J.; Jin, F.; Wang, J.; Wang, Y.; Cong, Y. The effects of CYP1A inhibition on alkyl-phenanthrene metabolism and embryotoxicity in marine medaka (*Oryzias melastigma*). *Environ. Sci. Pollut. Res.* **2016**, *23*, 11289–11297. [CrossRef] [PubMed]
24. Shirmohammadi, M.; Chupani, L.; Salamat, N. Responses of immune organs after single-dose exposure to phenanthrene in yellowfin seabream (*Acanthopagrus latus*): CYP1A induction and oxidative stress. *Chemosphere* **2017**, *186*, 686–694. [CrossRef] [PubMed]
25. Wu, J.Y.; Yan, Z.G.; Liu, Z.T.; Liu, J.D.; Wang, W.L. Development of water quality criteria for phenanthrene and comparison of the sensitivity between native and non-native species. *Environ. Pollut.* **2015**, *196*, 141–146. [CrossRef] [PubMed]
26. Fontana, R.J.; Lown, K.S.; Paine, M.F. Effects of a chargrilled meat diet on expression of CYP3A, CYP1A, and P-glycoprotein levels in healthy volunteers. *Gastroenterology* **1999**, *117*, 89–98. [CrossRef]
27. Abel, L. Chemical Neurobiology: An Introduction to Neurochemistry: H. F. Bradford. Freeman, New York, 1976. *Neurochem. Int.* **1987**, *10*, 594–595. [CrossRef]
28. Costa, J.; Ferreira, M.; Rey-Salgueiro, L.; Reis-Henriques, M.A. Comparision of the waterborne and dietary routes of exposure on the effects of Benzo(a)pyrene on biotransformation pathways in *Nile tilapia* (*Oreochromis niloticus*). *Chemosphere* **2011**, *84*, 1452–1460. [CrossRef]
29. Sun, H.J.; Zhang, J.Y.; Wang, Q.; Zhu, E.; Chen, W.; Lin, H.; Chen, J.; Hong, H. Environmentally relevant concentrations of arsenite induces developmental toxicity and oxidative responses in the early life stage of zebrafish. *Environ. Pollut.* **2019**, *254*, 113022. [CrossRef]
30. Burkina, V.; Zlabek, V.; Zamaratskaia, G. Effects of pharmaceuticals present in aquatic environment on Phase I metabolism in fish. *Environ. Toxicol. Pharmacol.* **2015**, *40*, 430–444. [CrossRef]
31. Zhang, J.; Shen, H.; Wang, X.; Wu, J.; Xue, Y. Effects of chronic exposure of 2,4-dichlorophenol on the antioxidant system in liver of freshwater fish *Carassius auratus*. *Chemosphere* **2004**, *55*, 167–174. [CrossRef]
32. Niu, Z.; Xu, W.; Na, J.; Lv, Z.; Zhang, Y. How long-term exposure of environmentally relevant antibiotics may stimulate the growth of Prorocentrum lima: A probable positive factor for red tides. *Environ. Pollut.* **2019**, *255*, 113149. [CrossRef] [PubMed]
33. Triebskorn, R.; KHler, H.R.; Honnen, W.; Schramm, M.; Adams, S.M.; Müller, E.F. Induction of heat shock proteins, changes in liver ultrastructure, and alterations of fish behavior: Are these biomarkers related and are they useful to reflect the state of pollution in the field. *J. Aquat. Ecosyst. Stress Recovery* **1997**, *6*, 57–73. [CrossRef]
34. Zanette, J.; Jenny, M.J.; Goldstone, J.V.; Woodin, B.R.; Watka, L.A.; Bainy, A.C.D.; Stegeman, J.J. New cytochrome P450 1B1, 1C2 and 1D1 genes in the killifish *Fundulus heteroclitus*: Basal expression and response of five killifish CYP1s to the AHR agonist PCB126. *Aquat. Toxicol.* **2009**, *93*, 234–243. [CrossRef]
35. Wenju, X.U.; Yuanyou, L.I.; Qingyang, W.U.; Wang, S.; Zheng, H.; Liu, W. Effects of phenanthrene on hepatic enzymatic activities in tilapia(*Oreochromis niloticus* × *O. aureus* male). *J. Environ. Sci.* **2009**, *21*, 854–857. [CrossRef]
36. Mu, J.L.; Wang, X.H.; Jin, F.; Wang, J.Y.; Hong, H.S. The role of cytochrome P4501A activity inhibition in three- to five-ringed polycyclic aromatic hydrocarbons embryotoxicity of marine medaka (*Oryzias melastigma*). *Mar. Pollut. Bull.* **2012**, *64*, 1445–1451. [CrossRef]
37. Correia, A.D.; Goncalves, R.; Scholze, M.; Ferreira, M.; Henriques, A.R. Biochemical and behavioral responses in gilthead seabream (*Sparus aurata*) to phenanthrene. *J. Exp. Mar. Biol. Ecol.* **2007**, *347*, 109–122. [CrossRef]
38. Whyte, J.J.; Jung, R.E.; Schmitt, C.J.; Tillitt, D.E. Ethoxyresorufin-O-deethylase (EROD) Activity in Fish as a Biomarker of Chemical Exposure. *CRC Crit. Rev. Toxicol.* **2000**, *30*, 347–570. [CrossRef]
39. Hayes, J.D.; Pulford, D.J. The Glut athione S-Transferase Supergene Family: Regulation of GST and the Contribution of the lsoenzymes to Cancer Chemoprotection and Drug Resistance Part I. *CRC Crit. Rev. Biochem.* **1995**, *30*, 521–600. [CrossRef]
40. Lu, G.H.; Wang, C.; Zhu, Z. The Dose–Response Relationships for EROD and GST Induced by Polyaromatic Hydrocarbons in *Carassius auratus*. *Bull. Environ. Contam. Toxicol.* **2009**, *82*, 194–199. [CrossRef]
41. Yin, Y.; Jia, H.; Sun, Y.; Yu, H.; Wang, X.; Wu, J.; Xue, Y. Bioaccumulation and ROS generation in liver of *Carassius auratus*, exposed to phenanthrene. *Comp. Biochem. Physiol. C-Toxicol. Pharmacol.* **2007**, *145*, 288–293. [CrossRef]
42. Olinga, P.; Elferink, M.G.L.; Draaisma, A.L.; Merema, M.; Castell, J.V.; Pérez, G.; Groothuis, G.M.M. Coordinated induction of drug transporters and phase I and II metabolism in human liver slices. *Eur. J. Pharm. Sci.* **2008**, *33*, 380–389. [CrossRef] [PubMed]
43. Sun, Y.; Li, Y.; Rao, J.; Liu, Z.; Chen, Q. Effects of inorganic mercury exposure on histological structure, antioxidant status and immune response of immune organs in yellow catfish (*Pelteobagrus fulvidraco*). *J. Appl. Toxicol.* **2018**, *6*, 843–854. [CrossRef] [PubMed]
44. Zheng, J.L.; Yuan, S.S.; Wu, C.W.; Li, W.Y. Chronic waterborne zinc and cadmium exposures induced different responses towards oxidative stress in the liver of zebrafish. *Aquat. Toxicol.* **2016**, *177*, 261–268. [CrossRef] [PubMed]

45. Costa, J.; Reis-Henriques, M.A.; Castro, L.F.C.; Ferreira, M. Gene expression analysis of ABC efflux transporters, CYP1A and GSTα in *Nile tilapia* after exposure to benzo(a)pyrene. *Comp. Biochem. Physiol., Part. C Toxicol. Pharmacol.* **2012**, *155*, 469–482. [CrossRef] [PubMed]
46. Bettim, F.L.; Galvan, G.L.; Cestari, M.M.; Yarnamoto, C.I.; de Assis, H.C.S. Biochemical responses in freshwater fish after exposure to water-soluble fraction of gasoline. *Chemosphere* **2016**, *144*, 1467–1474. [CrossRef]
47. Anna, L.; Malin, C.C.; Lars, F. Effects of medetomidine on hepatic EROD activity in three species of fish. *Ecotoxicol. Environ. Saf.* **2008**, *69*, 74–79. [CrossRef]
48. Goksoyr, A.F. The cytochrome P-450 system in fish, aquatic toxicology and environmental monitoring. *Aquat. Toxicol.* **1992**, *22*, 287–311. [CrossRef]
49. Henson, K.L.; Gallagher, E.P. Glutathione S-Transferase Expression in Pollution-Associated Hepatic Lesions of Brown Bullheads (*Ameiurus nebulosus*) from the Cuyahoga River, Cleveland, Ohio. *Toxicol. Sci.* **2004**, *80*, 26–33. [CrossRef]

International Journal of
Environmental Research and Public Health

MDPI

Article

Mitigating the Health Effects of Aqueous Cr(VI) with Iron-Modified Biochar

Zhihong Zheng [1,2] and Xiaohan Duan [1,*]

[1] School of Water Conservancy, North China University of Water Resources and Electric Power, Zhengzhou 450046, China; zzh@ncwu.edu.cn
[2] Henan Vocational College of Water Conservancy and Environment, Zhengzhou 450008, China
* Correspondence: b2019081505@stu.ncwu.edu.cn

Abstract: A large amount of chromium (Cr) has entered the natural environment from the wastewater and waste residues, and the hexavalent (Cr(VI)) is highly poisonous, threatening the ecological environment and human health directly. In this study, iron-modified biochar was prepared using honeysuckle residue as raw material and the ferric chloride impregnation method. Batch Cr(VI) adsorption experiments were carried out using the modified honeysuckle-derived biochar (MHDB) as an adsorbent. The results indicate that a pH of 2 was best for the adsorption removal of Cr(VI) in the initial pH range of 2–10. The adsorption kinetic data fitted the pseudo-second-order model best out of the two models, and the Langmuir model was better than the Freundlich model to describe the adsorption process. Thermodynamic analysis indicated that the adsorption process of Cr(VI) on MHDB had an endothermic and spontaneous nature, and the increasing temperature was conducive to the adsorption. The main mechanisms of Cr(VI) adsorption might be the physical adsorption (electrostatic interactions) and chemical adsorption (ion exchange, the reduction of Cr(VI) to Cr(III)). The efficient adsorption of Cr(VI) makes MHDB a potential material for Cr(VI)-containing wastewater treatment. This study provides a feasible adsorption material for mitigating the environmental hazards of chromium, which has a certain reference value for protecting environmental health.

Keywords: adsorption; environmental health; chromium; ferric chloride; traditional Chinese medicine residual

Citation: Zheng, Z.; Duan, X. Mitigating the Health Effects of Aqueous Cr(VI) with Iron-Modified Biochar. *Int. J. Environ. Res. Public Health* **2022**, *19*, 1481. https://doi.org/10.3390/ijerph19031481

Academic Editor: Cheng Yan

Received: 16 December 2021
Accepted: 25 January 2022
Published: 28 January 2022

1. Introduction

Heavy metal pollutants have raised worldwide concerns because of their toxicity and persistence, causing an indefinite threat to the sustainable development of environmental health and human safety [1–4]. Among these heavy metals, chromium (Cr) is a commonly known pollutant derived from various industries, such as steel manufacturing, dye manufacturing, leather tanning, oil color, and electroplating [5–10]. Cr usually exists in the forms of trivalent (Cr(III)) and hexavalent (Cr(VI)) [11]. Cr(III) and Cr(VI) can be converted to each other under certain conditions [12]. Cr(III) is a micronutrient with low solubility and mobility that can be easily removed from water in the form of stable and insoluble precipitates [13]. In contrast, Cr(VI) is several hundred times more toxic than Cr(III) due to its high toxicity, higher migration, and non-biodegradability [14,15]. Cr(VI) has much higher mobility in water and soil and can be transformed into various reactive and toxic intermediates, eventually endangering human health because of the accumulation characteristics [16,17]. Cr(VI) can cause cell mutations and lead to skin injury, intestinal diseases, kidney diseases, cancer, and other diseases [18–21]. Consequently, Cr(VI) is a significant threat to the quality of water, soil environments, and human health and has been considered a top-priority pollutant to be addressed in water remediation [22]. To minimize the environmental risk caused by Cr(VI) and ensure human health, the maximum levels of chromium allowed in drinking water must not outnumber 0.05 ppm according to the

World Health Organization (WHO) [23]. Therefore, it is of great importance to remove Cr(VI) from water safely and efficiently.

Various techniques, including chemical precipitation, ion exchange, ultrafiltration, membrane separation, adsorption, and so on, have been developed to remove Cr(VI) from water [24–26]. Among the techniques, adsorption is a promising method for Cr(VI) removal due to its low cost, simple operation, and high efficiency [27,28]. Biochar, considered one of the most promising adsorbents because of its low cost and excellent adsorption effect, has been used for the removal of chromium-containing wastewater successfully to protect environmental health [29,30]. Biochar (BC) is obtained by the pyrolysis of biomass that is rich and diverse. Biomass made up of bagasse, corn stalk, peanut shell, oak wood, sludge, and rice straw has been used for the preparation of biochar successfully [31–35].

With the development of traditional Chinese medicine in recent years, plenty of traditional Chinese medicine residuals (TCMRs) have been created. Up to 70 Mt of TCMRs are produced per year in China. Owing to the limited availability of effective recycling methods, TCMRs are mainly disposed of as solid waste in landfill [36]. Thus, using TCMRs as a raw material with which to prepare biochar is a promising method for its reclamation. According to the statistical data, honeysuckle is a widely used traditional Chinese medicine, which showed the best adsorption effect of Cr(VI) among ten kinds of TCMRs, including astragalus membranaceus, radix Isatidis, thorowax, and others in our pre-experiment.

In order to enhance the adsorption performance of biochar, researchers often use surface modification methods to improve the pore structure and the surface functional groups, such as chemical oxidation, chemical reduction, and metal impregnation [37,38]. Ferric chloride ($FeCl_3$) is a common chemical reagent that can provide hydrogen ions in aqueous solutions. Previous studies have indicated that modification using $FeCl_3$ could increase the mesopores of carbon materials and that Fe oxides have great selectivity and affinity for Cr(VI) [39]. Hence, $FeCl_3$ was chosen as a modifier to promote the Cr(VI) adsorption capacity in this study and, thus, to better reduce the harm of Cr(VI) in the environment.

Therefore, in order to prepare an effective biochar material to solve the problem of Cr(VI) in the environment, honeysuckle residue was used as a raw material to prepare honeysuckle-derived biochar (HDB), and the ferric chloride impregnation method was used to obtain modified honeysuckle-derived biochar (MHDB) to improve the adsorption capacity. The influencing factors of Cr(VI) adsorption were investigated to explore the adsorption mechanism. The study could be a valuable reference for the resource recovery of TCMRs and provide a resolvent for chromium-containing wastewater while alleviating the environmental pollution problem and protecting human health.

2. Material and Method

2.1. Preparation of the Biochar and Modified Biochar

Honeysuckle residue (HR) was obtained from a local pharmacy in Zhengzhou, China. The HR was dried at 105 °C for 2 h and smashed and passed through a 60 mesh sieve. The HR particles were calcinated in a tube furnace at 400 °C for 3 h in a N_2 environment. The black matter was taken out, ground, and screened with a 100 mesh sieve in sequence.

Ferric chloride was added to a conical flask along with HDB. The optimum concentrations of Fe^{3+} and the solid–liquid ratio were determined to be 1 mol/L and 1:20 (W/V) in our pre-experiment, respectively. After vigorous stirring for 10 min, the mixture was oscillated in a shaker at 150 rpm for 8 h. The product was washed with deionized water until a constant pH was achieved. It was then dried at 60 °C for 12 h and sifted with a 100 mesh sieve to obtain the MHDB.

2.2. Characterization Techniques

The surface morphology of the HDB and MHDB before and after Cr(VI) adsorption was performed using a scanning electron microscope (SEM, Zeiss Genimi500, Zeiss, Barkhausenbau, Germany). The porosities and surface areas of the HDB and MHDB were measured using a surface area analyzer with a N_2 atmosphere (BELSORP-max, MicrotracBEL, Osaka, Japan). The structures and crystallinities of the HDB and MHDB before and after Cr(VI) adsorption were characterized using an X-ray diffractometer (XRD, ZSX Primus II, Rigaku Corporation, Tokyo, Japan). The surface functional groups of HDB and MHDB before and after Cr(VI) adsorption were detected using Fourier transform infrared spectroscopy (FTIR, Thermo Fisher IS50, Thermo Fisher, Waltham, MA, USA). The elements and morphologies of the sample surface were analyzed using X-ray photoelectron spectroscopy (XPS, Thermo Scientific Escalab 250Xi+, Thermo Fisher, Waltham, MA, USA). The zeta potential of the MHDB was determined using a zeta potential meter after sonication (Particle Metrix GmbH, Particle Metrix, Munich, Germany).

2.3. Synthesis of Cr(VI)-Containing Wastewater

Cr(VI) solutions with different concentrations were synthesized by dissolving $K_2Cr_2O_7$ into deionized water. The pH value of the wastewater was adjusted using 0.1 mol/L of NaOH or HCl. All chemicals used in the study were analytically pure.

2.4. Batch Adsorption Experiment

A total of 0.05 g biochar was added into 25 mL of Cr(VI) solution, and then the conical flasks were put into a temperature-controlled shaker and reacted at 120 rpm. At the end of the experiment, the mixture was filtered with a 0.45 μm filter membrane, the residual Cr(VI) was determined using diphenylcarbazide spectrophotometry at 540 nm. Each set was replicated three times, and the average value was used for subsequent analysis. The adsorption capacity was calculated by Equation (1), as follows:

$$q = \frac{(C_0 - C_e)V}{m} \quad (1)$$

where q (mg/g) is the adsorption capacity; C_0 and C_e (mg/L) are the initial and final concentration of Cr(VI), respectively; m (g) is the mass of biochar used for the adsorption; V (L) is the solution volume.

3. Results and Discussion

3.1. Sample Characterization

3.1.1. Scanning Electron Microscope (SEM) Images

The SEM images of (A) HDB, (B) MHDB, and (C) MHDB after Cr(VI) adsorption are shown in Figure 1. The surface of the MHDB was rougher than that of the HDB due to the modification of ferric chloride ($FeCl_3$); more surface pores in circular and irregular shapes appeared. Figure 1C shows that more irregular particles were dispersed on the surface after adsorption, which indicates that the Cr(VI) was successfully adsorbed onto the surface of the MHDB.

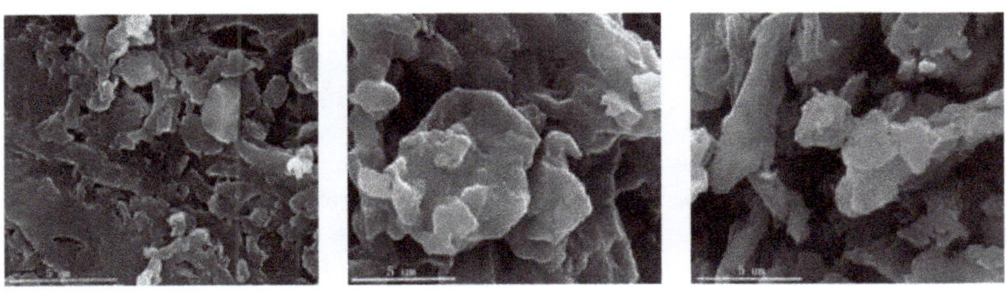

(A) (B) (C)

Figure 1. SEM images of (**A**) HDB, (**B**) MHDB, and (**C**) MHDB after adsorption of Cr(VI).

3.1.2. Brunauer–Emmett–Teller (BET) Analysis

Figure 2 shows the N_2 sorption–desorption isotherms of HDB and MHDB, which fitted the type IV isotherm, indicating the adsorbents are mesoporous materials. The average pore diameters of the HDB and MHDB were 12.45 nm and 8.33 nm, respectively. The hysteresis loop was at $p/p_0 > 0.4$, as shown in Figure 2, indicating that the capillary condensation occurred in the biochar channels at relatively high pressures. The mesopore size distribution of the HDB was unimodal, with a peak pore size of 3.718 nm, as shown in Figure 2C. In addition, the peak pore size of the MHDB was 3.75 nm, and multiple peaks appeared, which indicates that the number of mesopores increased after modification. Therefore, the diffusion resistance decreased and the Cr(VI) diffused more easily into the pores of the adsorbent. The BET-specific surface areas of the HDB and MHDB were 0.81 m^2/g and 1.13 m^2/g, respectively, and the small specific surface area was mainly due to the raw material characteristics and the low sintering temperature. Considering the material properties, the mechanism of Cr(VI) adsorption might mainly be electrostatic adsorption, and the modification with iron enhanced the adsorption capacity of the adsorbent effectively.

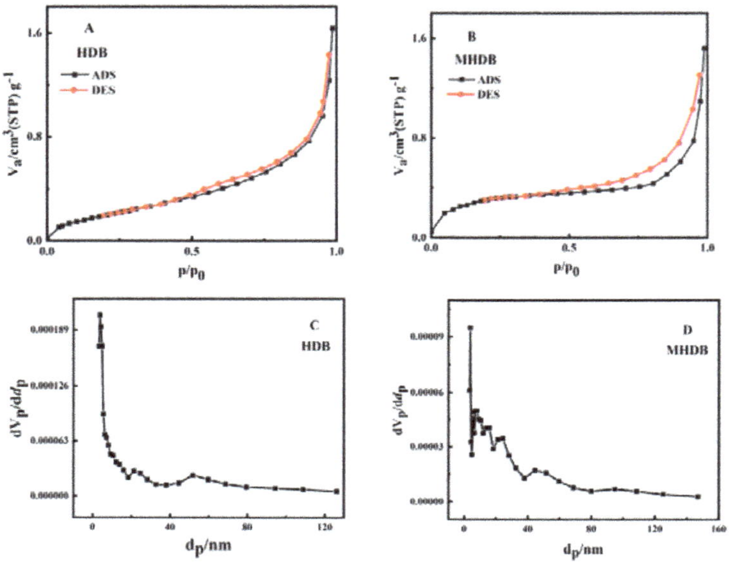

Figure 2. Nitrogen physisorption–desorption isotherms of HDB (**A**) and MHDB (**B**); BJH mesopore size distributions of HDB (**C**) and MHDB (**D**).

3.1.3. X-ray Diffractometer (XRD) Analysis

For further study, the HDB and MHDB were characterized by XRD, and the results are shown in Figure 3A. The XRD patterns of the biochars show a large number of characteristic peaks, and the main peak is at 26.49°, indicating that the surface of the material contained crystalline material for SiO_2. No obvious sharp peak appeared, indicating that the iron of the MHDB mainly existed with low crystallinity [40]. The small peaks of the MHDB at $2\theta = 29.45°$, 42.77°, and 60.05° indicate that the Fe_3C particles may have been present in the MHDB, which was consistent with the XPS analysis results [41,42]. In addition, the modification of the ferric chloride ($FeCl_3$) caused changes in the materials, according to Figures 1 and 2. It is quite clear that the surface of the original HDB was smoother than that of the MHDB, and the average pore diameter increased after modification.

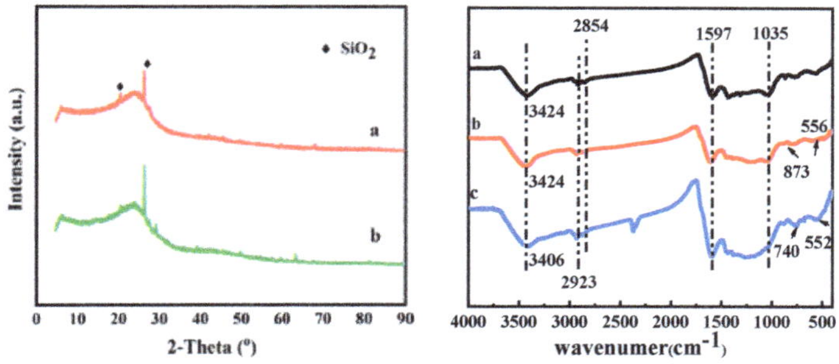

Figure 3. (**A**) The XRD patterns of (a) HDB and (b) MHDB; (**B**) FTIR spectra of (a) HDB, (b) MHDB; (c) MHDB after adsorption of Cr(VI).

3.1.4. Fourier Transform Infrared (FTIR) Analysis

Figure 3B shows that the wavenumber at 3424 cm^{-1} corresponds to the stretching oscillation of -OH, with a wide adsorption peak [43]. The wavenumbers at 2923 cm^{-1} and 2854 cm^{-1} could link to the symmetric and asymmetric stretching vibrations of C–H and C–C double bonds. The wavenumber at 1597 cm^{-1} is attributed to the stretching oscillation of C=C and C=O double bonds. The peak at 1035 cm^{-1} is assigned to the stretching vibrations of C–O. Except for the common functional groups, the oxygen-containing functional groups of the MHDB could have provided more chemical adsorption sites that improved the adsorption capacity, such as Ar-H, based on the wavenumber at 873 cm^{-1}. Meanwhile, the wavenumber at 556 cm^{-1} is attributed to Fe–O, indicating that the adsorbent contained Fe [44]. Figure 3B shows that the peaks for the functional groups changed after Cr(VI) adsorption, as the peak around 3424 cm^{-1} (-OH) shifted to 3406 cm^{-1}, indicating that the functional groups were favorable for Cr(VI) removal. A new wavenumber at 2363 cm^{-1} appeared after adsorption, which corresponds to the formation of H_3O^+ [45]. In addition, the peak near 552 cm^{-1} (Fe–O) obviously increased after Cr(VI) adsorption. In summary, the change in the infrared spectrum after adsorption resulted from chemical denaturation, which indicates that the adsorption process was accompanied by a chemisorption reaction. The adsorption effect was closely related to the functional groups on the surface of the adsorbent, and the addition of functional groups by modification could have enhanced the adsorption.

3.1.5. X-ray Photoelectron Spectroscopy (XPS) Analysis

The full XPS spectra of the (a) HDB, (b) MHDB, and (c) MHDB after adsorption of Cr(VI) are shown in Figure 4A. The C1s spectra of the HDB resolved into two peaks at 284 eV and 287.82 eV, representing the C–H and C=O, respectively [46,47]. The O1s spectra of the HDB resolved into two peaks at 532.29 eV and 533.1 eV, representing the -OH and C-O, respectively [48,49]. The binding energies and intensities of the peaks changed after modification, suggesting that the structure of the MHDB changed due to the iron modification. As shown in Figure 4B, the major peaks in the Fe 2p spectra are at 711.1 eV (Fe 2p3/2) and 724.3 eV (Fe 2p1/2) [50]. The distinctive satellite peaks at 705.9eV and 719.8eV are assigned to Fe_3C, which is in agreement with a previous study [51]. After the adsorption of Cr(VI), the main peaks shifted to 711.4 eV and 725.1 eV, respectively, indicating that the iron was involved in the adsorption. As shown in Figure 4C, the binding energies at 576.8 eV (Cr 2p3/2) and 586.7 eV (Cr2p1/2) correspond to Cr(III), and the peaks of 579.0 eV (Cr 2p3/2) and 588.3 eV (Cr2p1/2) correspond to Cr(VI) [52,53]. Hence, the MHDB could have effectively reduced the Cr(VI) concentration in the water, converting the Cr(VI) to the stable Cr(III). At this point, the toxicity and harm of Cr(VI) to human health declines sharply after adsorption.

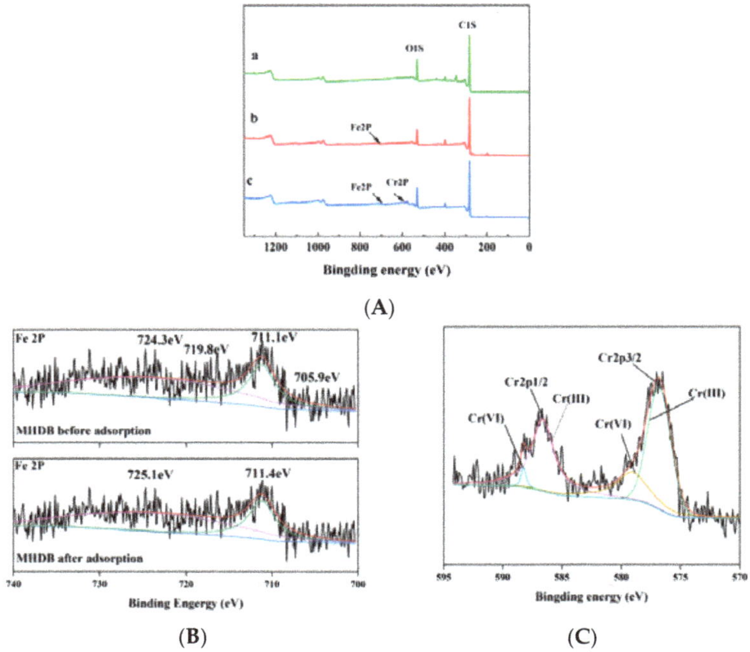

Figure 4. (**A**) The full XPS spectra of (a) HDB, (b) MHDB, (c) MHDB after adsorption; (**B**) detailed graph of the Fe 2p before and after adsorption; (**C**) detailed graph of the Cr 2p after adsorption.

In sum, the sample characterization results suggest that the MHDB changed after the Cr(VI) adsorption, and the main mechanisms of the Cr(VI) adsorption might be physical adsorption (electrostatic interactions) and chemical adsorption (ion exchange, the reduction of Cr(VI) to Cr(III)).

3.2. Adsorption Study

3.2.1. Effect of Initial Solution pH

The effect of the initial aqueous pH on the adsorption of Cr(VI) was studied by varying the initial pH from 2 to 10. Figure 5A showed the adsorption capacity of Cr(VI) by the MHDB decreased with the increasing pH, which is consistent with the results of a previous study [54]. The Cr(VI) adsorption capacity decreased slowly from 8.10 mg/g to 7.39 mg/g as the initial pH increased slowly from 2 to 6, and then the adsorption capacity decreased to 5.23 mg/g quickly as the initial pH rose to 10.

Previous studies have shown that the pH causes changes in the surface charge of the biochar and the degree of protonation of the functional groups on the biochar, affecting the form of Cr(VI) in the adsorption system significantly [55]. The main forms of chromium were CrO_4^{2-} and $HCrO_4^-$ in the pH range of 2–8. Further research has shown that $HCrO_4^-$ was the main form, with pH values of 2–6.5, and the form of CrO_4^{2-} mainly presented in water at pH values of >6.5 (Figure 5B) [56]. In addition, the free energy of the $HCrO_4^-$ was low, meaning it could be removed from the system more easily.

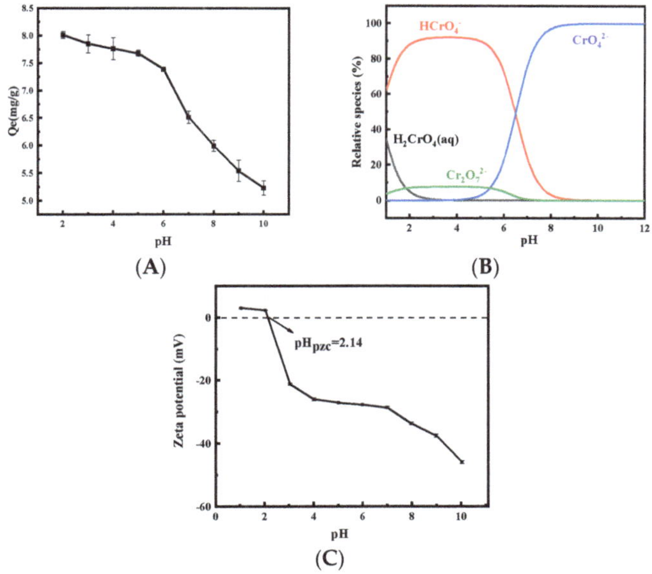

Figure 5. (**A**) Effect of initial solution pH on adsorption performance (C_0 = 50 mg/L, time = 6 h, and temperature = 25 °C); (**B**) the existing forms of Cr(VI) in the solution at different pH levels; (**C**) zeta potential of MHDB.

The dominant chemical reactions of the reduction of Cr(VI) to Cr(III) in the solution are expressed as Equations (2) and (3) [57]. As seen in Figure 5C, the pH_{PZC} value of the MHDB was measured to be 2.14, which indicates that the surface charge of the MHDB was positive at solution pH values lower than its pH_{PZC}. The highest removal efficiency at a pH of 2 was due to the possible protonation of the adsorbent, which led to the weakening of the electrostatic repulsion and the enhancement of the electrostatic adsorption of Cr(VI) [58]. The protons in the solution were depleted, and the positive charge of the adsorbent surface turned into a negative charge as the initial pH increased; hence, the electrostatic attraction with Cr(VI) decreased. With the initial pH increase, the alkaline environment meant an increased presence of OH^-, which indicates that the equilibrium in Equations (2) and (3) shifted to the left. That is, the alkaline environment suppressed the conversion reaction of Cr(VI) to Cr(III) and was detrimental to the removal of Cr(VI). The experiment results indicate that a lower pH was beneficial for Cr(VI) removal, and a pH of 2 was best for Cr(VI) adsorption by MHDB in the pH range of 2–10. Therefore, the severe threat of Cr(VI) to human health is mitigated by MHDB more effectively in an acidic environment.

$$Cr_2O_7{}^{2-} + 14H^+ + 6e^- \rightarrow 2Cr^{3+} + 7H_2O \quad (2)$$

$$HCrO_4{}^- + 7H^+ + 3e^- \rightarrow Cr^{3+} + 4H_2O \quad (3)$$

3.2.2. Effect of Adsorption Time and Adsorption Kinetics

The adsorption of Cr(VI) was affected by the adsorption time, and the result is illustrated in Figure 6A. It can be seen that Cr(VI) adsorption increased with the increase in reaction time, and the Cr(VI) adsorption took place in two stages. The adsorption capacities in the first 12 h rapidly reached 91.65% and 76.65% of the equilibrium adsorption capacity with the initial Cr(VI) concentrations of 10 and 50 mg/L, respectively. In the second stage, the removal rate showed slow growth, and the adsorption of Cr(VI) reached equilibrium at 48 h. The active sites of the adsorbent were unbound in the early stage, meaning it could combine with Cr(VI) quickly. Most of the active sites were occupied along with the reaction time, which led to slow adsorption and the equilibrium of the adsorption at 48 h.

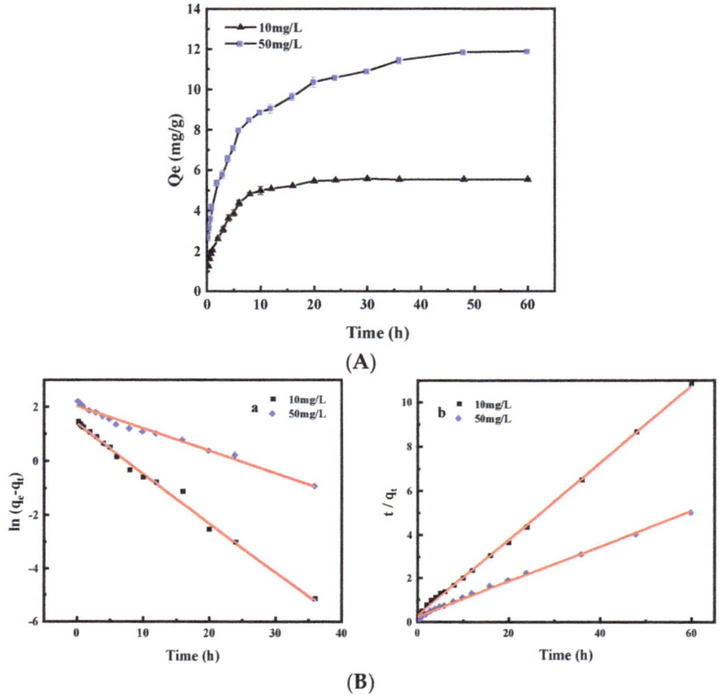

Figure 6. (**A**) Effect of adsorption time on adsorption performance (C_0 = 10 and 50 mg/L, pH = 2, and temperature = 25 °C); (**B**) Linear kinetic plots of pseudo-first model (**a**) and pseudo-second model (**b**) for Cr(VI) adsorption by MHDB.

Pseudo-first-order and pseudo-second-order models were used to fit the data obtained from the experiment [59]. The two Equations (4) and (5) are expressed as follows:

$$ln(q_e - q_t) = lnq_e - k_1 t \quad (4)$$

$$\frac{t}{q_t} = \frac{t}{q_e} + \frac{1}{k_2 q_e^2} \quad (5)$$

where q_t and q_e (mg/g) are the Cr(VI) uptakes at time t and at equilibrium time, respectively; k_1 (h^{-1}) is the equilibrium rate constant, and k_2 [g/(mg·h)] is the equilibrium rate constant, respectively.

Figure 6B shows the linear kinetic plots of the two models, and the kinetics parameters for the adsorption are listed in Table 1. It can be seen that the regression coefficient of the pseudo-second-order equation is higher out of the two equations (R^2 > 0.993). Moreover, the adsorption amounts calculated based on the pseudo-second-order equation at a different initial concentration of Cr(VI) were 5.74 and 12.33 mg/g, which are in good agreement with the experimental data at 5.54 and 11.84 mg/g, respectively. In sum, the pseudo-second-order equation was better to describe the adsorption process of Cr(VI) by MHDB, and the human health impact of Cr(VI) was weakened after adsorption.

Table 1. The kinetics parameters for the adsorption of Cr(VI) by MHDB.

C_0 mg/L	Pseudo-First-Order			Pseudo-Second-Order		
	q_e	k_1	R^2	q_e	k_2	R^2
10	4.05	0.1838	0.9925	5.74	0.1741	0.9991
50	7.78	0.0826	0.9789	12.33	0.0811	0.9962

3.2.3. Adsorption Isotherms

As the initial Cr(VI) concentration increased from 10 to 100 mg/L, the uptake of Cr(VI) by MHDB increased from 4.96 to 9.93 mg/g, from 5.13 to 11.68 mg/g, and from 5.17 to 13.70 mg/g at 15, 25, and 35 °C, respectively (Figure 7A). The results might be attributed to the fact that the concentration difference between the adsorbent and the solution obviously increased when the initial Cr(VI) concentration increased. In this respect, the difference accelerated the mass transfer rate and promoted the adsorption reaction of Cr(VI).

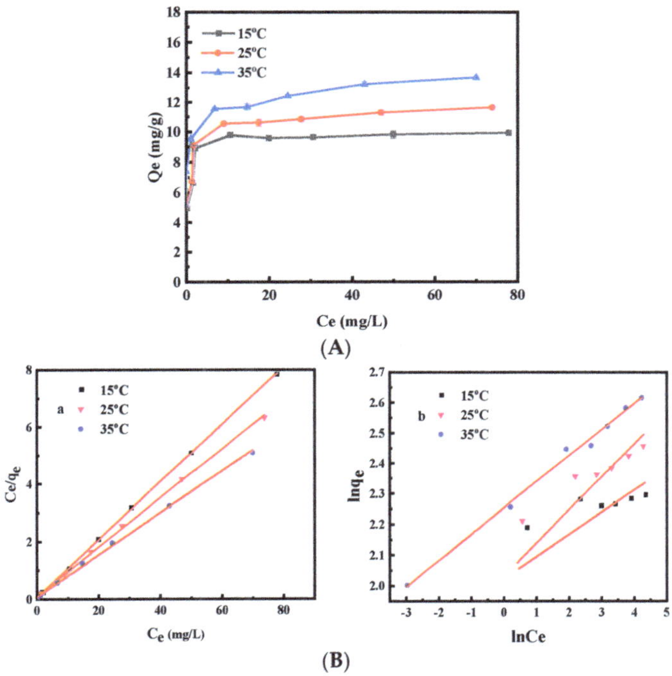

Figure 7. (**A**) Isotherms of Cr(VI) adsorption removal by MHDB (C_0 = 10–100 mg/L, pH = 2, time = 48 h, and temperature = 15, 25 and 35 °C); (**B**) adsorption isotherms of Cr(VI) by fitting the Langmuir isotherm model (**a**) and Freundlich isotherm model (**b**).

Figure 7A shows the relationship between the Cr(VI) concentration and the adsorption capacity of the MHDB at equilibrium. In addition, the adsorption isotherm data were fitted by the Langmuir and Freundlich models [60]. The linearized forms of the two models are expressed by Equations (6) and (7) below:

$$\frac{C_e}{q_e} = \frac{1}{Q_m K_L} + \frac{C_e}{Q_m} \quad (6)$$

$$lnq_e = lnK_f + \frac{1}{n} lnC_e \quad (7)$$

where C_e (mg/L) and q_e (mg/g) are the concentration and adsorption capacity at equilibrium, respectively; K_L (L/mg) is the Langmuir constant related to the binding energy, and Q_m (mg/g) is the maximum adsorption capacity; K_f is the Freundlich constant related to the adsorption capacity, and n is the adsorption intensity.

Figure 7B shows the fitting results of the Langmuir and Freundlich models, and the isotherm parameters are listed in Table 2. The regression coefficients R^2 of the Langmuir equation were higher than those of the Freundlich equation at temperatures of 15, 25, and 35 °C, indicating that the Langmuir equation was more suitable for describing the adsorption process ($R^2 > 0.997$), and the adsorption of Cr(VI) by MHDB had a monolayer adsorption nature. Though other isotherm models might offer further insight, considering that the Langmuir equation was significantly suitable for the adsorption process, other isotherm models were no longer used for analysis in this study. Based on the Langmuir isotherm model, the calculated maximum adsorptions were 9.95, 11.70, and 13.70 mg/g at 15, 25, and 35 °C, respectively. In sum, the impact of Cr(VI) on human health declined more easily by MHDB at higher temperatures.

Table 2. The parameters of Langmuir equation and Freundlich equation.

Reaction Temperature °C	Langmuir Equation			Freundlich Equation		
	Q_m	K_L	R^2	K_f	$1/n$	R^2
15	9.95	2.039	0.9998	7.551	0.0724	0.5089
25	11.70	0.972	0.9989	7.605	0.1085	0.7089
35	13.70	0.979	0.9979	9.511	0.0859	0.9924

The maximum adsorption capacity of the MHDB was about 13.70 mg/g at 35 °C, which was higher than those of multiple other types of biochars from previous studies. For instance, the adsorption capacities of Cr(VI) by the pineapple peel-derived biochar, the natural zeolite-derived biochar, the nano-magnetite-modified biochar material, and the cornstalk-derived biochar were 7.44 mg/g, 6.08 mg/g, 9.92 mg/g, and 9.25 mg/g, respectively [61–64]. Thus, the modified honeysuckle-derived biochar is a promising adsorbent for Cr(VI) removal and could be a valuable reference for the resource recovery of TCMRs, alleviating the environmental pollution problem from the chromium-containing wastewater.

3.2.4. Adsorption Thermodynamics

The thermodynamics of Cr(VI) adsorbed by MHDB were studied at 288 K, 298 K, and 308 K, and the parameters were calculated using Equations (8)–(10), as follows [65]:

$$\Delta G^0 = \Delta H^0 - T\Delta S^0 \tag{8}$$

$$lnK_d = \frac{\Delta S^0}{R} + \frac{\Delta H^0}{RT} \tag{9}$$

$$K_d = \frac{q_e}{C_e} \tag{10}$$

where K_d is the distribution coefficient (L/g); the K_d value is obtained by plotting $\ln(q_e/C_e)$ against C_e and extrapolating to zero C_e; ΔS^0 is the standard entropy change (kJ/mol·K); ΔH^0 is the standard enthalpy change (kJ/mol); R is the ideal gas constant (8.314J mol/k); T is the reaction temperature (K); ΔG^0 is the standard Gibbs free energy change (kJ/mol); C_e is the concentration of Cr(VI) at equilibrium (mg/L); q_e is the adsorption capacity of MHDB on Cr(VI) at equilibrium (mg/g).

Figure 8 shows the relationship between the $1/T$ and $\ln K_d$. The values of ΔS^0, ΔH^0, and ΔG^0 are shown in Table 3. It can be seen that the adsorption process had a spontaneous character since all three values of ΔG^0 at 288, 298, and 308 K were negative. Meanwhile, the spontaneous nature of the adsorption process increased since the value of ΔG^0 decreased gradually as the temperature increased. The higher temperature was favorable for the adsorption of Cr(VI) by MHDB due to the fact that the high reaction temperature decreased the solution viscosity and the diffusion resistance of Cr(VI) ions, which could speed up the adsorption process. Thus, the higher temperature was beneficial for lessening the harmful effects of Cr(VI) on human health.

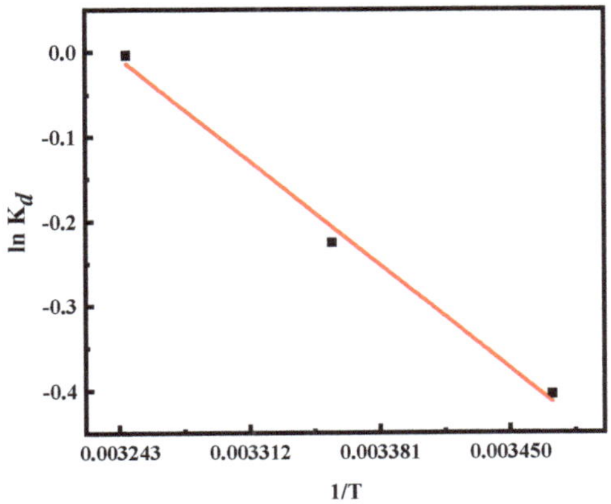

Figure 8. Thermodynamics plot for adsorption of Cr(VI) by MHDB.

Table 3. Thermodynamic parameters for adsorption of Cr(VI) by MHDB.

Reaction Temperature	ΔG^0	ΔH^0	ΔS^0
288 K	−0.15		
298 K	−0.67	14.74	51.73
308 K	−1.19		

3.2.5. Effect of Co-Existing Ions

In general, wastewater may contain other anions that could influence the adsorption of Cr (VI). Therefore, the impact of co-existing ions on the adsorption of Cr(VI) was investigated in this study, and the results are shown in Figure 9. It can be seen that the adsorption capacity of Cr(VI) decreased with the presence of anions, and the effect of PO_4^{3-} was the most significant among the three anions. In addition, the adsorption capacity of Cr(VI) decreased with the increase of the anion concentration. As shown in Figure 9, the PO_4^{3-} with a high concentration (0.1 mol/L) significantly inhibited Cr(VI) removal, and the adsorption capacity of Cr(VI) decreased to 4.68 mg/g—a decrease of 61.20%.

The competition between anions and Cr(VI) for adsorption might mainly be due to their similar ionic nature; the anions occupied the active sites on the surface of the adsorbent and formed complexes [66,67]. The interference was directly proportional to the charge on the anions; the anionic charges provided by the anions Cl^-, SO_4^{2-}, and PO_4^{3-} were −1, −2, and −3, respectively, and hence, the negative effects on the Cr(VI) removal efficiency followed the order of $Cl^- < SO_4^{2-} < PO_4^{3-}$. Overall, these results suggest that the presence of anions could affect the removal of Cr(VI) by MHDB, and the impact was positively related to the charges and concentrations of the anions. Therefore, the anions reduced the adsorption capacity of the MHDB for Cr(VI); in other words, they increased the harmful effects of Cr(VI) on human health.

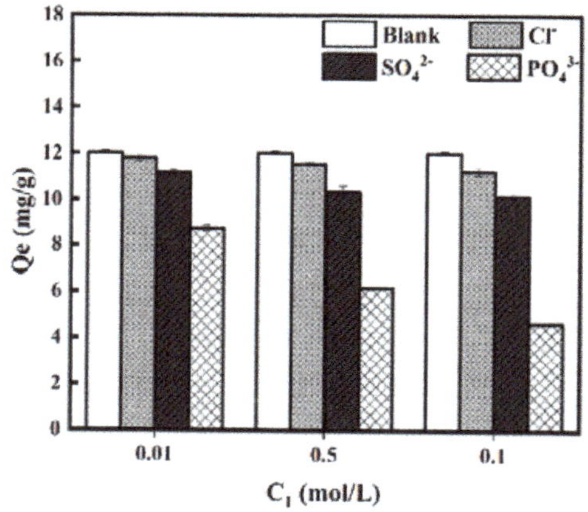

Figure 9. Effect of co-existing ions on adsorption performance (C_0 = 150 mg/L, time = 6 h, pH = 2, temperature= 25 °C).

4. Conclusions

Hexavalent chromium (Cr(VI)) is a severe threat to human health. In order to solve the hazard of chromium in the environment, the removal of Cr(VI) using Fe^{3+}-modified honeysuckle-derived biochar (MHDB) was researched in this study. The highest Cr(VI) removal was observed at a pH of 2 in the initial pH range of 2–10. The adsorption of Cr(VI) took place in two stages; the removal efficiency increased rapidly in the first 12 h, and then increased slowly until the equilibrium was reached. The adsorption process followed pseudo-second-order kinetics and the Langmuir isotherm. The thermodynamic study showed that the adsorption process was spontaneous and endothermic. The anions had negative effects on Cr(VI) adsorption in the order of $Cl^- < SO_4^{2-} < PO_4^{3-}$. Based on the experimental results, the MHDB reduced the Cr(VI) concentration in water distinctly, demonstrating a high adsorption capacity and an efficient ability to convert the Cr(VI) to the Cr(III). In general, this study provides a new material for the adsorption of Cr(VI) and has important reference value for mitigating the harm of chromium pollution and protecting environmental health.

Author Contributions: Conceptualization, Z.Z. and X.D.; methodology, Z.Z. and X.D.; investigation, X.D.; supervision, Z.Z.; writing—original draft, X.D.; writing—review and editing, Z.Z. and X.D. All authors have read and agreed to the published version of the manuscript.

Funding: The authors are thankful for the financial support from the Major Science and Technology Project in Henan Province of China for the integration and demonstration of sewage treatment technology for typical villages and towns of Henan Province (161100310700).

Institutional Review Board Statement: Not applicable.

Informed Consent Statement: Not applicable.

Acknowledgments: The authors thank the Director of laboratory in North China University of Water Resources and Electric Power for offering the experiment sites and experiment materials.

Conflicts of Interest: The authors declare no conflict of interest.

References

1. Hu, X.; Zhao, H.P.; Liang, Y.; Chen, R. Energy level mediation of $(BiO)_2CO_3$ via Br doping for effificient molecular oxygen activation and ciproflfloxacin photodegradation. *Appl. Catal. B Environ.* **2019**, *258*, 117966. [CrossRef]
2. Mao, W.; Zhang, L.X.; Liu, Y.; Wang, T.Y.; Bai, Y.C.; Guan, Y.T. Facile assembled N, S-codoped corn straw biochar loaded Bi_2WO_6 with the enhanced electron-rich feature for the efficient photocatalytic removal of ciprofloxacin and Cr(VI). *Chemosphere* **2021**, *263*, 127988. [CrossRef] [PubMed]

3. Yuan, R.R.; Yue, C.L.; Qiu, J.L.; Liu, F.Q.; Li, A.M. Highly effificient sunlight-driven reduction of Cr(VI) by TiO_2@NH_2-MIL-88B(Fe) heterostructures under neutral conditions. *Appl. Catal. B Environ.* **2019**, *251*, 229–239. [CrossRef]
4. Seleiman, M.F.; Santanen, A.; Mäkelä, P.S.A. Recycling sludge on cropland as fertilizer-Advantages and risks. *Resour. Conserv. Recycl.* **2020**, *155*, 104647. [CrossRef]
5. Dong, H.R.; He, Q.; Zeng, G.M.; Tang, L.; Zhang, C.; Xie, Y.K.; Zeng, Y.L.; Zhao, F.; Wu, Y.A. Chromate removal by surface-modified nanoscale zero-valent iron: Effect of different surface coatings and water chemistry. *J. Colloid Interface Sci.* **2016**, *471*, 7–13. [CrossRef]
6. Kong, Q.P.; Wei, J.Y.; Hu, Y.; Wei, C.H. Fabrication of terminal amino hyperbranched polymer modified graphene oxide and its prominent adsorption performance towards Cr(VI). *J. Hazard. Mater.* **2019**, *363*, 161–169. [CrossRef]
7. Antony, G.S.; Manna, A.; Baskaran, S.; Puhazhendi, P.; Ramchary, A.; Niraikulam, A.; Ramudu, K.N. Non-enzymatic reduction of Cr(VI) and it's effective biosorption using heat-inactivated biomass: A fermentation waste material. *J. Hazard. Mater.* **2020**, *392*, 122257. [CrossRef]
8. Shan, R.; Shi, Y.Y.; Gu, J.; Bi, J.W.; Yuan, H.R.; Luo, B.; Chen, Y. Aqueous Cr(VI) removal by biochar derived from waste mangosteen shells: Role of pyrolysis and modifification on its absorption process. *J. Environ. Chem. Eng.* **2020**, *8*, 103885. [CrossRef]
9. Lu, Z.; Zhang, H.; Shahab, A.; Zhang, K.; Zeng, H.T.; Bacha, A.; Nabi, I.; Ullah, H. Comparative study on characterization and adsorption properties of phosphoric acid activated biochar and nitrogen-containing modified biochar employing Eucalyptus as a precursor. *J. Clean. Prod.* **2021**, *303*, 127046. [CrossRef]
10. Seleiman, M.F.; Ali, S.; Refay, Y.; Rizwan, M.; Alhammad, B.A.; El-Hendawy, S.E. Chromium resistant microbes and melatonin reduced Cr uptake and toxicity, improved physio-biochemical traits and yield of wheat in contaminated soil. *Chemosphere* **2020**. [CrossRef]
11. Shi, Y.; Shan, R.; Lu, L.; Yuan, H.; Jiang, H.; Zhang, Y.; Chen, Y. High-efficiency removal of Cr(VI) by modified biochar derived from glue residue. *J. Clean.* **2020**, *254*, 119935. [CrossRef]
12. Richard, F.C.; Bourg, A.C.M. Aqueous geochemistry of chromium: A review. *Water Res.* **1991**, *25*, 807–816. [CrossRef]
13. Su, C.L.; Wang, S.; Zhou, Z.Y.; Wang, H.J.; Xie, X.J.; Yang, Y.Y.; Feng, Y.; Liu, W.F.; Liu, P. Chemical processes of Cr(VI) removal by Fe-modified biochar under aerobic and anaerobic conditions and mechanism characterization under aerobic conditions using synchrotron-related techniques. *Sci. Total Environ.* **2021**, *768*, 144604. [CrossRef] [PubMed]
14. Zhao, J.J.; Boada, R.; Cibin, C.; Palet, C. Enhancement of selective adsorption of Cr species via modifification of pine biomass. *Sci. Total Environ.* **2021**, *756*, 143816. [CrossRef]
15. Zhu, D.C.; Shao, J.G.; Li, Z.Q.; Yang, H.P.; Zhang, S.H.; Chen, H.P. Nano nickel embedded in N-doped CNTs-supported porous biochar for adsorption-reduction of hexavalent chromium. *J. Hazard. Mater.* **2021**, *416*, 125693. [CrossRef]
16. Khalil, U.; Shakoor, M.B.; Ali, S.; Alyemeni, M.N.; Wijaya, L. Adsorption-reduction performance of tea waste and rice husk biochars for Cr(VI) elimination from wastewater. *J. Saudi Chem. Soc.* **2020**, *24*, 799–810. [CrossRef]
17. Lv, Z.M.; Tian, X.L.; Wang, C.H.; Alsaedi, A.; Hayat, T.; Chen, C.L. Metal-organic frameworks-derived 3D yolk shell-like structure Ni@carbon as a recyclable catalyst for Cr(VI) reduction. *Chem. Eng. J.* **2020**, *389*, 123428. [CrossRef]
18. Wang, X.S.; Tang, Y.P.; Tao, S.R. Kinetics, equilibrium and thermodynamic study on removal of Cr (VI) from aqueous solutions using low-cost adsorbent Alligator weed. *Chem. Eng. J.* **2009**, *148*, 217–225. [CrossRef]
19. Kahraman, H.T.; Pehlivan, E. Cr^{6+} rovemal using oleaster (Elaeagnus) seed and cherry (Prunus avium) stone biochar. *Powder Technol.* **2017**, *306*, 61–67. [CrossRef]
20. Zhong, J.W.; Yin, W.Z.; Li, Y.T.; Li, P.; Wu, J.H.; Jiang, G.B.; Gu, J.J.; Liang, H. Column study of enhanced Cr(VI) removal and longevity by coupled abiotic and biotic processes using Fe0 and mixed anaerobic culture. *Water Res.* **2017**, *122*, 536–544. [CrossRef]
21. Rafique, M.I.; Usman, A.R.; Ahmad, M.; Al-Wabel, M.I. Immobilization and mitigation of chromium toxicity in aqueous solutions and tannery waste-contaminated soil using biochar and polymer-modified biochar. *Chemosphere* **2021**, *266*, 129198. [CrossRef] [PubMed]
22. He, B.; Yun, Z.J.; Shi, J.B.; Jiang, G.B. Research progress of heavy metal pollution in China: Sources, analytical methods, status, and toxicity. *Chin. Sci. Bull.* **2013**, *58*, 134–140. [CrossRef]
23. WHO. *Guidelines for Drinking-Water Quality: Fourth Edition Incorporating the First Addendum*; World Health Organization: Geneva, Switzerland, 2017.
24. Basumatary, A.K.; Kumar, R.V.; Ghoshal, A.K.; Pugazhenthi, G. Cross flow ultrafiltration of Cr (VI) using MCM-41, MCM-48 and Faujasite (FAU) zeolite-ceramic composite membranes. *Chemosphere* **2016**, *153*, 436–446. [CrossRef] [PubMed]
25. Pakade, V.E.; Tavengwa, N.T.; Madikizela, L.M. Recent advances in hexavalent chromium removal from aqueous solutions by adsorptive methods. *RSC Adv.* **2019**, *9*, 26142–26164. [CrossRef]
26. Tiadi, N.; Dash, R.R.; Mohanty, C.R.; Patel, A.M. Comparative Studies of Adsorption of Chromium(VI) Ions onto Different Industrial Wastes. *J. Hazard. Toxic Radioact. Waste* **2020**, *24*, 04020021. [CrossRef]
27. Rodrigues, E.; Almeida, O.; Brasil, H.; Moraes, D.; Reis, M.A.L. Adsorption of chromium(VI) on hydrotalcitehydroxyapatite material doped with carbon nanotubes: Equilibrium, kinetic and thermodynamic study. *Appl. Clay Sci.* **2019**, *172*, 57–64. [CrossRef]
28. Jiang, R.Q.; Yu, G.W.; Ndagijimana, P.; Wang, Y.; You, F.T.; Xing, Z.J.; Wang, Y. Effective adsorption of Direct Red 23 by sludge biochar based adsorbent: Adsorption kinetics, thermodynamics and mechanisms study. *Water Sci. Technol.* **2021**, *83*, 2424–2436. [CrossRef]

29. Roy, P.; Dey, U.; Chattoraj, S.; Mukhopadhyay, D.; Mondal, N.K. Modeling of the adsorptive removal of arsenic(III) using plant biomass: A bioremedial approach. *Appl. Water Sci.* **2017**, *7*, 1307–1321. [CrossRef]
30. Dong, H.R.; Deng, J.M.; Xie, Y.K.; Zhang, C.; Jiang, Z.; Cheng, Y.J.; Hou, K.J.; Zeng, G.M. Stabilization of nanoscale zero-valent iron (nZVI) with modified biochar for Cr(VI) removal from aqueous solution. *J. Hazard. Mater.* **2017**, *332*, 79–86. [CrossRef]
31. Yang, C.; Ge, C.Z.; Li, X.L.; Li, L.; Wang, B.; Lin, A.J.; Yang, W.J. Does soluble starch improve the removal of Cr(VI) by nZVI loaded on biochar? *Ecotoxicol. Environ. Saf.* **2021**, *208*, 111552. [CrossRef]
32. Yin, Z.H.; Liu, Y.G.; Liu, S.B.; Jiang, L.H.; Tan, X.F.; Zeng, G.M.; Li, M.F.; Liu, S.J.; Tian, S.R.; Fang, Y. Activated magnetic biochar by one-step synthesis: Enhanced adsorption and coadsorption for 17β-estradiol and copper. *Sci. Total Environ.* **2018**, *639*, 1530–1542. [CrossRef] [PubMed]
33. Zhao, Y.L.; Zhang, R.Y.; Liu, H.B.; Li, M.X.; Chen, T.H.; Chen, D.; Zou, X.H.; Frost, R.L. Green preparation of magnetic biochar for the effective accumulation of Pb(II): Performance and mechanism. *Chem. Eng. J.* **2019**, *375*, 122011. [CrossRef]
34. Lyu, H.H.; Tang, J.C.; Huang, Y.; Gai, L.S.; Zeng, E.Y.; Liber, K.; Gong, Y.Y. Removal of hexavalent chromium from aqueous solutions by a novel biochar supported nanoscale iron sulfide composite. *Chem. Eng. J.* **2017**, *322*, 516–524. [CrossRef]
35. Zhang, Y.T.; Jiao, X.Q.; Liu, N.; Lv, J.; Yang, Y.D. Enhanced removal of aqueous Cr(VI) by a green synthesized nano scale zero-valent iron supported on oak wood biochar. *Chemosphere* **2020**, *245*, 125542. [CrossRef] [PubMed]
36. Guo, F.Q.; Dong, Y.P.; Zhang, T.H.; Dong, L.; Guo, C.W.; Rao, Z.H. Experimental Study on Herb Residue Gasification in an Air-Blown Circulating Fluidized Bed Gasifier. *Ind. Eng. Chem. Res.* **2014**, *53*, 13264–13273. [CrossRef]
37. Zeng, H.T.; Zeng, H.Z.; Zhang, H.; Shahab, A.; Zhang, K.; Lu, Y.Q.; Nabi, I.; Naseem, F.; Ullah, H. Efficient adsorption of Cr (VI) from aqueous environments by phosphoric acid activated eucalyptus biochar. *J. Clean. Prod.* **2021**, *286*, 124964. [CrossRef]
38. Lee, Y.G.; Shin, J.W.; Kwak, J.W.; Kim, S.W.; Son, C.J.; Kim, G.Y.; Lee, C.H.; Chon, K.M. Enhanced Adsorption Capacities of Fungicides Using Peanut Shell Biochar via Successive Chemical Modifification with $KMnO_4$ and KOH. *Separations* **2021**, *8*, 52. [CrossRef]
39. Feng, Z.Y.; Chen, N.; Feng, C.P.; Gao, Y. Mechanisms of Cr(VI) removal by $FeCl_3$-modified lotus stem-based biochar ($FeCl_3$@LS-BC) using mass-balance and functional group expressions. *Colloids Surf. A Physicochem. Eng. Asp.* **2018**, *551*, 17–24. [CrossRef]
40. Zhong, Z.X.; Yu, G.W.; Mo, W.T.; Zhang, C.J.; Huang, H.; Li, S.; Gao, M.; Lu, X.J.; Zhang, B.P.; Zhu, H.P. Enhanced phosphate sequestration by Fe(III) modified biochar derived from coconut shell. *RSC Adv.* **2019**, *9*, 10425. [CrossRef]
41. Cho, D.; Yoon, K.; Kwon, E.E.; Biswas, J.K.; Song, H. Fabrication of magnetic biochar as a treatment medium for As(V) via pyrolysis of $FeCl_3$-pretreated spent coffee ground. *Environ. Pollut.* **2017**, *229*, 942–949. [CrossRef]
42. Hou, T.Y.; Sun, X.H.; Xie, D.L.; Wang, M.J.; Fan, A.; Chen, Y.Y.; Shu, C.; Zheng, C.M.; Hu, W.B. Mesoporous Graphitic Carbon-Encapsulated Fe_2O_3 Nanocomposite as High-Rate Anode Material for Sodium-Ion Batteries. *Chem.—A Eur. J.* **2018**, *24*, 14786–14793. [CrossRef] [PubMed]
43. Liang, M.; Ding, Y.M.; Zhang, Q.; Wang, D.Q.; Li, H.H.; Lu, L. Removal of aqueous Cr(VI) by magnetic biochar derived from bagasse. *Sci. Rep.* **2020**, *10*, 21473. [CrossRef] [PubMed]
44. Zhu, S.S.; Ho, S.; Huang, X.C.; Wang, D.W.; Yang, F.; Wang, L.; Wang, C.Y.; Cao, X.D.; Ma, F. Magnetic Nanoscale Zerovalent Iron Assisted Biochar: Interfacial Chemical Behaviors and Heavy Metals Remediation Performance. *ACS Sustain. Chem. Eng.* **2017**, *5*, 9673–9682. [CrossRef]
45. Yang, L.; Chen, J.P. Biosorption of hexavalent chromium onto raw and chemically modified Sargassum sp. *Bioresour. Technol.* **2008**, *99*, 297–307. [CrossRef]
46. Guo, J.H.; Yan, C.Z.; Luo, Z.X.; Fang, H.D.; Hu, S.G.; Cao, Y.L. Synthesis of a novel ternary HA/Fe-Mn oxide-loaded biochar composite and its application in cadmium(II) and arsenic(V) adsorption. *J. Environ. Sci.* **2019**, *85*, 168–176. [CrossRef] [PubMed]
47. Zhang, R.C.; Leivisk, T.; Taskila, S.; Tanskanen, J. Iron-loaded Sphagnum moss extract residue for phosphate removal. *J. Environ. Manag.* **2018**, *218*, 271–279. [CrossRef]
48. De, S.H.; Mishra, S.; Poonguzhali, E.; Rajesh, M.; Tamilarasan, K. Fractionation and characterization of lignin from waste rice straw: Biomass surface chemical composition analysis. *Int. J. Biol. Macromol.* **2020**, *145*, 795–803. [CrossRef]
49. Benis, K.Z.; Soltan, J.; Mcphedran, K.N. Electrochemically modified adsorbents for treatment of aqueous arsenic: Pore diffusion in modified biomass vs. Biochar. *Chem. Eng. J.* **2021**, *423*, 130061. [CrossRef]
50. Zhang, B.A.Q.; Zhao, Z.W.; Chen, N.; Feng, C.P.; Lei, Z.F.; Zhang, Z.Y. Insight into effificient phosphorus removal/recovery from enhanced methane production of waste activated sludge with chitosan-Fe supplementation. *Water Res.* **2020**, *187*, 116427. [CrossRef]
51. Liang, Y.; Liu, P.; Xiao, J.; Li, H.B.; Wang, C.X.; Yang, G.W. A microfifibre assembly of an iron-carbon composite with giant magnetisation. *Sci. Rep.* **2013**, *3*, 3051. [CrossRef]
52. Wang, Z.W.; Wu, T.; Geng, X.; Ru, J.J.; Hua, Y.X.; Bu, J.J.; Xue, Y.; Wang, D. The role of electrolyte ratio in electrodeposition of nanoscale Fe-Cr alloy from choline chloride-ethylene glycol ionic liquid: A suitable layer for corrosion resistance. *J. Mol. Liq.* **2021**. [CrossRef]
53. Zhou, M.; Yang, X.N.; Sun, R.H.; Wang, X.Z.; Yin, W.Q.; Wang, S.S.; Wang, J. The contribution of lignocellulosic constituents to Cr(VI) reduction capacity of biochar-supported zerovalent iron. *Chemosphere* **2021**, *263*, 127871. [CrossRef] [PubMed]
54. Zhao, G.X.; Huang, X.B.; Tang, Z.W.; Huang, Q.F.; Niu, F.L.; Wang, X.K. Polymer-based nanocomposites for heavy metal ions removal from aqueous solution: A review. *Polym. Chem.* **2018**, *9*, 3562–3582. [CrossRef]

55. Qu, J.H.; Tian, X.; Jiang, Z.; Cao, B.; Akindolie, M.S.; Hu, Q.; Feng, C.C.; Feng, Y.; Meng, X.L.; Zhang, Y. Multi-component adsorption of Pb(II), Cd(II) and Ni(II) onto microwave-functionalized cellulose: Kinetics, isotherms, thermodynamics, mechanisms and application for electroplating wastewater purification. *J. Hazard. Mater.* **2020**, *387*, 121718. [CrossRef]
56. Liu, K.; Li, F.B.; Tian, Q.W.; Nie, C.R.; Ma, Y.B.; Zhu, Z.L.; Fang, L.P.; Huang, Y.Y.; Liu, S.W. A highly porous animal bone-derived char with a superiority of promoting nZVI for Cr(VI) sequestration in agricultural soils. *J. Environ. Sci.* **2021**, *104*, 27–39. [CrossRef] [PubMed]
57. Mahanty, B.; Mondal, S. Synthesis of Magnetic Biochar Using Agricultural Waste for the Separation of Cr(VI) From Aqueous Solution. *Arab. J. Sci. Eng.* **2021**, *46*, 10803–10818. [CrossRef]
58. Zhao, Z.G.; Wei, J.X.; Li, F.X.; Qu, X.L.; Liang, S.; Zhang, H.D.; Yu, Q.J. Synthesis, characterization and hexavalent chromium adsorption characteristics of Aluminum and sucrose-incorporated tobermorite. *Materials.* **2017**, *10*, 597. [CrossRef]
59. Xu, H.; Gao, M.X.; Hu, X.; Chen, Y.H.; Li, Y.; Xu, X.Y.; Zhang, R.Q.; Yang, X.; Tang, C.F.; Hu, X.J. A novel preparation of S-nZVI and its high efficient removal of Cr(VI) in aqueous solution. *J. Hazard. Mater.* **2021**, *416*, 125924. [CrossRef]
60. Rezak, N.; Bahmani, A.; Bettahar, N. Adsorptive removal of P(V) and Cr(VI) by calcined Zn-Al-Fe ternary LDHs. *Water Sci. Technol.* **2021**, *83*, 2504–2517. [CrossRef]
61. Wang, C.H.; Gu, L.F.; Liu, X.Y.; Zhang, X.Y.; Cao, L.; Hu, X.X. Sorption behavior of Cr(VI) on pineapple-peel-derived biochar and the inflfluence of coexisting pyrene. *Int. Biodeterior. Biodegrad.* **2016**, *111*, 78–84. [CrossRef]
62. Deveci, H.; Kar, Y. Adsorption of hexavalent chromium from aqueous solutions by bio-chars obtained during biomass pyrolysis. *J. Ind. Eng. Chem.* **2013**, *19*, 190–196. [CrossRef]
63. Song, X.; Zhang, Y.; Cao, N.; Sun, D.; Zhang, Z.; Wang, Y.; Wen, Y.; Yang, Y.; Lyu, T. Sustainable Chromium (VI) Removal from Contaminated Groundwater Using Nano-Magnetite-Modified Biochar via Rapid Microwave Synthesis. *Molecules* **2021**, *26*, 103. [CrossRef] [PubMed]
64. Tang, W.; Zanli, B.L.G.L.; Chen, J. O/N/P-doped biochar induced to enhance adsorption of sulfonamide with coexisting Cu^{2+}/Cr(VI) by air pre-oxidation. *Bioresour. Technol.* **2021**, *341*, 125794. [CrossRef] [PubMed]
65. Qhubu, M.C.; Mgidlana, L.G.; Madikizela, L.M.; Pakade, V.E. Preparation, characterization and application of activated clay biochar composite for removal of Cr(VI) in water: Isotherms, kinetics and thermodynamics. *Mater. Chem. Phys.* **2020**, *260*, 124165. [CrossRef]
66. Zhu, S.H.; Wang, S.; Yang, X.; Tufail, S.; Chen, C.; Wang, X.; Shang, J.Y. Green Sustainable and Highly Efficient Hematite Nanoparticles Modified Biochar-clay Granular Composite for Cr(VI) Removal and Related Mechanism. *J. Clean. Prod.* **2020**, *276*, 123009. [CrossRef]
67. Zhao, N.; Zhao, C.F.; Tsang, D.C.W.; Liu, K.Y.; Zhu, L.; Zhang, W.H.; Zhang, J.; Tang, Y.T.; Qiu, R.L. Microscopic mechanism about the selective adsorption of Cr(VI) from salt solution on O-rich and N-rich biochars. *J. Hazard. Mater.* **2021**, *404*, 124162. [CrossRef] [PubMed]

MDPI
St. Alban-Anlage 66
4052 Basel
Switzerland
Tel. +41 61 683 77 34
Fax +41 61 302 89 18
www.mdpi.com

International Journal of Environmental Research and Public Health Editorial Office
E-mail: ijerph@mdpi.com
www.mdpi.com/journal/ijerph

www.ingramcontent.com/pod-product-compliance
Lightning Source LLC
Chambersburg PA
CBHW061616210926
51850CB00009B/2343
9783036558516